ACTIVE CARBON

ROOP CHAND BANSAL
Panjab University
Chandigarh, India

JEAN-BAPTISTE DONNET
Centre National de la
Recherche Scientifique
Mulhouse, France

FRITZ STOECKLI
University of Neuchâtel
Neuchâtel, Switzerland

MARCEL DEKKER, INC.   **New York and Basel**

Library of Congress Cataloging-in-Publication Data

Bansal, Roop Chand
  Active carbon.

  Includes indexes.
  1. Carbon, Activated.   I. Donnet, Jean-Baptiste.
II. Stoeckli, Fritz.   III. Title.
TP245.C4B35 1988   622'.93   88-3571
ISBN 0-8247-7842-1

MARCEL DEKKER, INC.
270 Madison Avenue, New York, New York 10016

Current printing (last digit):
10 9 8 7 6 5 4 3 2 1

PRINTED IN THE UNITED STATES OF AMERICA

Preface

Activated carbons are unique and versatile adsorbents because of
their extended surface area, microporous structure, universal adsorption effect, high adsorption capacity, and high degree of surface
reactivity. They are extensively used to purify, decolorize, deodorize,
dechlorinate, and detoxicate potable waters; for solvent recovery
and air purification in inhabited spaces such as restaurants, food-
processing, and chemical industries; in the purification of many
chemical and foodstuff products; and in a variety of gas phase applications. They are also increasingly being used in hydrometallurgy
for the recovery of gold and silver and other inorganics and in the
treatment of domestic and industrial wastewaters. Their use in medicine
for certain types of bacterial ailments is well known. Thus active
carbons are of interest in many economic sectors and concern industries
as diverse as food processing, pharmaceuticals, chemical, petroleum,
mining, nuclear, automobile, and vacuum manufacturing.

Some of these applications are very demanding with regard to the
surface chemistry and the surface characteristics of these adsorbent
carbons. In the past, many users of active carbon were able to get
along with any grade of carbon and were concerned only with the price.
The current emphasis, however, is toward unit operations which tend
to utilize the entire adsorption space of the carbon, and this puts a
high premium on quality, reproducibility, adsorption capacity, and
surface reactivity. It was assumed until some years ago that investigation of the fine structure of carbon surfaces was necessary only
in particular fields of technological research and in the manufacture

of carbons. Today it is realized that characterization of the carbon surface and its porous structure with respect to the chemical composition of the surface, pore size distribution, and surface area is of vital importance whenever quantitative data for processes occurring or starting at the surface of the carbon, such as surface reactions, adsorption, degasing, and surface treatment, are required. A more precise knowledge of the surface chemistry of carbon surface is also essential for the proper development and improvement of active carbon for specific applications. As a maximum in the specific surface area now appears to have been realized, there is need, therefore, to modify carbon surfaces and their porous structure by different surface treatments to develop newer carbons with adequate porosity and with modified properties to cope with new problems and recent developments in the area of wastewater treatment, where large quantities of active carbons are likely to be used. It is with these views that the present book has been conceived.

The book describes briefly the basic steps involved in the manufacture of activated carbons, the selectivity and suitability of different raw materials, and the probable mechanism of the physical and chemical activation processes. The chapter on chemical structure presents an exhaustive survey of the nature and the characteristics of the carbon–oxygen surface structures on acidic and basic carbons and their estimation by physical, chemical, and physicochemical techniques, which include ESCA and the latest innovations in infrared spectroscopy. The chapter on porous structure includes the classification of pores, their characterization and contribution to surface area, distinction between internal and external surfaces, their contribution to the adsorption of gases and vapor, and the thermodynamic consequences of Dubinin's micropore volume theory. The characterization of adsorbent carbons by immersion calorimetry, by adsorption of polar and nonpolar vapors, and by adsorption from solutions is discussed. The importance of active sites and their characterization and of active surface area and its measurement in determining the reactivity of carbons is emphasized. The procedures for the modification of carbon surfaces by surface impregnation, oxidation, halogenation, hydrogenation, and sulfurization are described, and the influence of these surface modifications on the surface characteristics and surface behavior of carbons is discussed.

The book presents a detailed survey of some of the applications utilizing bulk amounts of carbons. The factors involved in the application of activated carbons for the removal of organics and inorganics from the aqueous phase are delineated; the use of activated carbons for the recovery of gold and silver from their cyanide solutions and their elution from the carbon surface are discussed; and the various theories of gold recovery are reviewed. The possibility of

using carbons as a catalyst for certain oxidation, combination, decomposition, halogenation, and dehalogenation reactions is also examined.
The book thus combines in one volume the manufacture, various facets
of the surface chemistry, characterization, and modifications, and the
important applications of adsorbent carbons so that carbon scientists
and technologists can take full advantage of carbon surfaces and
their modifications promoting a relationship between carbon surface
structure, carbon surface properties, and their applications and
can have access to the relevant literature. The unified approach
promotes further research toward improvement and development of
newer carbon adsorbents.

We express our thanks to our colleagues Dr. P. Ehrburger, Dr.
J. Lahaye, and Dr. Derai for making Professor Bansal comfortable dur-
ing his stay in Mulhouse; to F. Muller for photography; and to Miss Caty
Bachmeyer for typing the manuscript. The authors are also grateful
to the Pergamon Press; Academic Press, London; Society of Chemical
Industry, London; Butterworths; Ann Arbor Science Publishing; the
Council of Mineral Science and Engineering, South Africa; Soutn
African Institute of Mining and Metallurgy; and various authors
for permission to reproduce figures and tables. Professor Bansal
is also thankful to the Centre National de la Recherche Scientifique
CNRS (French Ministry for University and Research) for the financial
support for his stay and the Panjab University, Chandigarh, India
for granting leave. Most important, the understanding, support,
and encouragement of our wives, Rajesh Bansal, Suzanne Donnet,
and Helen Stoeckli-Evans, made the completion of this project
possible.

Roop Chand Bansal
Jean-Baptiste Donnet
Fritz Stoeckli

Introduction

The term *activated carbon* in its broadest sense includes a wide range
of amorphous carbon-based materials prepared to exhibit a high degree
of porosity and an extended interparticulate surface area. These
are obtained by combustion, partial combustion, and thermal decomposition of various carbonaceous substances. These materials may
be granular or in powdered from. The granular form is characterized
by a large internal surface and small pores, whereas the finely
divided powdered form is associated with larger pore diameters but
a smaller internal surface.

The use of activated carbon in the form of carbonized wood dates
back many centuries. The Egyptians used it around 1500 B.C. as
an adsorbent for medicinal purposes and also as a purifying agent.
The ancient Hindus in India filtered their drinking water through
charcoal. The basis for the industrial production of active carbons,
however, was established in 1900—1901 in order to replace bone
char in the sugar refining process. This active carbon was prepared
by carbonizing a mixture of materials of vegetable origin in the
presence of metal chlorides or by the action of carbon dioxide or
steam on charred materials. Activated carbons with better decolorizing
power were prepared by the action of zinc chloride on wood and
other materials of high carbon content. The manufacture of better
quality gas-adsorbent carbons received fresh impetus during World
War I, when they were used in gas masks for protection against
poisonous gases. In the late 1930s the activated carbons were also
manufactured, from sawdust by chemical activation with zinc chloride,
for volatile solvent recovery and for the removal of benzene from
tower gas.

Activated carbons are excellent adsorbents and thus are used to purify, decolorize, deodorize, dechlorinate, detoxicate, filter, or remove or modify the salts, separate, and concentrate in order to permit recovery; they are also used as catalysts and catalyst supports. These applications of active carbons are of interest to most economic sectors and concern areas as diverse as the food, pharmaceutical, chemical, petroleum, mining, nuclear, automobile, and vacuum industries as well as the treatment of drinking water, industrial and urban wastewater, and air and gas. Nearly 80% (220,000 tons/yr) of the total active carbon is consumed for liquid phase applications where both the granulated and the powdered forms of active carbon are used. The use of powdered carbon is more ancient and generally involves processing of food and drinking water. The total consumption of active carbon in gas phase applications is around 60,000 tons/yr, which includes exclusively the granular form of the active carbon, which may be extruded or pounded, the principal uses being in the purification of air, recovery of gold, and cigarette filters. The consumption of active carbons is high in the United States and Japan, which together consume two to four times more active carbon than the Western European countries. The per capita consumption of active carbon per year is 0.5 kg in Japan, 0.4 kg in the United States, 0.2 kg in Europe, and 0.03 kg in the rest of the world.

The adsorbent properties of activated carbons are essentially attributed to their large surface area, a high degree of surface reactivity, universal adsorption effect, and favorable pore size, which makes the internal surface accessible, enhances the adsorption rate, and enhances mechanical strength. The most widely used commercial active carbons have a specific surface area of the order of $800-1500 \ m^2/g$. This surface area is contained predominantly within micropores, which have effective diameters smaller than 2 nm. In fact, a particle of active carbon is made up of a complex network of pores which have been classified into micropores (diameters < 2 nm), mesopores (diameter between 2 and 50 nm), and macropores (diameters > 50 nm). The macropores do not contribute much toward surface area but act as conduits for the passage of the adsorbate into the interior mesopore and the micropore surface where most of the adsorption takes place. The pore size distribution in a given carbon depends on the type of the raw material and the method of manufacture of the carbon (Fig. 1).

The large surface area of the active carbon is the result of the activation process in which a carbonaceous char with little internal surface is oxidized in an atmosphere of air, carbon dioxide, or steam at a temperature between 800 and 900°C. This causes the oxidation of some of the regions within the char in preference to others so that as combustion proceeds a preferential etching occurs, resulting in the development of a large internal surface area, which in some cases may be as high as $2500 \ m^2/g$.

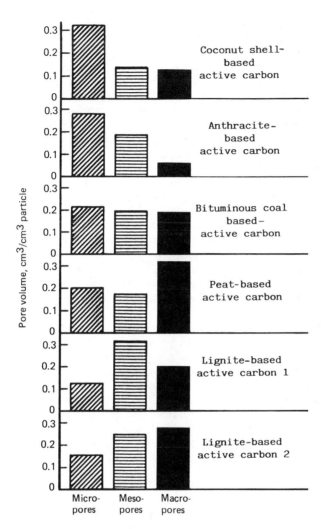

Figure 1 Pore size distribution in some active carbons obtained using different precursors.

The structure of activated carbons has been repeatedly compared with that of graphite and considered to be microcrystalline. However, in recent years transmission electron microscopy of carbonaceous materials has shown that the analogy with graphite is very poor and that the active carbon structure can be visualized as stacks of flat aromatic sheets crosslinked in a random manner (see Fig. 1 of Chapter 3). Activation by carbon dioxide or steam in the range 800–900°C reduces the number of these aromatic sheets in the original stacks, leaving in some cases single and in general nonplanar layers.

Electron spin resonance studies have revealed that the aromatic sheets in active carbons contain free radical structures or structures with unpaired electrons. These unpaired electrons are resonance stabilized and are trapped during the carbonization process as a result of the breaking up of bonds at the edges of the aromatic sheets thus creating edge carbon atoms. These edge carbon atoms have unsatisfied valencies and can thus interact with heteroatoms such as oxygen, hydrogen, nitrogen, and sulfur, giving rise to different types of surface functional groups. Thus the elemental composition of a typical activated carbon was found to be 88% C, 0.5% H, 0.5% N, 1% S, and 6–7% O, the balance representing inorganic ash constituents. The oxygen content of an activated carbon can, however, vary between 1 and 25%, depending on the type of raw material and the conditions of the activation process. The oxygen content of an activated carbon decreases with an increase in temperature of activation, the contents being minimum when the activation temperature is 1000°C or higher. The oxygen is picked up by the carbon from the raw material or from the oxidizing gases (CO_2, O_2, or steam) used for their activation.

The activation temperature significantly influences the nature of the carbon–oxygen surface complex. At low temperature of activation the carbon–oxygen surface complexes formed are less stable and are removed as carbon dioxide on heat treatment in vacuum or in nitrogen. On the other hand, when the temperature of activation is high, the functional groups formed are more stable and could be removed only as carbon monoxide, also at higher temperatures.

Extensive investigations using varied chemical and physicochemical techniques have been reported from several laboratories. The low-temperature–activated carbons have been found to develop acidic surface functional groups which are hydrophilic in character and show negative zeta potential. These acidic surface groups have been identified as phenolic, carboxylic, lactonic, cyclic peroxides, and carbonyl groups. The carboxylic and lactone groups render the carbon surface polar in character and tend to decrease the adsorption of nonpolar aromatic compounds, whereas the carbonyl groups in the form of quinones and hydroquinones tend to enhance the adsorption of aromatic compounds through the formation of an electron acceptor–donor complex. The carbon activated at higher temperature (800–1000°C) develops basic oxides and exhibits positive zeta potential. These basic surface oxides have been suggested to have chromene and pyronelike structures.

Besides oxygen, the activated carbons are invariably associated with hydrogen, which is present partly in the surface oxygen functional group and partly combined with the carbon atoms and dispersed in the granules of the carbon particles. The hydrogen is bonded more strongly than oxygen and cannot be removed completely even on evacuating at 1200°C, when about 30% of the total hydrogen is left behind in the carbon. Infrared studies of some anthracites have shown that the hydrogen is present both in the aliphatic and the aromatic forms. The aromatic hydrogen is bonded covalently to the carbon atoms at the periphery of the aromatic sheets while the aliphatic hydrogen may be present as aliphatic chains and alicyclic groups attached to the peripheral aromatic rings.

The characterization of activated carbons is carried out on the basis of several physical and chemical properties, commonly including their surface area, pore size distribution, impact hardness, ability to adsorb several selected substances such as benzene, carbon tetrachloride, nitrogen from the gaseous phase as well as iodine, molasses, phenol, and methylene blue from the aqueous phase. The nitrogen BET value, for example, expresses the surface that can be covered by nitrogen in a monomolecular layer. Typical nitrogen BET surface area values are found to be between 400 and 1500 m^2/g, the former representing low-activity carbons and the latter, high-activity carbons. However, surface area measurements alone are not sufficient to characterize a carbon product since the nitrogen molecule is very small and can penetrate into pores which are not available for larger molecules. The accessibility of larger molecules that are involved in the actual use of activated carbons may be small compared to adsorption of nitrogen. Furthermore, the adsorption of nitrogen being carried out at very low temperature (-195°C), the nitrogen adsorption cannot measure some of the extremely ultrafine microcapillary pores. Thus the BET surface area should be used with caution. It is essential, therefore, that characterization of carbons be carried out by their adsorption capacity toward larger molecular species such as phenol, iodine, methylene blue, and molasses, under standard experimental conditions. The adsorption capacity of activated carbons toward these molecular species gives the distribution of their internal accessible volume among pores of different sizes. Thus the adsorption capacity for iodine, which is generally called the iodine number, measures the number of pores above 10 Å whereas the molasses number indicates the number of pores larger than 30 Å in diameter. A more sophisticated approach to the measurement of pore size distribution is carried out by forcing mercury under pressure into the pores of the carbon in a mercury porosimeter, while a complete characterization involves gas adsorption and calorimetric measurements as described later in this book. Typical values of some of the properties used to characterize activated carbons are given in Table 1 of Chapter 1.

This book is divided into six chapters covering manufacture, surface structure (which includes chemical as well as porous structure), surface characterization, surface modifications, and applications of active carbons. The various factors involved in the manufacture of activated carbons using several raw materials are closely guarded secrets. Chapter 1 thus describes the basic steps involved in the manufacture of active carbons, the selectivity and suitability of different raw materials, and the probable mechanisms of the physical and chemical activation processes.

The surface structure of the activated carbons is discussed in two chapters. Chapter 2 deals with the chemical structure, presenting a comprehensive survey of the nature and the characteristics of the carbon—oxygen surface structures, the physical, chemical, and physicochemical methods of their measurements, which include ESCA (XPS), and the latest innovations in infrared spectroscopy. The surface chemistry of basic carbons has also been included. The porous structure of active carbons briefly describes the classification of pores, their characterization and contribution to surface area, distinction between internal and external surfaces, and their contribution to the adsorption of gases and vapors. A brief description of various theories dealing with physical adsorption, and the thermodynamic consequences of the Dubinin's micropore volume theory are also discussed in Chapter 3.

The characterization of active carbons by immersion calorimetry, adsorption of polar and nonpolar vapors, and adsorption from solutions are considered in Chapter 4. The importance of active surface area, its measurement, and measurement of active sites and their characterization are also discussed.

The modification of active carbons by surface impregnation, halogenation, nitrogenation, and sulfurization are dealt with in Chapter 5. The influence of different types of surface structures (surface complexes) on the surface characteristics and surface behavior of carbons is discussed.

Chapter 6 describes some applications of active carbons. The factors involved in the application of active carbons for the removal of organics and inorganics from the aqueous phase are delineated. The use of activated carbon for the recovery of gold and silver from their cyanide solution is discussed and the various theories of gold recovery are reviewed. The possibility of using active carbon as a catalyst for certain oxidation, combination, decomposition, halogenation, and dehalogenation reactions is also discussed.

Contents

1

Manufacture of Active Carbons

1.1 MANUFACTURE

The manufacture of activated carbons involves two main steps: the carbonization of the carbonaceous raw material at temperatures below 800°C in the absence of oxygen and the activation of the carbonized product. Thus all carbonaceous materials can be converted into activated carbon, although the properties of the final product will be different, depending on the nature of the raw material used, the nature of the activating agent, and the conditions of the activation process. During carbonization most of the noncarbon elements such as oxygen and hydrogen are eliminated as volatile gaseous products by the pyrolytic decomposition of the starting material. The residual carbon atoms group themselves into sheets of condensed aromatic ring systems with a certain degree of planar structure. The mutual arrangement of these aromatic sheets is irregular and therefore leaves free interstices between them which may become filled with the tarry matter or the products of decomposition or at least blocked partially by the disorganized carbon. These interstices give rise to pores, which make active carbons excellent adsorbents.

Simple coking or carbonization does not give rise to products that have high adsorption capacity because of their less developed pore structure and low surface area. This pore structure is enhanced during the activation process, which converts the carbonized raw material into a form that contains the greatest possible number of randomly distributed pores of various shapes and sizes giving rise to an extended and extremely high surface area of the product.

Figure 1 is the most general flow diagram for the manufacture of activated carbon from any raw material. In the figure sizing involves breaking down the raw material into lumps or granules of the proper size, which can be effectively handled in the subsequent operations. Reconstitution involves pulverizing the raw material and then agglomeration by extrusion, briquetting, or tableting. During carbonization the raw material is heated under a time schedule with a certain rate of heating in order to eliminate the volatiles and to form a fixed carbon mass with a rudimentary pore structure. This pore structure can be developed during activation. Activation is an oxidation reaction at elevated temperatures where the oxidizing agent is usually steam and carbon dioxide and only sometimes air. Activation and carbonization steps are sometimes carried out simultaneously using chemical activating agents such as phosphoric acid, zinc chloride, and sulfuric acid. These activating agents act as dehydrating agents as well as oxidants so that carbonization and activation take place simultaneously.

Most of the raw materials can be processed into activated carbon by the most direct and economical route as shown in Figure 1. The type of activated carbon produced depends on the type of raw material. However, the nature of the product can be varied by reconstitut-

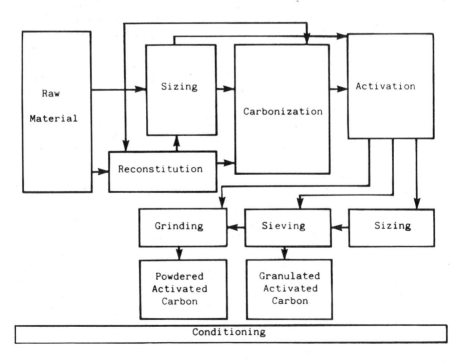

Figure 1 Basic process flow sheet.

ing the raw material or by partial carbonization and then reconstitution or by compressing the raw material during carbonization or immediately after activation, specifically when zinc chloride is the activating
agent (1). The rate of heating during carbonization, the final temperature, and the length of the activation period are some of the
other factors that change the pore volume, surface area, and mean
pore diameter of the final product.

1.2   RAW MATERIALS

Any cheap material with a high carbon content, low in inorganics,
can be used as a raw material for the production of activated carbon.
In early production procedures, preference was given to younger
fossil materials such as wood, peat, and wastes of vegetable origin,
which included fruit stones, nutshells, and sawdust. The chars
obtained from them could be activated easily and produced reasonably
high-quality activated carbons. The current trend, however, is toward an increasing use of various kinds of natural coal, which are
cheap and readily available. Of considerable interest also is the
utilization of different wastes such as waste lignin, sulfite liquors,
and wastes from the processing of petroleum and lubricating oil industries. The following criteria are considered when chosing a carbonaceous raw material:

1. Potential for obtaining high-quality activated carbon.
2. Presence of minimum inorganics.
3. Volume and cost of the raw material.
4. Storage life of the raw material.
5. Workability of the raw material.

In the case of raw materials from vegetable or fossil origin, the
amount of cinder should not exceed 3% and should be preferably 2%
for coal and wood char and between 1 and 2% for peat. In practice
five different types of carbonaceous materials are being used for
industrial-scale production of activated carbons. These raw materials
in the order of their importance and in terms of activated carbon production capacity are (2) the following:

Wood	130,000 tons/yr
Coal	100,000
Lignite	50,000
Coconut shell	35,000
Peat	35,000
Others	10,000

In the case of wood as a raw material, pine represents by far the largest production of activated carbons, about 70,000 tons/yr, while all other woods contribute up to 60,000 tons/yr. The tender and hard wood together make about 50,000 tons in the form of sawdust. The coals used for the production of activated carbons are essentially more than 90% of bituminous and subbituminous variety.

The properties of some of the raw materials and the most general uses of the activated carbon obtained from them are seen in Table 1. The property requirements of an activated carbon, and hence of the raw material, are also changing with their new applications. Low inorganics are essential to keep ash content low in the final product because the ash content in terms of percentage of the final product can increase many times after activation. In addition, high density and sufficient volatile content of the raw material are of considerable importance. High density contributes to enhanced structural strength of the carbon so that it can withstand excessive particle crumbling during use. Volatile content together with density considerably modify the manufacturing processes. Low-density materials like wood and lignin, which also contain high volatile content, produce activated carbons with large pore volume but with low density. These carbons, therefore, are generally used for some liquid phase applications and are not very suitable for vapor adsorption applications. However, the quality of these carbons can be improved by carrying on the processing in such a way that there is less loss of carbon during carbonization and by densification of the carbon by reconstitution or compression during carbonization. Coconut shells, fruit pits, and other nut shells which have higher densities than wood and have high volatile contents produce hard, granular carbons with large micropore volumes and are suitable for vapor as well as solution phase applications. Lignite, although like wood, also produces hard carbons but with small micropore volume. These carbons are generally preferable for water treatment. Soft coals have to be reconstituted before carbonization. The carbons obtained from soft coals have density and hardness between carbons obtained from coconut shells and lignite and therefore can be used both for vapor phase and liquid phase applications. Petroleum cokes produce activated carbons similar to those obtained from soft coals, whereas semihard and hard coals produce carbons similar to coconut shell carbons.

1.3   CARBONIZATION

Carbonization involves thermal decomposition of the carbonaceous material, eliminating noncarbon species and producing a fixed carbon mass and a rudimentary pore structure. The process is usually carried out in rotary kilns or multiple hearth furnaces at temperatures below 800°C in a continuous stream of an inert gas. The char produced in this manner is sometimes further calcinated at 1000°C in the absence of any

Table 1 Properties of Some Raw Materials Used in the Manufacture of Activated Carbon

Raw material	Carbon (%)	Volatiles (%)	Density (kg/L)	Ash (%)	Texture of activated carbon	Application of activated carbon
Soft wood	40–45	55–60	0.4–0.5	0.3–1.1	Soft, large pore volume	Aqueous phase adsorption
Hard wood	40–42	55–60	0.55–0.80	0.3–1.2	Soft, large pore volume	Aqueous phase adsorption
Lignin	35–40	58–60	0.3–0.4	—	Soft, large pore volume	Aqueous phase adsorption
Nutshells	40–45	55–60	1.4	0.5–0.6	Hard, large micropore volume	Vapor phase adsorption
Lignite	55–70	25–40	1.00–1.35	5–6	Hard, small pore volume	Wastewater treatment
Soft coal	65–80	20–30	1.25–1.50	2–12	Medium hard, medium micropore volume	Liquid and vapor phase adsorption
Petroleum coke	70–85	15–20	1.35	0.5–0.7	Medium hard, medium pore volume	Wastewater treatment
Semihard coal	70–75	10–15	1.45	5–15	Hard, large pore volume	Gas vapor adsorption
Hard coal	85–95	5–10	1.5–1.8	2–15	Hard, large pore volume	Gas vapor adsorption

gas. The important parameters that determine the quality and the yield of the carbonized product are the rate of heating, the final temperature, the soaking time at the final temperature, and the nature and physical state of the raw material. Low heating rate during purolysis results in lower volatilization and higher char yield (3) because of increased dehydration and better stabilization of the polymeric components. However, the char microporosity was found to be independent of the precursor composition and the pyrolyzing heating rate. The basic microstructure was formed by 500°C, although some of these pores were blocked by the pyrolysis products and could be available only when high-temperature treatment was given (3).

These factors also had a marked influence on activation and on the quality of the final product (4). When the chars were prepared at temperatures lower than the activation temperature they underwent further pyrolytic decomposition during activation, resulting in weight loss independent of the activating gas. The oxidative reactivity of the activating gas depended largely on the heating rate below 500°C, duration of exposure to temperature near 900°C during pyrolysis, and the nature of the oxidizing atmosphere. Thus the low-temperature chars gasify at a much faster rate in the initial stages of the activation process. When the weight loss was above 20−30% the low-temperature chars gasified almost at the same rate as the high-temperature chars. However, in the range of gasification chars prepared at low heating rates gasified more slowly.

In the case of wood carbonization (5), the material is being dried at temperatures up to 170°C. The partial degradation of the material begins above this temperature with the evolution of CO_2, CO, and $C_2H_4O_2$. When the temperature reaches 270−280°C the exothermal decomposition takes over and releases a considerable amount of tar, methanol, and other substances. The carbonization is almost complete in the temperature range 400−600°C. The carbon content of the product, during this process, attains the usual value of about 80%. In this the carbonization is generally carried out at a rate sufficiently fast to minimize the contact between the carbonized product and the volatile products. The rate of pyrolysis is significantly influenced by the moisture content and the temperature of carbonization.

The carbonization involves two important stages that markedly determine the properties of the final product. The first stage is the softening period, during which the temperature control has an important bearing on the type of char obtained. After the softening period the char begins to harden and shrink. The shrinkage of the char also plays a role in the development of porosity in the char. In the case of soft coal, the temperature rise during the softening stage should be very slow so that the gases can escape through the pores in the granules without a collapse or deformation. However, in the case of wood, lignin, coconut, and petroleum coke the softening stage does not produce any particular problems but the low rate of heating can result in denser and harder chars. In the case of low-density materials such as wood, compression can be applied to the chars (1) during the softening stage to obtain an activated carbon with micropore volume comparable to that of coconut carbon (Table 2). However, the low rate of heating can promote shrinkage, which reduces pore volume. In the case of wood char, even the low heating rate gives sufficient pore volume, which can later be developed during the activation stage. But the problem is more acute in the case of semihard and hard coals where the shrinkage can reduce the pore volume tremendously. Therefore, in these cases it is advantageous to bypass the slow carbonization step and activate the chars immediately to enlarge the pores before shrinkage occurs.

1.4   ACTIVATION

The objective of the activation process is to enhance the volume and to enlarge the diameters of the pores which were created during the carbonization process and to create some new porosity. The structure of the pores and their pore size distribution are largely predetermined by the nature of the raw material and the history of its carbonization. As mentioned earlier, the activation removes disorganized carbon, exposing the aromatic sheets to the action of activation agents in the first phase and leads to the development of a microporous structure. In the latter phases of the reaction the significant effect is the widening of the existing pores or the formation of the large-sized pores by the complete burnout of the walls between the adjacent pores. This

Table 2   Physical and Adsorptive Properties of Wood-Based Activated before and after Compression and Extrusion

Raw material	Bulk density (kg/L)	Total pore volume (cm^3/g)	CCl$_4$ adsorption capacity (%)
Wood	0.27	1.13	36.5
Compressed wood	0.47	0.81	81
Extruded wood	0.48	0.77	88
Coconut shell	0.48	0.75	72

Source: Juhola (1).

results in an increase in the transitional pores and macroporosity, whereas the volume of the micropores decreases. Thus the extent of burnoff of the carbon material is a measure of the degree of activation. According to Dubinin and Zaverina (6), a microporous active carbon is produced when the degree of burnoff is less than 50% and a macroporous active carbon when the extent of burnoff is greater than 75%. When the degree of burnoff is between 50 and 75%, the product has a mixed porous structure and contains all types of pores.

Although the exact mechanism of the activation process is not completely understood, it can be visualized as an interaction between the activating agent and the carbon atoms which form the structure of the intermediate carbonized product. These carbon atoms differ from each other in their reactivity depending on their spatial arrangement. The carbon atoms which are located at the edges and periphery of the aromatic sheets or those located at defect positions and dislocations or discontinuities are associated with unpaired electrons or have residual valencies and are rich in potential energy. Consequently, these carbon atoms are more reactive and have a tendency to form surface compounds by oxidation. These later break down and peel off the oxidized carbon from the surface as gaseous oxides leaving behind new unsaturated carbon atoms for further reaction with the activating agent.

The exact procedures for the activation of carbons are secrets closely guarded by the manufacturers. Nevertheless, the methods most commonly employed are broadly divided into two main types: chemical activation and physical activation. In the chemical activation process, the carbonization and activation are carried out in a single step by carrying out thermal decomposition of the raw material impregnated with certain chemical agents. The physical activation involves gasification of the char in the mass of the active carbon by oxidation with water vapor or carbon dioxide in the temperature range 850–1100°C. In this case the carbonization and activation are carried out in two different steps.

The activated carbons are commonly characterized by the mode of the activation process because this is the process in the course of which one of the reactants—the carbonaceous material—is transformed into one of the products—the activated carbon. Thus the activated carbons obtained by chemical activation are called chemical carbons and those obtained by physical activation are called physical carbons.

1.4.1   Chemical Activation

Chemical activation usually is carried out when the raw material is of wood origin. The starting material is impregnated with the activating agent in the form of concentrated solution usually by mixing and kneading. This results in the degradation of the cellulosic material. The chemical impregnated material is then extruded and pyrolyzed in a

rotary kiln between 400 and 600°C in the absence of air. The pyrolyzed product is cooled and washed to remove the activating agent,
which is recycled. On calcination, the impregnated chemicals dehydrate
the raw material, which results in charring and aromatization of the
carbon skeleton and the creation of a porous structure. The most
widely used activating agents are phosphoric acid, zinc chloride, and
sulfuric acid, although potassium sulfide, potassium thiocyanate hydroxides and carbonates of alkali metals, chlorides of calcium, magnesium,
and ferric iron have also been suggested (5). The common feature
of these activating agents is that they are dehydrating agents which
influence the pyrolytic decomposition and inhibit the formation of tar.
They also decrease the formation of acetic acid, methanol, etc., and
enhance the yield of carbon.

Chemical activation is usually carried out at temperatures between
400 and 800°C. When zinc chloride is the activating agent, the optimum temperature is about 600−700°C (7). These temperatures are
lower than needed in the physical activation process and therefore the
development of a porous structure is better in the case of the chemical
activation. The pore size distribution in the final carbon product is
determined largely by the degree of impregnation. The larger the
degree of impregnation, the larger is the pore diameter of the carbon.
However, activation by zinc chloride is being abandened.

In the case of sawdust, the activating agent most commonly used
is phosphoric acid. The dried sawdust is mixed with a concentrated
solution of phosphoric acid into a paste, which is then calcined between
350 and 500°C (Fig. 2). The calcined product is cooled by immersion
in a dilute phosphoric acid solution where the dehydrated acid gets
rehydrated and dissolved. The phosphoric acid is recovered by filtration and washing with water. The active carbon is dried and ground
finely. The phosphoric acid recovered is concentrated and recycled.
The characteristics of the final product very with:

Degree of impregnation, i. e., ratio of P_2O_5 to wood.
Temperature of heating of wood−acid mixture.
Temperature at which the wood−acid mixture is kept in the oven.
Composition of the combustion gas used for heating.

The chemical carbons obtained in the powder form are products of
high activity and adsorption capacity and are used for the adsorption
of larger molecules (decolorization). Kadlec et al. (8) observed that
the shapes of the pores in activated carbons were different when obtained by chemical and physical activation methods. The pores were
usually bottle shaped in the case of chemical activation and cone shaped in physical activation. The bottle-shaped pores in the former were
attributed to the fact that during the course of chemical activation at
temperatures around 500°C, the carbonized carbonaceous material was

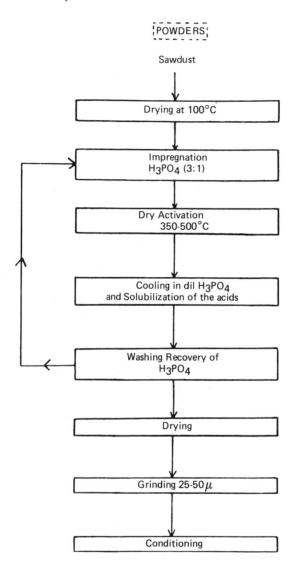

Figure 2 Process flow sheet for sawdust. [From Caron (2).]

in a plastic state. The gases released on thermal decomposition created hollows in the plastic matter from which they escape through small
passages. This may be the cause of creation of bottle-shaped pores.
During activation of carbonized wood raw materials by water vapor at
850—950°C, there is a water vapor concentration gradient between the
entrance to and the center of the pores. The oxidation process, which
is the cause of the activation, occurs at a higher rate in the neighbor-
hood of the pore entrances than in the pore centers. Thus in this
case pores of conical shape with wide entrances are more likely to be
created.

Several types of raw material and chemical agents are now being
mentioned for the production of chemical carbons. Wennerberg and
O'Grady (9) prepared active carbons of high surface area from coal,
coke, petroleum coke, and their mixtures by impregnating them with
hydrated potassium hydroxide. The impregnating material was first
calcined between 300 and 450°C and then further heated in another
calcinator between 700 and 850°C for times up to 4 hr. The resultant
product after cooling in inert atmosphere and washing with water to
remove the impregnant was found to be highly microporous and had a
cagelike structure. The surface area of carbons obtained by this
method was of the order of 3000—4000 m^2/g for coke and 1800—3000
m^2/g for coal carbons. Schafer (10) prepared active carbons from
brown coal treated with hydrochloric acid to remove cations and then
with an aqueous solution of potassium chloride at pH 8.3 to convert
coal into potassium coal. The pH was controlled by the addition of
potassium hydroxide or ammonia. The potassium coal was made into
pellets and then pyrolyzed in a stream of nitrogen by raising the temperature slowly to 900°C and then maintaining it at 900°C overnight.
The active carbon obtained after washing with hot water to leach out
the potassium salt had a surface area of 1100—1500 m^2/g. The yield
of the carbon was 43%. Nishino et al. (11) observed that when coking
coals mixed with potassium salts such as potassium hydroxide, potassium carbonate, potassium bicarbonate, or potassium sulfate were subjected to carbonization at about 700°C for 4—5 hr, they produced a
coke which on further activation with steam at 800—1000°C yielded an
activated carbon. The adsorption capacity of the final product was
a function of the grain size of the coal as well as the amount of potassium salt added (Table 3).

Ehrburger and co-workers (12,13) carried out the pyrolysis of
coals in the presence of alkaline compounds such as potassium hydroxide and sodium hydroxide and examined the influence of alkaline additives on the carbonization mechanism as well as investigating the physicochemical reactions that result in the creation of microporosity. The
micropore volume of the carbons obtained by these workers is given
in Table 4. It was found that the micropore volume first decreased
upon pyrolysis of coal with small additions of the alkalis (up to 10%

Table 3 Adsorption Capacities of Activated Carbon Prepared from Coal by Impregnation with Potassium Salts

Grain size of coal (µm)	Amount of potassium added (%)	Benzene adsorption capacity (%)	Acetone adsorption capacity (%)
1600	3	14.9	13.0
800—1600	3	21.8	18.8
600—800	3	29.3	25.0
250—600	3	35.1	29.2
600	2.0	35.3	28.1
600	1.2	36.3	30.2
600	0.8	32.0	27.3
600	0.4	22.0	18.1
600	0.2	15.8	13.3
600	0.0	11.2	10.0

Source: Ehrburger et al. (12).

Table 4 Micropore Volume of Carbons Obtained by the Carbonization of Coal in the Presence of Alkali Hydroxide

Hydroxide content (wt. %)	Micropore volume (cm^3/g)	
	Potassium hydroxide	Sodium hydroxide
0	0.169	0.169
10	0.078	0.049
20	0.177	0.067
30	0.386	0.142
70	0.627	—

Source: Ehrburger et al. (12).

by weight) but increased appreciably as the amount of alkali additive was enhanced. The increase in micropore volume was considerably higher in the case of potassium hydroxide. At potassium hydroxide content of 70% by weight of the coal, a highly activated carbon with a surface area of the order of 1600 m^2/g and a micropore volume of 0.627 cm^3/g was obtained. Enthalpies of immersion using liquids of different molecular dimensions showed that the widening of the micropores occurred simultaneously with increase in the micropore volume, indicating that carbonization in the presence of potassium hydroxide also caused activation of the carbon.

Kwok and Miller (14) suggested a method for preparing chemically activated carbon from petroleum coke by treating it with a polyphosphoric acid which contained an equivalent P_2O_5 content of at least 74% by weight. The phosphoric acid—treated coke was activated in steam between 700 and 900°C to a weight loss of about 30%. The surface area of the activated carbon so obtained was about 700 m^2/g. The particle size of the coke was important in determining the properties of the final carbon product. Cokes with particle size between 4 and 8 mesh were preferable. Das (15) described a method for the production of activated carbon from bituminous coal which was first leached with a solution of HF (4%) and HNO_3 (18%) and remaining H_2O at 85%. Oxygen was bubbled through the leach solution during the leaching process to facilitate removal of impurities. The coal was separated from the leach solution by filtration and then dried at 100°C until the coal contained a small amount of the leaching solution. This leaching solution caused swelling of the coal and promoted the development of pores. The swollen coal material was calcined at about 600°C to obtain an activated carbon with surface area of 280 m^2/g. Wood materials such as sawdust were also converted into active carbons by first carbonizing them in HCl or HBr vapor between 100 and 700°C and then activating the carbonized material with steam or CO_2 in the temperature range 700—900°C (16).

1.4.2   Physical Activation

Physical activation is the process through which the carbonized product develops an extended surface area and a porous structure of molecular dimensions, as previously mentioned. This step is generally carried out at temperatures between 800 and 1100°C in the presence of suitable oxidizing gases such as steam, carbon dioxide, air or any mixture of these gases. The heating is carried out by the combustion of coke oven or natural gas, when available, because it is most economical as both the required heat and the activation agent are supplied simultaneously. However, in directly fired activators, which could be rotary kilns or multiple hearth furnaces, extra steam is added to moderate the temperature. The active oxygen in the activating agent basically burns away the more reactive portions of the carbon skeleton as CO and CO_2, the extent of burnoff (gasification) depending on the nature of the gas employed and the temperature of activation. The burning out of the carbon skeleton also occurs at different rates at different parts of the exposed surface. Figure 3 is a schematic representation of the manufacture of activated carbon by physical activation.

Gasification of the carbonized material with steam and carbon dioxide occurs by the following endothermic reactions:

$$C + H_2O \longrightarrow CO + H_2 - 29 \text{ kcal} \tag{1}$$

$$C + CO_2 \longrightarrow 2CO - 39 \text{ kcal} \tag{2}$$

The reaction of steam with carbon is accompanied by the water–gas formation reaction, which is catalyzed by the carbon surface as

$$CO + H_2O \longrightarrow CO_2 + H_2 + 10 \text{ kcal} \tag{3}$$

Since the reactions of carbon with steam and with carbon dioxide are both endothermic, the activation process leads itself to accurate control of conditions in the kiln. External heating is required to drive reactions (1) and (2) and maintain the reaction temperature.

The rate of reaction with steam is retarded by the product H_2, which is strongly adsorbed on the active centers of the carbon surface and reduces the rate of activation. The activation with CO_2 is also retarded both by the H and the CO. The CO acts either by chemisorption on the active centers or by increasing the rate of the backward reaction. Rand and Marsh (17), however, observed that the presence of CO helps in making the gasification rate more uniform. These workers used a mixture of CO_2 and CO for the gasification of a polyfurfuryl alcohol carbon carbonized at 850°C and observed that the addition of CO in the gas steam (at a higher flow rate) resulted in the development of a better microporous structure as well as in decreasing the rate of gasification. In the case of the iron-catalyzed reaction as well, the development of microporosity was better when a mixture of CO and CO_2 was used. The rate of gasification was retarded only when iron catalyst was rendered inactive (Table 5).

Activation with CO_2 involves a less energetic reaction than that with steam and therefore requires a higher temperature. In actual industrial processes, the activating agent used is generally flue gas to which a certain amount of steam is added so that a combined activation with steam and CO_2 can occur.

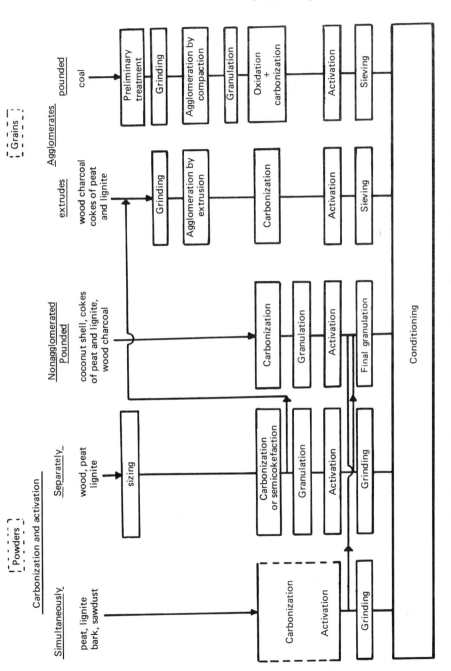

Figure 3 Physical activation manufacture process. [From Caron (2).]

Table 5 Gasification of Polyfurfuryl Alcohol Carbon (PFA, 850°) at 25% Burnoff

Particle size (B. S. mesh)	Reactant gas composition (P = cm Hg)	Gasification temperature (C°)	Rate of gasification (%/hr)	Micropore volume (cm^3/g; N$_2$ at 77 K)	Particle (mercury) density (g/cm^3)
- 30 + 44	1 atm (CO_2)	800	0.31	0.340	1.08
- 120	1 atm (CO_2)	800	0.35	0.445	1.02
- 30 + 44	$P_{CO_2} = 45.6$ $P_{CO} = 0$ $P_{N_2} = 30.4$	840	0.36	0.33	1.08
- 30 + 44	$P_{CO_2} = 45.6$ $P_{CO} = 15.2$ $P_{N_2} = 15.2$	840	0.094	0.41	1.04

- 30 + 44	P_{CO_2} = 45.6 P_{CO} = 5.6 P_{N_2} = 24.8	840	0.16	0.38	—
+ Fe - 30 + 44	P_{CO_2} = 45.6 P_{CO} = 0 P_{N_2} = 30.4	840	0.6	0.16	1.26
+ Fe - 30 + 44	P_{CO_2} = 45.6 P_{CO} = 5.6 P_{N_2} = 24.8	840	0.17	0.23	1.22

Source: Rand and Marsh (17).

In the case of activation with oxygen both the reactions

$$C + O_2 \longrightarrow CO_2 + 92.4 \text{ kcal}$$

and

$$C + O_2 \longrightarrow 2CO + 53.96 \text{ kcal}$$

are exothermic, so that there is excessive burning and the reaction is difficult to control. Since there is always some local overheating, the product obtained is not uniform. As the reaction is very aggressive, the burning is not just restricted to pores but also occurs on the surface of the grains, causing excessive weight loss. Thus the method is rarely used.

1.4.3 Mechanism of Activation

An activated carbon with high adsorption capacity can be obtained only by activating the carbonized material under such conditions that the activating agent reacts with the carbon. The activation reaction occurs in two steps. In the first step the disorganized carbon is burned out preferentially when the burnoff does not exceed 10%. This results in the opening of the blocked pores. In the second stage, the carbon of the aromatic ring system starts burning, producing active sites and wider pores. Activation with carbon dioxide promotes external oxidation and development of larger pores compared to activation with steam. The relative amount of external and internal oxidation depends on how well developed the pores are in the carbonized material. The activation of chars with no developed pore structure only results in decrease in the carbon granule size.

Kalback et al. (18) reacted cylinders of a very pure graphitized porous carbon with carbon dioxide at 1030°C to 10–30% burnoff and measured the pore size distributions as a function of the burnoff. These workers observed that the major part of early pore development occurred by preferential burning of the single aromatic sheets, which constitute the active carbon structure, and yielded pores of approximately the same size as the original carbon sheets. The pore development in this region reached a maximum and then ceased probably because the exposed properly oriented sheets were exhausted. This was followed by the burning away of the walls of the layer planes, resulting into the formation of larger pores. Marsh and Rand (19), while reacting a polyfurfuryl alcohol carbon with carbon dioxide, observed that the activation caused opening of previously inaccessible pores and widening of existing pores.

McEnancy and Dovaston (20) prepared a series of microporous carbons from cellulose triacetate by heat treatment in the temperature

range 1230–2275 K and activated them with carbon dioxide to 30%
burnoff. The changes in porosity were measured by nitrogen adsorption and mercury density measurements. The activation at lower temperatures predominated in the development of mesopores and macropores.
Heat treatment in the temperature range 1500–1700 K converted open
porosity into closed porosity.

Tourkow et al. (21) studied the activation of two brown coals carbonized at 900°C with water vapor, carbon dioxide, and oxygen and
observed that each activation produced a different distribution of
porosity. All the three activating agents at low weight loss (burnoff)
produced solely micropores and their volume was the highest with oxygen activation. With water vapor activation there was an indication
that the activation resulted in the development of mesoporosity to a
higher extent than with either carbon dioxide or oxygen.

At higher burnoffs the differences in porosity created by the different activating agents become more pronounced. Activation with
water vapor resulted in a progressive development and widening of
all sized pores until at a burnoff of 70%, the activated product contained a well-developed porous system with a wide pore size distribution in which almost all pore sizes were represented. Activation between 50 and 70% burnoff caused an increase in the total adsorption
volume from 0.6 to 0.83 cm^3/g. But as this was associated mainly
with widening of the pores, the surface area was not affected and for
both burnoffs it was almost the same (920 m^2/g).

Activation with carbon dioxide developed mainly microporosity over
the entire range of burnoffs. The micropores accounted for about 73%
of the total adsorption pore volume and for over 90% of the total surface area. The micropores contributed only 33% toward total pore
volume and 63% toward surface area in the case of the steam-activated
carbons (Fig. 4). Thus activation with carbon dioxide produces a
more uniform porosity. The activated carbons produced by carbon
dioxide activation had lower total pore volume (0.49 cm^3/g) than those
of the corresponding samples obtained by activation with steam, but
the effective surface in both the cases was almost the same ($\cong$ 900
m^2/g). This was due to the contribution of micropores to surface area.

The activation with oxygen showed a different behavior. A very
low burnoff resulted in the development of strong microporosity, which
changed very little with further burnoff (Fig. 4). The micropore volume of the sample increased from 0.002 cm^3/g for the nonactivated
carbon to 0.2 cm^3/g when the burnoff was as small as 1%. However,
it increased only very slightly to 0.23 cm^3/g and 0.27 cm^3/g when
the activation was carried to 8 and 25% burnoff. On increasing the
burnoff further to 70%, there was a slight decrease in the pore volume. Such a small decrease in microporosity at higher burnoffs did
occur in the case of activation with water vapor, but the decrease in
microporosity was compensated by the increase in the macroporosity,

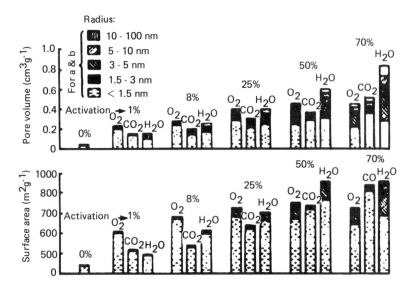

Figure 4 Pore distribution according to activation process. [From Tourkow et al. (21).]

which clearly indicated that micropores were being converted into meso-pores and macropores during activation by water vapor. However, no such decrease in microporosity was observed at higher burnoffs in the case of activation by carbon dioxide. Thus, in the case of activation by oxygen, the activation occurs only in the initial stages of the pro-cess. In the latter stages of activation, the micropores are blocked by the surface oxygen structures formed at their entrances causing the interior of the micropores to be inaccessible for further activa-tion. In addition, burning of the outer parts can take place. Thus the oxygen-activated carbon at 70% burnoff has the lowest total adsorp-tion volume (0.45 cm^3/g) of these activated carbons and because of the presence of very fine micropores also has the lowest effective sur-face area ($\cong 650$ m^2/g).

Caron (2) carried out a detailed study of the activation of car-bonized pure wood char with water vapor at 950°C following the acti-vation as a function of weight loss (degree of gasification) by mea-suring the BET (N_2) surface area, the benzene index, the methylene blue index, and the molasses index. The variation of these para-meters per unit mass of the active carbon obtained and of the original carbonized material used as a function of weight loss is shown in Fig-ures 5 and 6. It is seen that the adsorption capacity of the activat-ed carbon for all molecules increases as the weight loss increases. But the increase in the adsorption capacity of the larger molecules is much faster than the increase in the adsorption capacity of smaller molecules. For example, when the weight loss increased from 10 to 80%, the BET area increased in the ratio of 1:2, the benzene index in the ratio 1:2.5, the methylene blue index in the ratio 1:6, and the molasses index increased in the ratio of 1:11 (Fig. 5). As the molecules adsorbed in the determination of these parameters have different molecular dimensions (nitrogen 3−4 Å, benzene 5−6 Å, methylene blue 8−9 A, and molasses 12−20 Å), the formation of different sized pores as a result of the activation is evident.

Figure 6 presents the values of these parameters calculated for unit mass of the original carbon material as a function of weight loss. This figure shows the changes in the efficiency of utilization of the initial porous potential of the carbon, which increases the degree of burnoff, passes through a maximum and then decreases. But the

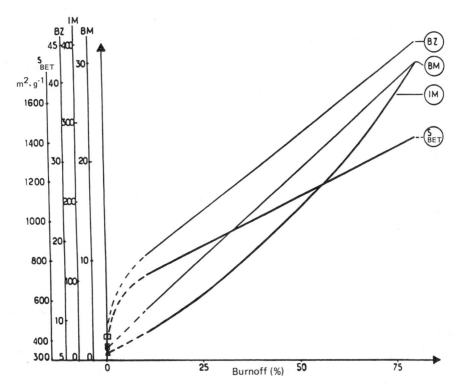

Figure 5 Activation process of a pine wood char with H_{20} at 950°C. (Surface area S, Benezene index BZ, methylene blue index BM, and molasses index IM as a function of the weight loss expressed for the obtained active carbon.) [From Caron (12).]

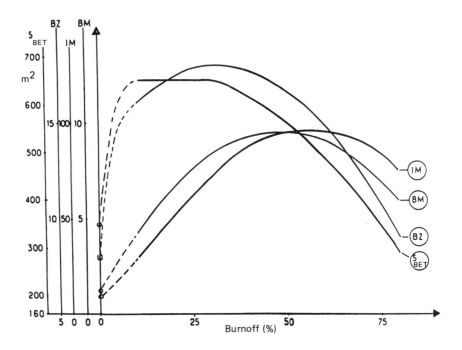

Figure 6 Activation of pine of wood char with H_2O at 950°C. (All indexes are expressed as a function of the weight loss of the coke starting material.) [From Caron (2).]

decrease is much faster when the molecule to be adsorbed is smaller. For example, the decrease in the BET surface area (where nitrogen is the adsorbate molecule) or the benzene index (where benzene is the adsorbate) is much faster compared to methylene blue index or the molasses index. These results show that in the initial stages of the activation process, the peripheral and more accessible single aromatic sheets are burned off, producing predominately micropores. The formation of these micropores is limited by the availability of these sheets, which continues to decrease. As the burnoff is continued, the larger sheets start burning out as the walls between the adjacent pores converting more and more of the micropores into macropores. Since the percentage contribution of the macropores to surface area is small, the surface area shows a steep decrease at higher degrees of burnoff.

Thus the most important factor in obtaining, from the same carbonized material, different qualities of activated carbon is the level of burnoff, i. e., the duration of activation and the time it stays in the oven. Thus for industrial-scale production of activated carbons a compromise has to be found between the capacity of production and the quality of the material produced.

1.5 CLASSIFICATION OF ACTIVATED CARBONS

Activated carbons are complex products which are difficult to classify on the basis of their behavior, surface charactertistics, and properties. For example, the BET surface area is in no manner a representative of the quality of the activated carbon because it does not indicate the surface accessible to larger molecules, which may be quite different. The only parameter that can give an idea of the quality of the carbon is its adsorption capacity, and that only in the case of a particular application. Activated carbons are therefore classified on the basis of their particle size and particle shapes into powdered, granulated, spherical, or pelleted activated carbons.

1.5.1 Powdered Activated Carbons

Powdered activated carbons have a fine granulometry less than 100 μm with average diameter between 15 and 25 μm. Thus they present a large external surface and a small diffusion distance. The rate of adsorption is very high and the problems related to mass transfer are very low. They are thus preferably used for adsorption from solution phase because of their low diffusion rates. Their application is simple. The carbon is added to the solution directly, agitated, left in contact for a short time (5–30 min), and then separated by filtration. To reduce the consumption of the carbon, the treatment is sometimes carried out in a countercurrent principle but the process involves heavy investment. This group of carbons includes decolorization and medicinal carbons. These carbons are generally prepared by chemical activation methods from sawdust.

1.5.2 Granulated Activated Carbons

Granulated activated carbons have a relatively larger size of carbon particles in the granules compared to carbon powders and consequently present smaller external surface. Diffusion of the adsorbate thus is an important factor. These carbons are therefore preferred for the adsorption of gases and vapors as their rates of diffusion are faster. Adsorption from the gas phase is carried out almost invariably under dynamic conditions by passing the gaseous mixture and the carrier gas through a bed of the carbon. The size of the carbon granules is an important factor. It should not be too small to prevent excessive pressure drop along the bed and carrying away of the carbon particles with the gaseous steam. In fact, the size of the grains is chosen depending upon the height of the bed to be used. The greater the height, the larger is the size of the granules. Granulated carbons are also now being used for water treatment and sometimes also for decolorization and separation of components of flow systems.

Granulated activated carbons can also be prepared by the physical activation methods using a variety of carbonaceous materials such as bituminous, subbituminous and lignite coals (22,23) from petroleum, heavy oil, or heavy residue (24), synthetic and natural rubber (25), and by the pyrolysis of waste rubber tires (26).

1.5.3 Spherical Activated Carbons

Katori et al. (27) and Nagai et al. (28) developed a process for the preparation of spherical carbons from pitch. The pitch is melted in the presence of naphthalene or tetralin and converted into spheres. These spheres are contacted with solvent naphtha, which extracts naphthalene and creates a porous structure. These porous spheres are then heated at temperatures between 100 and 400°C in the presence of an oxidizing gas containing about 30% of oxygen by weight. The pitch spheres chemisorb some oxygen, which should be around 10%. The oxidized spheres are then heated between 150 and 700°C in the presence of ammonia to introduce nitrogen into the spheres, which are then activated in steam or CO_2. The activated carbon spheres so obtained were found to have a high mechanical strength and excellent SO_2 and NO_2 adsorption capacities. Spherical activated carbons in the form of bellow spheres (29) and in the form of granulated spherical pellets (30) have also been obtained.

1.5.4 Impregnated Carbons

Carbons containing several types of inorganic impregnant such as iodine (31), silver (32), cations such as aluminum, manganese, zinc, iron, lithium, and calcium (33), and organic impregnants such as pyridines (34), ketones (35), and tertiary amines (36) have also been prepared. The iodine-impregnated carbons can be used as catalysts, in the removal of SO_2 and H_2S from a gas steam and in removing methyl iodide gas from the effluent of a reactor.

1.5.5 Polymer-Coated Carbon

Fennimore et al. (37) described a process by which a porous carbon can be coated with a biocompatible polymer to give a thin, smooth, and permeable coating without blocking the pores. The resulting carbon is useful for hemoperfusion.

REFERENCES

1. Juhola, A. J., *Kewie-Kemi*, *11*:543 (1977).
2. Caron, J., Private communication, 1985.
3. Mackay, D. M. and Roberts, P. V., *Carbon*, *20*:95 (1982).
4. Mackay, D. M. and Roberts, P. V., *Carbon*, *20*:105 (1982).
5. Smisek, M. and Cerny, S., *Active Carbon*, Elsevier, Amsterdam, 1970.
6. Dubinin, M. M. and Zaverina, E. D., *Dokl. Akad. Nauk. SSSR*, *65*:295 (1949).
7. Yamada, D., *Bull. Fac. Eng. Yokohama Natl. Univ.*, *8*:125 (1959).
8. Kadlec, O., Varhanikova, A., and Zukal, A., *Carbon*, *8*:321 (1970).
9. Weunerberg, A. N. and O'Grady, T. M., U. S. Patent 4,082,694, Apr. 4, 1978.
10. Schafer, H. N. S., U. S. Patent 4,039,473, Aug. 2, 1977.
11. Nishino, H., Kubo, H., and Ichukawa, H., U. S. Patent 2,764, 561, Oct. 9, 1973.
12. Ehrburger, P., Addoun, A., Addoun, F., and Donnet, J. B., Submitted for presentation at FUNCAT COGAS conference.
13. Kraehenbuehl, F., Stoeckli, H. F., Ehrburger, P., Addoun, A., and Donnet, J. B., *Carbon* (accepted) (1986).
14. Kwok, J. and Miller, A., U. S. Patent 3,767,592, Oct. 23, 1973.
15. Das, S. K., U. S. Patent 4,083,801, Apr. 11, 1978.
16. Shindo, L., Souna, I., and Nakanishi, Y., U. S. Patent 3,557, 020, Jun. 19, 1971.
17. Rand, B. and Marsh, H., *Carbon*, *9*:79 (1971).
18. Kalback, W. M., Brown, L. F., and Wert, R. E., *Carbon*, *8*:117 (1970).
19. Marsh, H. and Rand, B., *Carbon*, *9*:47 (1971).
20. McEnancy, B. and Dovaston, N., *Carbon*, *13*:515 (1975).
21. Tourkow, K., Siemieniewska, T., Czeckowski, F., and Gankowska, A., *Fuel*, *56*:121 (1977).
22. Johnson, B. C., Sinha, R. K., and Urbanic, J. E., U. S. Patent 4,014,817, Mar. 29, 1977.
23. Murthy, H. N., U. S. Patent 4,032,476, June 28, 1977.
24. Yokogawa, A., Mitooka, M., and Shima, K., U. S. Patent 3,940, 344, Feb. 24, 1976.
25. Devong, G. J., U. S. Patent 3,886,088, May 27, 1975.
26. Watanabe, Y. and Miyajina, T., U. S. Patent 4,002,587, June 11, 1977.
27. Katori, K., Nagai, H., and Shüki, Z., U. S. Patent 4,045,368, Aug. 30, 1977.
28. Nagai, H., Katori, K., Shüki, Z., and Amagi, Y., U. S. Patent 3,909,449, Sept. 30, 1975.
29. Kobayashi, K., Watari, S., Kato, T., Shiraishi, M., and Kawana, Y., U. S. Patent 3,891,574, June 24, 1975.

30. Voet, A. and Lamond, T. G., U. S. Patent 3,533,961, Oct. 31, 1970.
31. Sterp, K., Wirth, H., Rottinger, G., and Hohmann, V., U. S. Patent 4,075,282, Feb. 21, 1978.
32. Piccione, S. and Urbanic, J. E., U. S. Patent 3,294,572, Dec. 27, 1966.
33. Zall, D. M., U. S. Patent 3,876,451, Apr. 8, 1975.
34. Dolian, F. E. and Hormats, S., U. S. Patent 2,963,441, Dec. 6, 1960.
35. Urbanic, J. E. and Sutt, R. F., U. S. Patent 3,778,387, Dec. 11, 1973.
36. Dietz, V. R. and Blachly, C. H., U. S. Patent 4,040,802, Aug. 9, 1977.
37. Fennimore, J., Ruder, G., and Simmonite, D., U. S. Patent 4,076,892, Feb. 28, 1978.

2

Surface Chemical Structures on Active Carbons

Active carbons are mainly and almost exclusively prepared by the pyrolysis of organic compounds at temperatures lower than 1000°C. During this pyrolysis process noncarbon elements such as oxygen, hydrogen, and nitrogen are eliminated as volatile gaseous products. The residual elementary carbon stoms are grouped into stacks of flat aromatic sheets crosslinked in a random manner. The mutual arrangement of these aromatic sheets is irregular and therefore leaves free interstices between the sheets, which may become filled with the tarry matter or the products of decomposition, or at least blocked partially by disorganized carbon. These interstices give rise to pores, which make active carbons excellent adsorbents. This porous structure in active carbons is further developed and enhanced during the activation process when the spaces between the aromatic sheets are cleared of various carbonaceous compounds and nonorganized carbon.

Besides the physical structure, active carbons have a chemical structure as well. The adsorption capacity of active carbons is determined by their physical or porous structure but is also strongly influenced by the chemical structure of their surface. In graphites, which have a highly ordered crystalline structure, the adsorption capacity is determined mainly by the dispersion component due to London forces, But the random ordering of the aromatic sheets in active carbons causes a variation in the arrangement of electron clouds in the carbon skeleton and results in the creation of unpaired electrons and incompletely saturated valencies, which would undoubtedly influence the adsorption behavior. Active carbons are also almost invariably associated with appreciable amounts of heteroatoms such as oxygen and hydrogen. In addition, they may be associated with atoms of chlorine, nitrogen, and sulfur. These heteroatoms are derived from the start-

ing material and become a part of the chemical structures as a result
of imperfect carbonization, or they become chemically bonded to the
surface during activation or during subsequent treatments. These
heteroatoms are bonded at the edges of the aromatic sheets and form
surface compounds (surface complexes or surface groups), or they
can be incorporated within the carbon layers forming heterocyclic
ring systems.

Thus all carbons can chemisorb oxygen even on mere exposure
to air or oxygen (1–28) preferably at 400–500°C. The oxygen is
fixed firmly and comes off only as oxides of carbon on high-temperature evacuations (29–43). Similary, it is well known that all active
carbons have chemically bonded hydrogen (33,38,44,45), the amount
depending on the history of their formation. The hydrogen is bonded so strongly that it is not given off completely even on outgassing
at 1000°C (47). These carbons can also bind nitrogen on treatment
with ammonia, sulfur on treatment with hydrogen sulfide, carbon
sulfide, or sulfur, and halogens on treatment with the halogen in
gaseous or solution phase (46–49). These treatments give rise to
carbon–nitrogen, carbon–sulfur, and carbon–halogen surface compounds respectively. These surface compounds and their influence
on the modification of carbon surface properties are discussed in
another chapter in the book.

2.1   CARBON–OXYGEN SURFACE GROUPS

Carbon–oxygen surface structures are by far the most important
structures in influencing the surface characteristics and surface behavior of carbons. Consequently, a considerable amount of work has
been carried out and is still being carried out on them. Although
the determination of the number and nature of these surface structures began more than half a century ago, the precise nature of the
functional groups is not entirely established. The estimates obtained
by investigators using varied techniques differ considerably because
the carbon surface is very complex and difficult to reproduce. The
surface groups cannot be treated as ordinary organic compounds.
The surface groups interact differently in different environments.
They behave as combined structures presenting numerous mesomeric
forms depending on their location on the same polyaromatic frame.
Recent electron spectroscopy for chemical analysis (ESCA) studies
have shown that irreversible transformation of surface structures
occurred when classical organic chemistry methods were used to identify and estimate them. It is, however, expected that the application
of more sophisticated methods of analysis such as Fourier transform
infrared (FTIR), ESCA, nuclear magnetic resonance (NMR) spectroscopy, and radiometric tracer studies will contribute significantly to
a more precise knowledge about the surface chemical groups on carbons.

Carbons have an acid—base character. This fact has encouraged many investigators to devote their research effort to understanding the cause and the mechanism by which a carbon acquires either and acid or a base character. Several theories—e.g., the electrochemical theory of Burstein and Frumkin (50,51), the oxide theory of Shilov and his school (52,53), the chromene theory of Garten and Weiss (54), and the pyrone theory of Boehm and Voll (146)—have been proposed to explain the basic character of carbons. These theories and related work have been elaborately reviewed and critically examined in several review articles (55—57). Rather than dwelling on these theories once again, we start from what has come out of these investigations regarding the acid or base character of active carbons.

It is now well accepted that the acid or base character of a carbon is developed as a result of surface oxidation. It is also well known that the acid or base character of a carbon will depend on the history of its formation and the temperature at which it was oxidized. Thus we classify active carbons into acid carbons and basic carbons and discuss them separately. As most of the work is related to the development and studies into acidic carbons, it is reasonable to discuss them first.

2.2   ACIDIC CARBONS

Acidic carbons are defined as those carbons that show acidic behavior and adsorb appreciable amounts of bases but very little of acids. These carbons are generally obtained when carbons outgassed at high temperatures in vacuum or in inert atmosphere are exposed to oxygen between 200 and 700°C. They can also be produced by oxidizing as-received carbons with oxidants in gaseous or solution phase. The optimum temperature for the development of maximum capacity to adsorb bases has been found to be around 400°C.

2.3   CHARACTERIZATION OF SURFACE CHEMICAL STRUCTURES

Attempts have been made to identify and estimate the surface oxygen chemical structures (functional groups) using several physical, chemical, and physicochemical techniques, which include neutralization of bases, desorption of the oxide layer, potentiometric, thermometric, and radiometric titrations, direct analysis of the oxide layer by specific chemical reactions, polarography, infrared (IR) spectroscopy, and X-ray photoelectron spectroscopy. As a result of these investigations, the existence of such functional groups as carboxyls, phenols, lactones, aldehydes, ketones, quinones, hydroquinones, anhydrides, and ethereal structures has been postulated. However, these methods have not yielded comparable results and the entire amount of combined oxygen has not been accounted for.

2.3.1 Thermal Desorption

The surface oxygen structures found on as-received carbons or formed as a result of interaction with oxygen or oxidizing gases or oxidizing solutions are generally quite stable even under vacuum at temperatures below their formation temperature. In general, these structures are stable at temperatures below 200°C irrespective of the temperature at which they are formed. However, when they are heated at higher temperatures they decompose to produce CO_2 and H_2O at lower temperatures and CO and H_2 at higher temperatures.

Puri and co-workers (29,32,33) evacuated a number of sugar and coconut shell charcoals and commercial-grade carbon blacks at gradually increasing temperatures up to 1200°C and observed that the products of decomposition were CO_2, water vapor, and CO in that order. Following Puri and co-workers, Bansal et al. (38) carried out vacuum pyrolysis of a number of active carbons obtained by the carbonization of a number of polymer precursors and measured the amount of oxygen evolved as CO_2, CO, and water vapors as a function of heat treatment temperature. The total of the three oxygens (obtained as CO_2, CO, and H_2O) agreed fairly with the total oxygen obtained by ultimate analysis (Table 1). Graphic representation of the data (Fig. 1) clearly shows that the different carbons are associated with different amounts of oxygen, the amount depending on the nature of the source material and the history of its formation. In general the chars prepared at lower temperatures are associated with larger amounts of oxygen and hydrogen. This has been attributed to the limited stabilities of the carbon--oxygen surface structures. Further, for the same temperature of carbonization the chars prepared from polyfurfuryl alcohol and urea formaldehyde contain a larger amount of chemisorbed oxygen because these polymers contained oxygen as part of their chemical structure, whereas others did not.

The disposition of oxygen as CO_2, CO, and water vapor on evacuation at gradually increasing temperatures (Figs. 2–4) shows that while oxygen evolved as CO_2 is given out in the 300–700°C range, that disposed as CO starts at comparatively higher tempertaures of 500–600°C and is completed only at about 1000°C. The oxygen given out as water vapors is evolved in the 200–600°C temperature range. Since the temperature was raised gradually in steps of 50°C and since the complete evolution of the gas at any temperature was ensured before raising the temperature after each step, the possibility of evolution of CO as a result of the secondary reaction between CO_2 and carbon and between water vapor and carbon was ruled out (33). It is evident from these and similar studies on carbon blacks (33) that the

Table 1 Gases Evolved on Outgassing Various Polymer Carbons at 1200°C

Sample identification	O_2 evolved on outgassing at 1200° C (g/100 g)				Oxygen by ultimate analysis (g/100 g)	Hydrogen evolved on outgassing at 1200° C (g/100 g)			Hydrogen by ultimate analysis (g/100 g)
	CO_2	CO	H_2O	Total		H_2O	H_2	Total	
PF-140°	7.00	13.05	2.05	22.10	22.80	0.26	3.60	3.86	4.90
PF-400°	4.90	5.10	1.30	11.30	10.40	0.16	2.85	3.01	4.20
PF-600°	1.40	4.60	Tr	6.00	3.90	Tr	2.00	2.00	2.90
PF-900°	0.50	1.60	Tr	2.10	1.40	Tr	1.05	1.05	1.30
PVDC-600°	3.50	2.00	0.20	5.70	3.60	0.03	1.49	1.52	0.80
PVC-850° (vac)	Tr	Tr	Tr	Tr	Tr	Tr	Tr	Tr	Tr
PVC-850° (N_2)	Tr	Tr	Tr	Tr	Tr	Tr	Tr	Tr	Tr
PVC-850° (CO_2)	1.01	0.48	0.18	1.67	1.81	0.02	0.30	0.32	0.65
UF-400°	2.80	4.80	2.40	10.00	10.62	0.30	2.95	3.25	4.27
UF-650°	2.59	4.08	Tr	6.67	6.05	Tr	2.25	2.25	3.50
UF-850°	0.25	1.71	Tr	1.96	2.10	Tr	0.30	0.30	0.50

Key: PF = polyfuryl alcohol carbon; PVDC = polyvinylidene chloride carbon; PVC = polyvinyl chloride carbon; UF = urea formaldehyde resin carbon. The number represents the temperature of carbonization.
Source: Bansal et al. (38).

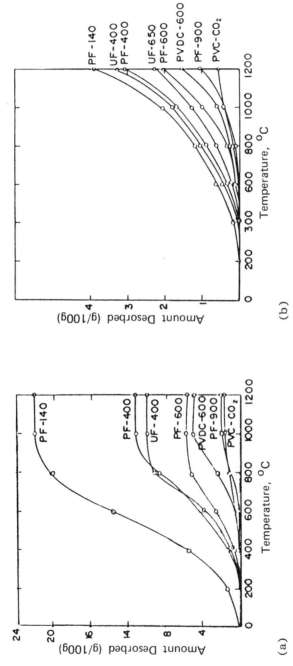

Figure 1 (a) Total oxygen evolved (as CO_2, CO, and H_2O) and (b) total hydrogen evolved (as H_2O and H_2) on outgassing polymer charcoals at different temperatures. [From Bansal et al. (38).]

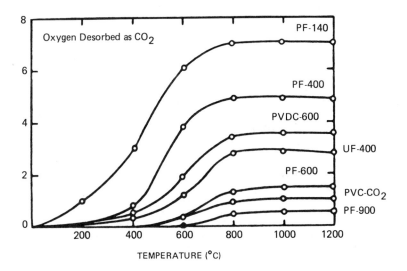

Figure 2 Oxygen evolved as CO_2 on outgassing polymer charcoals at different temperatures.

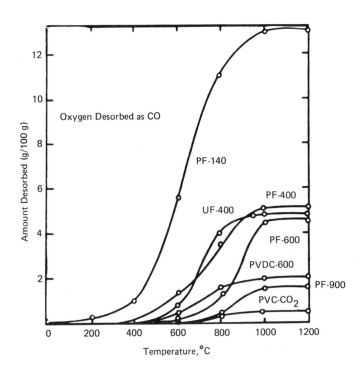

Figure 3 Oxygen evolved as CO on outgassing polymer charcoals at different temperatures. [From Bansal et al. (38).]

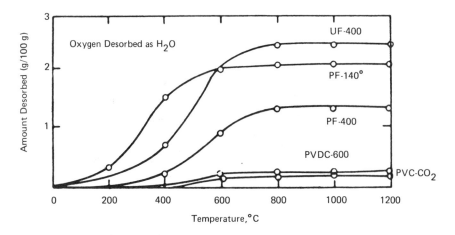

Figure 4   Oxygen evolved as H_2O on outgassing polymer charcoals at different temperatures.   [From Bansal et al. (38).]

chemisorbed oxygen constitutes different surface structures, which involve different sites associated with varying energies. The composition of the evolved gas in a particular temperature range seems to depend on the nature of the surface structurs (group) or structures decomposing in that range. Bansal et al. (40) also studied the decomposition of carbon−oxygen surface compounds formed on oxidation of ultraclean surfaces of an activated graphon sample using a mass spectrometer and observed that both CO_2 and CO were primary products obtained by the decomposition of different oxygen functional groups from different sites.

Van Driel (58) carried out thermogravimetric analysis of an active carbon and analyzed the gases evolved using a gas chromatograph. The adsorption products were mainly CO_2 and CO. The desorption of CO_2 showed two maxima at 600°C and at 700°C. Comparison of the results after oxidation of the carbon showed that the amount of CO_2 desorbed increased with oxidation. The oxidized carbon showed two maxima for CO as well, one at 750°C and the other at 900°C. These results were attributed to the existence of different surface structures with varying stabilities.

The mechanism and the nature of the gaseous species formed on thermal desorption of carbon−oxygen surface compounds were also studied by Lang and Magnier (59), using IR and gas chromatographic analysis, by Bonnetain et al. (60,61) using a chemical separation technique, by Tucker and Mulcahy (62) using a thermogravimetric technique, and by Dollimore et al. (39) using mass spectrometer. The main observations of these workers were that a major part of surface structures decomposed in the temperature range 600—800°C and almost completely at 1000°C. The amount of oxygen chemisorbed could be completely accounted for by the evolution of CO_2 and CO. The thermal desorption products separately obeyed the Elovich equation for desorption in the range 400—900°C. The activation energy for desorption increased with decreasing surface coverage, indicating that these gases were evolved by the decomposition of different surface species.

Trembley et al. (36) used linear programed thermal desorption analysis of carbon—oxygen surface compounds to measure their energies. The rate of production of CO from the decomposition of the surface compounds formed by low-temperature chemisorption of oxygen on activated graphon indicated that at least two types of surface group must exist. The desorption energies of these surface groups were a function of their coverage, indicating that the surface complex consisted of several types of functional group, which decomposed in different temperature ranges.

Matsumoto and co-workers (35,63) carried out thermal desorption of oxidized samples of vitreous carbon, diamond, and graphite at temperatures up to 950°C using mass spectrometer. The desorption spectra of the samples (Fig. 5) show different maxima for CO_2 and CO as a function of temperature. This indicates that these gases are being evolved by the decomposition of different surface structures.

Thus there is overwhelming evidence from thermal desorption studies that there are two types of surface chemical structure, one that evolves CO_2 on decomposition and the other that evolves CO. The surface chemical structure that evolve CO_2 are less stable and decompose at temperatures as low as 300°C and could be carboxylic or lactonic functions. The other chemical structures, which evolve CO, are fairly stable, and decompose only above 500°C, could be postulated as phenols or quinones.

2.3.2   Neutralization of Bases

Titration with alkalis is one of the earliest and simplest methods used to determine the nature of the acidic surface groups on carbons and graphites. However, the standard conditions under which comparable results can be obtained have been realized during the last few decades only. It is now recognized that base neutralization capacity of a carbon should be measured after degassing the sample at about 150 °C so as to free it from any physically adsorbed gases and vapors. The alkali solution used should be sufficiently strong (0.1—0.2 N) and the contact period should be long enough (24—72 hr) to obtain reproducible neutralization values. This contact time can be reduced to a few hours if the carbon and the alkali suspension are heated under reflux. These standard conditions are now being followed by many of the workers.

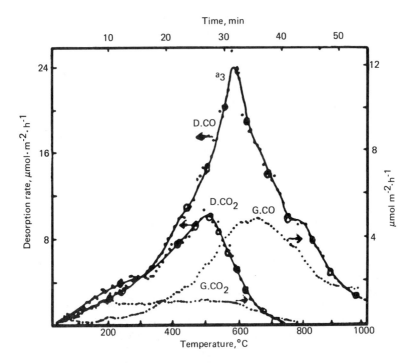

Figure 5 Desorption of chemisorbed oxygen from oxidized diamond (D) and graphite (G) as a function of temperature. [From Matsumoto and Setaka (35).]

Kruyt and De Kadt (66,67) observed that oxidation of charcoals with oxygen at 400°C developed surface acidic structures which neutralized strong alkalis and which were eliminated on heat treatment at 800°C. These groups were postulated to be carboxylic because they were decarboxylated at 800°C. The presence of carboxylic groups on the surface of ink and channel blacks was also suggested by Schweitzer and Goodrich (68), Studebaker et al. (69), and Hoffman and Ohlerich (70) on the basis of titration with alkalis. Garten and Weiss (55), on the basis of the shapes of the titration curves they obtained on sugar carbons and Villars (71) obtained on ink and carbon blacks, suggested that the acidity of both types of carbon was due to phenolic hydroxyl groups attached to the edges of the layer planes. However, later, on the basis of the reaction of carbons with diazomethane and subsequent hydrolysis, these workers (55) suggested that the surface acidic groups could also be lactones of the type present in fluorescein and phenolphthalein dyestuffs.

These lactones, which behaved as weak acids, neutralized alkalis, were methylated by diazomethane, and subsequently were hydrolyzed by a mineral acid, were termed f-lactones. The difference between the surface acidity measured by alkali neutralization and that due to lactones and phenols was compensated by postulating the existence of yet another lactone group (n-lactone), which reacted with alkalis but not with diazomethane. The presence of these two types of lactone was deduced on the basis of their IR adsorption bands at 1760 and 1710 cm^{-1}. Bruin and Van der Plas (72) attempted to identify lactones in carbon blacks by reacting them with hydrobromic acid dissolved in glacial acetic acid. There was an increase in the bromine content but without a corresponding increase in the carboxyl content as determined by titration with sodium bicarbonate (73,74) or treatment with calcium acetate (75,76). This indicated the absence of lactone groups in carbon blacks.

Rivin (77,78) combined acidimetry and vacuum pyrolysis techniques to determine the distribution of functional groups on several carbon blacks and attributed surface acidity to carboxylic, phenolic, neutral lactones and to quinone groups. These results were based on determination of total acidity by titration with lithium aluminum hydride, selective neutralization techniques using bases of different strengths (56), reduction of neutral groups to acidic groups, and analysis of vacuum pyrolysis products of the surface oxide layer.

Puri and co-workers (29,32,79) examined a large number of sugar and coconut charcoals before and after outgassing and extensive oxidation treatments in oxygen (80) as well as in oxidizing solutions such as potassium persulfate, nitrate, iodate (81), aqueous chlorine (64), hydrogen peroxide (82), and tried to correlate the base neutralization capacity of the charcoal with the oxygen evolved as CO_2 on evacuation at 1200°C. It was found (Tables 2,3) that in each case the amount of alkali neutralized was close to the amount of CO_2 evolved on evacuation (they termed this CO_2 complex). As the amount of the CO_2 complex decreased on outgassing or increased on oxidation, the base-neutralized capacity of the charcoal decreased correspondingly. When the entire amount of the complex was eliminated on outgassing around 750°C, the carbon lost almost completely the property to neutralize alkali even though it still contained appreciable amounts of combined oxygen (Tables 2,3). This work was later extended using a number of commercial carbon blacks representing furnance, channel, and color blacks by Puri and Bansal (83). Their surface acidity, as determined by neutralization of sodium and barium hydroxides, was found to be close to each other as well as to the amount of CO_2 complex (Tables 2,3) contained in each sample. The base-neutralization capacity of a carbon black decreased gradually with outgassing of the samples at gradually increasing temperatures and the decrease at any temperature corresponded with the decrease in

Table 2 Relationship between Amount of Alkali Neutralized and CO_2 Evolved on Evacuation at 1200° for Various Charcoals

Carbon sample	CO_2 evolved on evacuation at 1200° C (mEq/100 g)	Sodium hydroxide neutralized (mEq/100 g)	Barium hydroxide neutralized (mEq/100 g)
Sugar charcoal			
Original	669	662	664
300°-outgassed	388	381	377
400°-outgassed	231	222	228
500°-outgassed	150	153	147
600°-outgassed	75	74	72
750°-outgassed	0	0	0
1000°-outgassed	0	0	0
Oxidized with H_2O_2	819	810	815
Oxidized with $K_2S_2O_8$	975	970	965
Coconut charcoal			
Original	394	395	400
300°-outgassed	181	165	170
500°-outgassed	38	35	20
600°-outgassed	15	12	16
750°-outgassed	0	0	0
1000°-outgassed	0	0	0
Oxidized with H_2O_2	591	584	582
Oxidized with $K_2S_2O_8$	623	627	619
Wood charcoal			
Original	531	532	535
300°-outgassed	400	399	405
500°-outgassed	50	50	49
600°-outgassed	22	17	25
750°-outgassed	0	0	0
1000°-outgassed	0	0	0
Cotton stalk charcoal			
Original	325	335	342
300°-outgassed	250	245	251
500°-outgassed	62	65	63
600°-outgassed	15	21	16
750°-outgassed	0	0	0
1000°-outgassed	0	0	0

Source: Puri et al. (29).

Table 3 Alkalis Neutralized by Various Samples of Carbon Blacks in Relation to CO_2 Evolved on Outgassing Them

Trade name	Type	Barium hydroxide neutralized (mEq/100 g)		Sodium hydroxide neutralized (mEq/100 g)		CO_2 evolved on outgassing at 1200° (mEq/100 g)
		By shaking for 60 hr	By heating for 6 hr	By shaking for 60 hr	By heating for 6 hr	
Pelletex	Furnace	2—7	3—3	5—1	4—3	3—2
Kosmos-40	Furnace	5—3	5—8	5—2	5—1	5—5
Statex-B	Furnace	7—5	6—3	6—5	7—6	7—2
Philblack-A	Furnace	10—6	9—5	9—1	9—6	11—7
Philblack-O	Furnace	15—9	15—8	12—8	14—4	13—1
Philblack-I	Furnace	21—2	20—7	19—5	19—2	21—8
Philblack-E	Furnace	23—9	25—2	22—7	26—9	25—1
Vulcan-SC	Furnace	26—6	28—4	23—7	27—9	26—8
Sterling-V	Furnace	17—1	18—9	16—9	18—5	18—8
Spheron-9	Channel	29—9	28—2	28—5	28—9	33—5
Spheron-6	Channel	31—9	29—4	26—6	28—9	31—2
Spheron-4	Channel	31—5	33—5	31—1	32—9	34—2
Spheron-C	Channel	37—7	37—8	38—8	35—7	35—9
ELF-O	Colo	70—5	69—3	68—2	72—2	73—5
Mogul-A	Colo	113—2	115—2	114—8	115—5	118—4
Mogul	Colo	130—3	138—5	129—6	134—7	137—8
CK-4	German	28—2	31—2	29—9	30—5	30—3

Source: Puri and Bansal (83).

the CO_2 complex at that temperature. The samples outgassed at 700–800°C, although retaining appreciable amounts of chemisorbed oxygen, lost the capacity to neutralize bases almost completely.

The charcoals and carbon blacks were oxidized with nitric acid to a weight loss of about 25–30%. This resulted in the fixation of considerable amounts of oxygen, most of which was desorbed as CO_2 on evacuation. The base neutralization capacity of the oxidized sample increased in each case in proportion to the amount of CO_2 complex formed. The removal of the CO_2 complex on evacuation caused a corresponding decrease in the base adsorption capacity.

Puri and Mahajan (84) reacted a number of charcoals and carbon blacks with dry ammonia at room temperature and observed that the amount adsorbed was close to the amount of CO_2 complex on the carbon sample. Anderson and Emmett (30) studied the adsorption of several amines on charcoals and found that the adsorption depended on the amount of the carbon–oxygen surface complex and its ability to form hydrogen bonds. Puri, Talwa, and Sandle (85), while studying the adsorption of n-butyl, dimethyl, diethyl, and triethyl amines from aqueous solutions by a series of charcoals and carbon blacks, observed that the process involved neutralization reactions with the acidic CO_2 complex sites. The amount adsorbed was close to the amount of CO_2 complex in the case of n-butylamine but appreciably lower in secondary and tertiary amines depending on the length and the number of chains in the amine. This was attributed to the orientation and the steric effect, which could block some of the CO_2 complex sites. The number of sites thus rendered unavailable was evidently greater in tertiary than in secondary amines, and in diamines the effect was greater in diethylamine than in dimethylamine.

Puri and co-workers therefore hold the view that in charcoals as well as in carbon blacks, the same surface group which evolves CO_2 on evacuation is involved in the neutralization of alkalis. This cannot be a carboxylic acid since no significant correlation between surface acidity and active hydrogen was found. This cannot be a lactonic group, as suggested by Garten and Weiss (55), because its reaction with alkalis and decomposition as represented by Garten and Weiss did not show equivalence between alkali neutralized and CO_2 evolved. However, these workers did not rule out the possibility of the existance of certain types of lactone structures which would hydrolyze to give a carboxylic group and a phenolic hydroxyl group, each capable of stoichiometric ionic adsorption. Alternatively these workers suggested that the acidic complex consists of two oxygen atoms attached to an "active" carbon atom lying at the edges of the basal planes, or with other active sites on the surface, behaving as a layer of frozen CO_2. The ionization of such a complex in water, analogous to ionization of acid complexes, as reported in the literature (16), may be represented as

$$)C\diagdown\begin{matrix}O\\ \text{|}\\ O\end{matrix} + 2H_2O \longrightarrow)C\diagdown\begin{matrix}O\text{----}OH^-\\ \\ O\text{----}OH^-\end{matrix} \Bigg| 2H^+$$

$$)CO_2 + H_2O \longrightarrow)CO_3^{2-} \ \Big| \ 2H^+$$

The hydrogen ions are directed toward the liquid phase while the equivalent negative charge is left on the carbon surface.

Boehm (56) observed neutralization values about twice as high with barium hydroxide as with sodium hydroxide, in contrast to Puri and co-workers, who observed no significant difference between the two values determined under similar conditions. According to Boehm (56), neutralization of barium hydroxide can occur in two ways:

$$\begin{matrix}\text{——— COOH}\\ \\ \text{——— COOH}\end{matrix} + Ba(OH)_2 \longrightarrow \begin{matrix}\text{——— } CO\bar{O}\\ \\ \text{——— } CO\bar{O}\end{matrix} + Ba^{2+} + 2H_2O$$

$$\begin{matrix}\text{——— COOH}\\ \\ \text{——— COOH}\end{matrix} + 2Ba(OH)_2 \longrightarrow \begin{matrix}\text{——— } CO\bar{O} \quad Ba^{2+}O\bar{H}\\ \\ \text{——— } CO\bar{O} \quad Ba^{2+}O\bar{H}\end{matrix} + 2H_2O$$

The neutralization is equivalent if the acid sites (COOH groups) are close to each other as in the case of most ion exchange resins. However, when the acid sites are some distance apart, equimolecular neutralization is expected to occur. A barium ion is bound by each acid group, the extra positive charge on the surface being balanced by a hydroxyl ion. However, with bases that have a univalent ion such as sodium hydroxide, only equivalent neutralization as represented by the first equation is possible, irrespective of whether the acid sites are close or distant.

Boehm (56) differentiated the acidic group present on oxidized charcoals and carbon blacks by selective neutralization techniques using bases of different strengths. In this method the carbon sample was agitated for at least 16 hr with 0.05 N solutions of $NaHCO_3$, Na_2CO_3, NaOH, and C_2H_5ONa (sodium ethoxide). The neutralization capacity of the four bases for most of the carbons was in the simple ratio 1:2:3:4 (Table 4), which according to Boehm is not by chance but implies that four different groups of charactertistic acidities occurred side by side on the carbon surface in equivalent amounts. Very likely they were part of a bigger complex. This stoichiometric ratio was observed only when the carbons were completely oxidized, which was achieved by heating smaller samples in an oxygen stream at 420–450°C after 5 hr when the burnoff was about 25–50%. Another important factor in achieving complete oxidation was slow cooling of

Table 4   Selective Neutralization of Acidic Surface Structures on Microcrystalline Carbons

Carbon sample	Neutralization (mEq/g) by			
	Sodium bicarbonate	Sodium carbonate	Sodium hydroxide	Sodium ethoxide
Sugar charcoal heat-treated in nitrogen at 1200°C	0.16	0.32	0.69	0.85
Sugar charcoal (activated)	0.21	0.43	0.72	0.89
Sugar charcoal activated and then heat treated	0.35	0.73	1.02	1.38
Eponite	0.16	0.34	0.63	1.06
CK-3	0.76	1.52	2.37	3.15
Philblack-O	0.57	1.09	1.64	2.34
Spheron-6	0.59	1.18	1.96	2.95
Spheron-C	0.64	1.28	1.88	2.56
Sugar charcoal oxidized with $KMnO_4$	0.39	0.64	0.88	0.96
Eponite oxidized with $KMnO_4$	0.78	1.15	1.61	2.21
Eponite oxidized with $(NH_4)_2S_2O_8$	0.88	1.34	1.74	2.35
Eponite oxidized with NaOCl	1.08	1.64	2.15	2.61
CK-3 oxidized with $(NH_4)_2S_2O_8$	0.20	0.31	0.35	0.58

Source: Boehm (56).

the sample under oxygen. A similar simple ratio of the neutralization capacities for the four bases was observed when the carbons were oxidized in aqueous suspension, but twice the amount of groups reacting with sodium bicarbonate was observed.

The strongly acidic groups neutralized by $NaHCO_3$ were postulated as COOH groups, whereas those neutralized by Na_2CO_3 but not by $NaHCO_3$ were believed to be lactones. The weakly acidic groups neutralized by NaOH but not by Na_2CO_3 were postulated as phenols. The reaction with C_2H_5ONa was not considered a true neutralization reaction since it did not involve exchange of H^+ by Na^+ ions. The groups reacting with C_2H_5ONa but not with NaOH were suggested to be carbonyls, which were created by the oxidation of disorganized aliphatic carbon (56). Puri (57), however, questioned the validity of the selective neutralization technique in determining the acidic groups of varying strengths. According to him the same acid group will neutralize different amounts of alkalis of varying strengths. For example, a weak acid like acetic acid can be neutralized only partially when titrated against $NaHCO_3$ or Na_2CO_3. However, the same acid can be completely neutralized by NaOH. In support of their viewpoint Puri and co-workers (57) carried out pH titration curves of a sugar charcoal using sodium carbonate, sodium bicarbonate, ammonium hydroxide, and barium hydroxide as the neutralizing bases. Titration curves with ammonium hydroxide and barium hydroxide showed fairly sharp endpoints, whereas the titration curves with sodium carbonate and sodium bicarbonate became asymptotic and did not show any inflections. The amounts of the various bases neutralized (the values for sodium carbonate and sodium bicarbonate correspond to asymptotic points) (Table 4) show slearly that titration with barium hydroxide alone could measure total acidity.

Barton and co-workers (37,41–43) studied acidic surface oxygen structures on a sample of graphite (42) and a carbon black Spheron-6 (43), by degassing these samples at different temperatures and determining quantitatively the amount of CO_2 evolved at each temperature using mass spectrometer and measuring the base neutralization capacity of the degassed samples. In the case of graphite every mole of CO_2 evolved reduced the base neutralization capacity by one equivalent, indicating that the acid group was monobasic. But in the case of Spheron-6 the presence of two types of acidic oxides both of which decompose to give CO_2, one decomposing at about 250°C and the other at 600°C (Fig. 6), was indicated. A plot of base uptake against CO_2 on the surface (Fig. 7) showed two intersecting straight lines. The first linear region, which corresponded to the desorption of CO_2 at temperatures around 600°C, showed a slope of 2, indicating that the acidic oxide being destroyed is dibasic. The second linear region, which corresponded to decomposition of the surface complex around 250°C, had a slope of 1, indicating that the acidic oxide was monobasic in character. Combining these studies with the changes

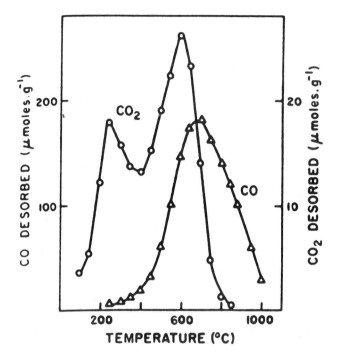

Figure 6 Differential desorption of CO_2 and CO from Spheron-6 as a function of degassing temperature. [From Barton et al. (43).]

in the surface acid groups on reaction with methyl magnesium iodide (37) and diazomethane (41), these workers suggested the presence in Spheron-6 of two types of acidic group, both of which were lactones but only one of which had active hydrogen associated with it. The one that had active hydrogen associated with it was present in graphite.

Bansal et al. (86) investigated the surface acidic structures on a number of polymer charcoals by titrating them with 0.2 N solutions of sodium hydroxide and sodium ethoxide. The base neutralization with sodium hydroxide was almost exactly equivalent to the amount of CO_2 evolved on evacuation at 1200°C in case of polyvinylidene chloride (PVDC), polyvinyl chloride (PVC), and Saran charcoals (Table 6) but was almost half of the amount of CO_2 evolved in the case of polyfurfuryl alcohol (PF) and urea formaldehyde (UF) charcoals. The base neutralization of a charcoal decreased on evacuation at gradually increasing temperatures and the decrease at any temperature

Table 5 Neutralization of Alkalis of Different Strength as Obtained from Titration Curves (mEq/100 g)

Charcoal	Sodium bicarbonate	Sodium carbonate	Ammonium hydroxide	Barium hydroxide	CO_2 complex
Sugar charcoal, original	204	270	295	650	669
Coconut charcoal, original	115	166	179	372	384

Source: Puri (57).

Table 6 Alkalis Neutralized by Various Polymer Carbons in Relation to CO_2 Evolved on Evacuation

Sample	Sodium hydroxide neutralized (mEq/100 g)	CO_2 evolved on evacuation (mEq/100 g)	Sodium ethoxide neutralized (mEq/100 g)
PVDC-600°	212	218	256
PVC-850° (CO_2)	62	63	71
Saran-600°	150	146	171
PF-600°	42	87	56
PF-900°	16	31	21
UF-650°	78	162	109
UF-850°	7	16	11

Source: Bansal et al. (86).

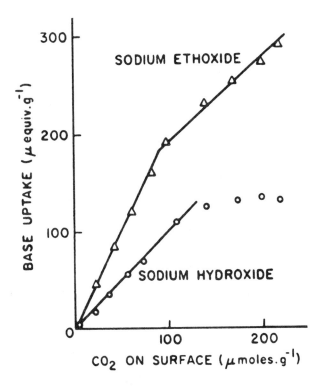

Figure 7   Relationship between base neutralization capacity of a
carbon and CO_2 desorbing complex on its surface from Spheron-G.
[From Barton et al. (43).]

corresponded to the amount of CO_2 evolved at that temperature (Table 7). Furthermore, the temperature interval over which the drop in
base neutralization capacity occurred appeared to be the same (Fig.
8) as the temperature interval over which CO_2 was eliminated from
the charcoal sample. At 600°c, when the desorption of CO_2 was complete, the charcoal lost its base neutralization capacity almost completely, although an appreciable amount (rather maximum amount) of CO
was being evolved from the sample at this temperature.   The CO-
desorbing surface structures, therefore, do not contribute to surface
acidity of carbons.   The relationship between the associated oxygen
desorbing as CO_2 on evacuation at gradually increasing temperatures
and the base neutralization capacity for PF-600° and PVDC-600° charcoals (Fig. 9) indicates that for every millimole of CO_2 evolved, the
base neutralization capacity decreased by 1 mEq in PF charcoal and
by 2 mEq in PVDC charcoal.   This showed that the acidic structures

Table 7 Decrease in the Amount of Sodium Hydroxide Neutralized by PVDC-600° after Outgassing at Various Temperatures in Relation to the Amount of CO_2 Evolved at Each Temperature

Temperature of outgassing (°C)	Sodium hydroxide neutralized (mEq/100 g)	Decrease in sodium hydroxide neutralized (mEq/100 g)	CO_2 evolved on outgassing (mEq/100 g)
30	212	0	0
100	212	0	0
200	210	2	0
300	204	8	10
400	176	36	31
600	100	112	119
800	12	200	212
1000	0	212	218

Source: Bansal et al. (86).

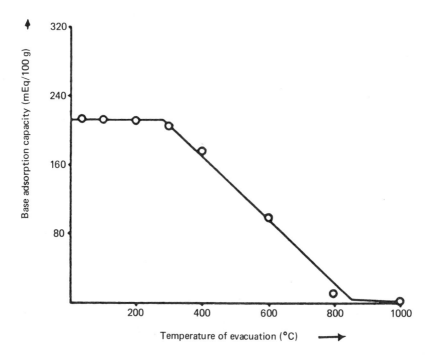

Figure 8 Base adsorption capacity in relation to evacuation temperature. [From Bansal et al. (86).]

were monobasic in the former and dibasic in the latter sample. The monobasic character of some carbons could be explained only if there is a phenolic group in conjunction with a carboxylic group. The phenolic group would not evolve CO_2 but will neutralize the base. Since the existence of isolated phenolic groups is still under dispute, it could be present as a part of a lactone. The possible structure for a lactone, which could explain most of the data obtained on base neutralization capacity and CO_2 desorbed on evacuation, could be an f-lactone, as suggested by Garten and Weiss (55) and later restated by Barton et al. (37,41). The lactone exists in two tautomeric forms. The keto form could explain the base neutralization capacity of PF and UF charcoals, whereas the enol form fitted well with the data on PVC, PVDC, and Saran charcoals (86).

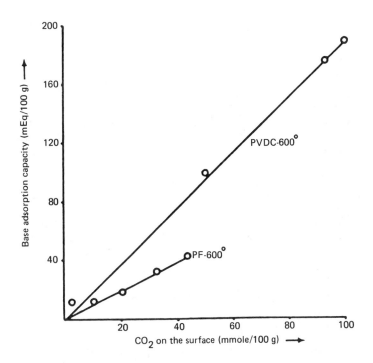

[From Bansal et al. (86).]

The existence of these two forms is quite reasonable in the case of different carbons since the presence of individual oxygen-containing structures and their proportions are very much dependent on the starting material from which the carbon is obtained.

Figure 9 Relationship between base adsorption and acidic oxide desorbing as CO_2. [From Bansal et al. (86).]

2.3.3   Electrochemical Methods

The most frequently used technique for determining base neutralization capacity to measure surface acidic structures has been equilibration of the carbon with an excess of alkali and back titration. However, this technique—although simple—requires large sample size
and gives accurate results only when the surface scidity value is
large. In comparison, electrochemical measurements can be carried
out by using very small sample size or samples with low acidity values
because of the sensitivity and accuracy with which current can be
measured. The procedure essentially consists of preparing a suspension of the carbon in CO_2-free distilled water and adding standard
alkali solutions in small amounts and measuring the current using a
precision instrument.

Puri and co-workers (87) reported pH titration curves of coconut
charcoals using barium hydroxide as the base. These curves resembled the titration curves of carbonic acid. The points of inflection
were sharp only as long as the acidity of the carbon was above a certain value. Puri and Bansal (83) used the same alkali in their potentiometric titration curves of a number of carbon blacks. The endpoints,
though not sharp (Fig. 10), could be located without much difficulty.
The base neutralization values determined at the inflections agreed
fairly well with those obtained by direct titrations. The shapes of
the titration curves indicated the presence of weaker acid structures
such as phenols because the pK values were between 9 and 10. However, when one of the sample, Mogul, was oxidized with nitric acid,
which extended its surface acidity from 115 to 530 mEq/100 g, the
shape of the titration curve indicated the presence of a much stronger
group like carboxyl, the pK value now being 5.5. When the oxidized
Mogul sample was subjected to gradual evacuation, the titration curves
of the residual products showed a gradual decrease in the strength
of the acid with decrease in the amount of CO_2 complex (Fig. 11).
However, it would be difficult to believe that a different group was
now being titrated. Puri and Bansal (83) agreed that it was the same
CO_2 complex that was responsible for surface acidity in each case.
As the concentration of this complex at the surface decreased, there
was a much lower amount of surface ionization, resulting in a lower
concentration of hydrogen ions per unit surface, which weakened the
acid character of the carbon.

Studebaker (88) carried out potentiometric titrations of a number
of carbon blacks in ethylene diamine using 0.1 N sodium aminoethoxide
as the titrant. The acidity values in nonaqueous media were appreciably higher than those obtained by titrating with sodium hydroxide
in aqueous media. The titration curves of carbon blacks with high
oxygen contents showed two inflections, which were attributed to
carboxylic and phenolic groups. The low oxygen—content carbon

blacks showed only one break, which was ascribed to phenols. Garten and Weiss (55), on the basis of the shapes of the titration curves for charcoals and carbon blacks, questioned the presence of acidic groups of such strengths as carboxylic.

Epstein et al. (89) observed that isotropic pyrolytic carbon electrodes subjected to electrochemical oxidations at widely varying current densities develop a reproducible surface functionality, which could be electrochemically reduced quantitatively and regenerated any number of times. They suggested that this oxidized and reduced

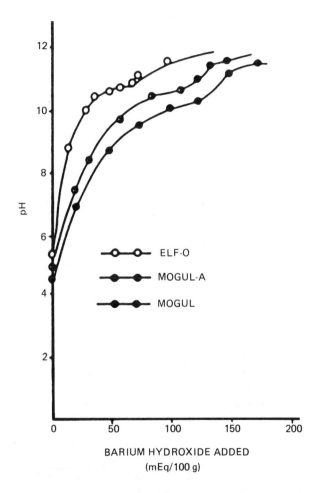

Figure 10   Titration curves of Mogul with barium hydroxide before and after different treatments.   [From Puri and Bansal (83).]

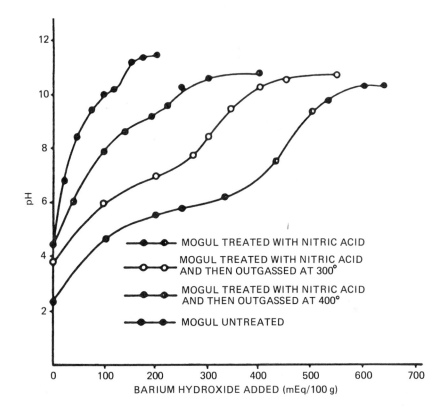

Figure 11 Titration curves of various carbon blacks with barium
hydroxide. [From Puri and Bansal (83).]

form of surface functionality corresponds to quinone and hydroquinone surface structures. Kinoshita and Bett (90,91) prepared Teflon-
bounded and platinum-supported electrodes using a porous carbon
black. The carbon balck was subjected to different degrees of heat
treatments. The surface oxygen structures were then studied using
a potentiodynamic sweep technique in 1M H_2SO_4. The surface structures observed by these workers were also quinone and hydroquinone. From the current−voltage curves, these workers calculated the
concentration of these structures to be 10^{-10} to 10^{-11} mol/cm^2, which
was in close agreement with those of earlier workers (83). Oxidations caused an increase in the surface oxides, whereas heat treatment resulted in a decrease.

Matsumura et al. (92,93) studied the titration curves of a number of carbon blacks after oxidation with nitric acid and hydrogen peroxide solutions using an automatic titrator and sodium hydroxide as the titrant. Oxidations for shorter periods both with nitric acid and hydrogen peroxide enhanced strong and weak acid groups, whereas longer oxidation time enhanced weak acid structures only. The stronger acids with pK lower than 7, which were postulated as carboxyls, occurred in amounts up to 0.475 mEq/g and the weaker acid structures, which were postulated as phenols, occurred in amounts up to 0.49 mEq/g (Table 8). The titration curves of the carbons after esterification with n-butanol followed by hydrolysis failed to show the existence of lactone structures.

Papirer and Guyon (94) titrated carbon blacks potentiometrically using bases of different strengths such as $NaHCO_3$, Na_2CO_3, and NaOH. The titration curves with $NaHCO_3$ indicated the presence of two different acid groups corresponding to pKa values 5.8 and 9.89 respectively. The difference observed with Na_2CO_3 as the titrant was very small. The titration curve with NaOH as the titrant showed two peaks: the first corresponded to physically adsorbed CO_2 (95) and the second to an acid group which could be decarboxylated at 1000°C and esterified with methanol in HCl. The potentiogram gave the pKa value of this acid group to be 5.8, which is typical of a carboxylic group.

Matsumura and Takahashi (96) carried out potentiometric titrations in which carbon blacks were reduced in aqueous medium with an excess of sodium borohydride ($NaBH_4$) and then backtitrated with iodine. The titration curves showed two types of reducible surface structures, reversibly and irreversibly reducible. The former were attributed to quinone structures and the latter to other reducible structures such as ketones and aldehydes. The total amounts of titrable quinones in the three carbon blacks were found to be 0.91, 0.98, and 1.18 mEq/g. However, these amounts determined by titrimetry were considerably lower than those obtained by using the methods of Studebaker and colleagues (44,97,102) and Suzuki and Matsumura (98) (hydrogen uptake by sodium borohydride; Table 9) and by Rivin (77,78) (increase in the active hydrogen content on treatment of the carbon blacks with lithium aluminum hydride). Thus these workers suggest caution in the determination of quinone groups by reduction methods since these would include other reducible groups such as aldehydes, ketones, and lactones.

Matsumara and Takahashi (96) and their colleagues are of the view that the titration method is advantageous in yielding quantitative results with more detailed information on the redox reactivities of carbon surfaces including quinones and other structures, although some interference with the side reactions and steric effects might still be included.

Table 8 Specific Surface Areas and the Acidities Obtained from the Titration Curves of Carbon Blacks

Carbon blacks	Surface area (m^2/g)	Initial pH of the suspension	Acids neutralized by NaOH ($\mu Eq/m^2$)	
			pKa 7.0	7.0 pKa 11.0
Mitsubishi-42	99.5	7.1	0	0.70
HNO₃ 60°C	111.6	3.9	0.29	0.98
HNO₃ 105°C	115.3	3.2	1.42	2.28
H₂O₂ 50°C	107.8	4.1	0.32	1.00
H₂O₂ 99°C	102.5	3.3	0.91	2.12
Air 400°C	354.4	3.1	0.44	0.81
Mitsubishi-100	94.8	4.9	0.21	0.66
Statex-R	75.7	6.8	0	0.54
Philblack-O	82.7	6.3	0.15	0.30
Diablack-H	82.0	7.7	0	0.32
Raven-35	94.6	4.2	0.23	1.06
Peerless-155	129.7	4.1	0.71	0.74
MA-11	84.0	3.7	0.45	0.41
MA-8	118.6	3.1	1.25	1.04
Neospectra-AG	824.5	3.0	0.58	0.59

Source: Matsumura et al. (92,93).

Table 9 Quinone Group Determined by Potentiometry and Borohydride Methods

| | Quinone groups | | | | |
| | Potentiometric backtitration with I_2 | | Potentiometric reducing titration with $NaBH_4$ (mEq/g) | Volumetric reduction with NBH_4 (mEq/g) | Total oxygen content (mEq/g) |
	Reversibly reducible (mEq/g)	Irreversibly reducible (mEq/g)			
Philblack-O	0.91	0.51	1.20	0.285	1.19
Peerless-155	1.18	0.41	3.52	0.366	1.97
Neospectra-AG	0.98	1.05	5.25	0.670	5.56

Source: Matsumura and Takahashi (96).

2.3.4   Thermometric Titrations

Indirect titrations with alkalis are time consuming and require long equilibration time, and potentiometric titrations sometimes do not give sharp endpoints. Thus accurate evaluations of the data are difficult in many cases. Consequently, a few attempts to estimate the surface acidic structures on carbons by thermometry have been reported. Since thermometry is related to the enthalpy of a reaction, it may be a reliable method when potentiometry fails.

Given and Hill (99) used direct nonaqueous titrations of several carbon blacks with thermometric detection of the endpoints. Several combinations of titrant and dispersant were used for the titrations. The carbon black sample varying from 0.15 to 0.5 g was completely dispersed in 10 ml of the dispersant using a magnetic stirrer. The slurry was then taken in a Dewar flask and an additional 30 ml of the dispersant was added. The tip of the titrant delivery tube, a nitrogen inlet, and the tip of the thermistor all extended into the slurry. Before adding the titrant, the slurry was stirred to get nearly a straight baseline. The titrant was added at the rate of 0.277 + 0.005 ml/min.

Plots for the thermometric titrations of five different carbon blacks (Fig. 12) show a slight cooling effect in the initial stages, a result of the slurries being at a temperature slightly higher than the ambient temperature because stirring was carried out over a magnetic stirrer. A slight curvature in the curves near the endpoint compared to the sharp break for benzoic acid solution was attributed to the variation in acid strengths of the surface structures. The amount of surface acidic structures determined by thermometry was independent of the titrant or the dispersant, indicating that steric effects are not significant in these neutralization reactions. On the basis of the comparison of the titration results for different carbon blacks obtained by thermometry with their phenolic contents determined by acetylation with C^{14}-labeled acetic anhydride (Table 10), Given and Hill (99) suggested that phenolic hydroxyl structures contributed between 5 and 100% of the total acid content of the carbon blacks.

Papirer and Guyon (94) used thermometric titration in aqueous media using 0.05 N sodium hydroxide as the titrant to determine surface acidity of a carbon black Carbolac I. The value 580 mEq/g obtained by thermometry was very much less than obtained by indirect titration in aqueous media. However, when the carbon black was allowed to stay in aqueous media for a prolonged duration before the thermometric titration, the acid value increased to 820 mEq/g, still much lower than the indirect titration value. This was attributed to the slow hydrolysis, which makes it difficult to determine weak acids even by thermometry in aqueous media (100). Papirer and Guyon (94) then used the Greenhow and Spencer method (101), i.e., an acrylonitrile polymerization reaction to indicate the endpoint of the

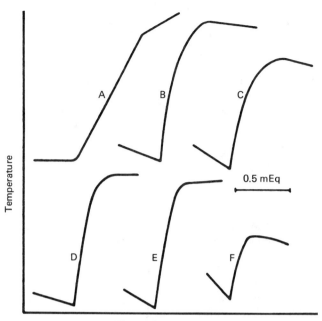

Figure 12 Thermometric titration curves of carbon blacks in DMF with potassium methoxide as titrant. (A) Benzoic acid in benzene–methanol solution; (B) Black Pearl 2; (C) Peerless-II; (D) Black Pearl 70; (E) Black Pearl A; and (F) Spheron-G. [From Given and Hill (99).]

Table 10 Acidity of Carbon Blacks as Determined by Thermometric Titrations

Sample	Total acid content (mEq/g)	Phenolic content by acetylation (mEq/g)
Black Pearls 2	1.44	1.25
Peerless-11	1.44	
Black Pearls 70	1.32	1.00
Black Pearls A	1.16	0.75
Excelsior		
No. 140 Powder		
Spheron-6	0.30	0.31
Raven-15		

Source: Given and Hill (99).

neutralization process. They used acrylonitrile as the monomer, a mixture of toluene, isopropanol, and acrylonitrile as a solvent, and an isoproponal solution of potassium hydroxide as the titrant. A correction was also applied for the amount of base required for the initiation of the polymerization reaction. The acid value (base up-take) so obtained agreed closely with that obtained by indirect titra-tion with aqueous alkali.

Thus direct thermometric titration appears to be a reliable tech-nique for determining total surface acidity of carbons.

2.3.5 Polarography

Polarography was first applied by Hallum and Drushel (102) to iden-tify quinone groups on carbon blacks, the evidence for which was indicated in their IR measurements. These workers carried out pol-arographic analysis of carbon black slurries in dimethyl formamide using tetra-n, butlylammonium iodide as supporting electrolyte. The samples were carefully extracted before polarographic analysis to ensure removal of extractables usually found in carbon blacks. The polarogram showed a distinct reduction wave rather than as expected for a 1,4-quinone. But the position of the wave was at the appropri-ate reduction potential (between -0.6 and -0.8 V) for quinonic groups. Treatment with lithium aluminum hydride caused the wave to disappear completely. The hydroquinone groups were detected by anodic polar-ography by the appearance of a wave at about + 1.00 V, which dis-appeared on treatment with hydrogen peroxide or diazomethane. These workers proposed the following structures for quinone and hydroqui-none groups in carbon blacks:

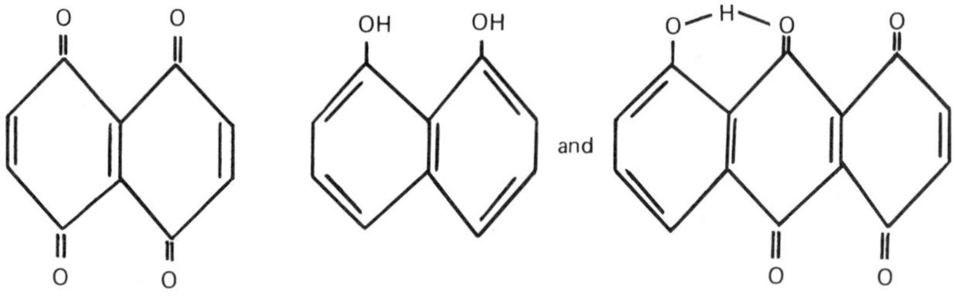

Donnet and co-workers (95) obtained a similar wave in the cath-odic polarogram of an oxidized carbon black which disappeared after treatment with isobutyronitrile. The disappearance of the wave was attributed to an addition reation of isobutyronitrile with quinone groups. However, no reaction with this reagent was observed after reduction of the carbon black with hydrogen iodide or after treat-ment with aniline or diazomethane, indicating that diazomethane was added to the quinones in carbon blacks. Jones and Kaye (98a) also obtained polarograms similar to that of Hallum and Drushel (102) on activated carbon samples (Fig. 13).

Given and Hill (99) also made polarographic studies of slurries of several carbon blacks using dimethyl formamide and acetonitrile, each 0.1 N in tetrabutyl ammonium iodide as electrolytes. The slurries were prepared by shaking degassed carbon black samples with oxygen-free electrolytes for 16 hr before transfer to the polarographic cell. Nitrogen was passed through the cell for about 20 min before taking the polarogram. None of the carbon black slurries produced a wave in the potential region from 0.0 to -1.0 V although most quinones exhibit at least one reduction wave in this region. The polarogram, however, showed steadily rising reduction current, which was attributed to the existence, in the carbon structures, of a whole series of quinone groups and aromatic nuclei having a range of half wave potentials.

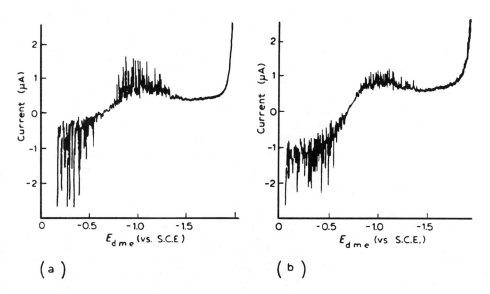

Figure 13 Polarograms of activated charcoal suspension in sodium hydroxide (a) after partial reduction with sodium dithionate and (b) after partial electrolytic reduction. [From Jones and Kaye (98a).]

2.3.6 Specific Chemical Reactions

Several workers have used more direct analysis of the surface oxygen structures on carbons by studying specific chemical reactions of organic chemistry. Most of these reactions have been carried out in conjunction with base neutralization studies to determine the types of surface structure on carbons. The more important groups for which specific chemical reactions have been used are carboxyls, phenols, quinones, and lactones. As the reactions used to identify and estimate these functional groups are discussed in most organic chemistry books and as the chemistry involved in this group analysis has been very nicely reviewed by several workers (55—57, 103), we merely offer a brief representation of the reactions with reference to the authors who have used these reactions for the study of surface structures in carbons in Tables 11 to 16.

Methylation of the carbon with diazomethane followed by hydrolysis of the methylated product with a mineral acid is the most commonly studied chemical reaction. The proportion of the methylated products which is unhydrolyzed is attributed to phenolic structures; the fraction which is hydrolyzed by the acid was attributed to carboxylic groups by Studebaker et al. (69) and to f-lactones by Garten and Weiss (55). The difference between the total methoxy content of a carbon and its sodium hydroxide neutralization capacity was attributed to n-lactone-type structures. Studebaker and Rinehart (97), however, determined lactonic structures in carbon blacks by titration with barium hydroxide as suggested by Puri and Bansal. The quinonic groups have generally been determined by reduction with sodium borohydride. The results of some of these studies obtained by different investigators for charcoals, activated carbons, and carbon blacks are presented in Tables 15 and 16.

Surface group estimation using direct chemical methods has not been able to account for the entire amount of the combined oxygen. Thus the validity of applying organic reactions for identification and estimation of surface oxygen structures is questionable. Boehm and co-workers, who have carried out considerable work on surface group determinations, are of the view that the groups on the surface of carbons are not likely to behave in the same way as those present in simple organic compounds because of possible association between adjacent structures and steric hindrances (56). Papirer and Guyon (94), on the basis of their group determinations using organic methods and spectroscopic analysis, have suggested that the surface structures on carbons should not be considered classical organic functional groups rather combined structures which may present numerous mesomeric forms largely favored by their location on the same polyaromatic frame.

Table 11 Carboxyl Groups R-COOH

Reactions of determination	Reference
Organomagnesium compound $R\text{-}COOH + CH_3MgI \quad RCOOMgI + CH_4$	Villars (71) Barton and Harrison (37) Matsumura et al. (93) Studebaker and Rinehart (124)
Diazomethane $R\text{-}COOH + CH_2N_2 \rightarrow RCOOCH_3 + N_2$	Studebaker (69) Papirer and Guyon (121) Barton et al. (41) Donnet and Marguier and Hueber (104)
Decarboxylation $R\text{-}COOH \xrightarrow{\text{heating above } 600°C} R\text{-}H + CO_2$	Rivin (78)
Hydride of aluminum and lithium $4R\text{-}COOH + 3LiAlH_4 \rightarrow LiAl(OCH_2R)_4$ $+ 2LiAlO_2 + 4H_2$ $LiAl\,(OCH_2R)_4 + 4H^+ \rightarrow Li^+ + Al^{3+}$ $+ 4RCH_2OH$	Test and Hansen (109) Rivin (78)
Isotopic exchange $R\text{-}COOH + {}^3H \rightarrow R\text{-}COO^3H + H$	De Bruin and Plas (72) Puri (72) Rivin (36) Test and Hansen (37) Boehm (43–45) Donnet (46) Dupuet and Bastick (47)
Formation of an acid chloride $R\text{-}COOH + SOCl_2 \rightarrow R\text{-}COCl + SO_2$ $+ HCl$ $R\text{-}COOH + PCl_5 \rightarrow RCOCl + POCl_3$ $+ HCl \;(POCl_3)$	Boehm et al. (74, 105) Donnet (110)
Schmidt reaction (formation of urethane) $R\text{-}COCl + NaN_3 \longrightarrow RCON_3 \longrightarrow R\text{-}NH\text{-}COOC_2HS +$ $N_2 + CH_3CH_2OH$	Boehm (56)
Dimethyl sulfate $2R\text{-}COOH + SO_2{\diagup}^{OCH_3}_{\diagdown OCH_3} \longrightarrow 2RCOOCH_3$ $+ SO_4H_2$	Donnet (110) Boehm (56)

Table 12 Hydroxyl Functions R-OH

Reactions of determination	Reference
Organomagnesium compound	Villars (71)
$R-OH + CH_3MgI \longrightarrow R-O-MgI + \overrightarrow{CH_4}$	Studebaker and Rinehart (97)
	Matumura et al. (93)
Diazomethane	Studebaker et al. (69)
$R-OH + CH_2N_2 \longrightarrow R-O-CH_3 + \overrightarrow{N_2}$	Barton et al. (41)
Hydride of aluminum and lithium	Test and Hansen (109)
$4\,R-OH + LiAlH_4 \longrightarrow LiAl(O-R)_4 + 4\,H_2$	Rivin (78)
Acetylation	Boehm et al. (74,105)
$R-OH + \begin{Bmatrix} CH_3COCl \\ CH_3COOH \\ \frac{1}{2}\,^{CH_3CO}_{CH_3CO}O \end{Bmatrix} \longrightarrow \begin{matrix} CH_3C=O \\ OR \end{matrix} + \begin{Bmatrix} HCl \\ H_2O \\ \frac{1}{2}\,H_2O \end{Bmatrix}$	Donnet et al. (111)
	De Bruin and Plas (72)
	Given and Hill (99)

2-4-Dinitrofluorobenzene

$$R-OH + F-\text{(2,4-dinitrophenyl)} \longrightarrow R-O-\text{(2,4-dinitrophenyl)} + HF$$

Boehm et al. (74,105)

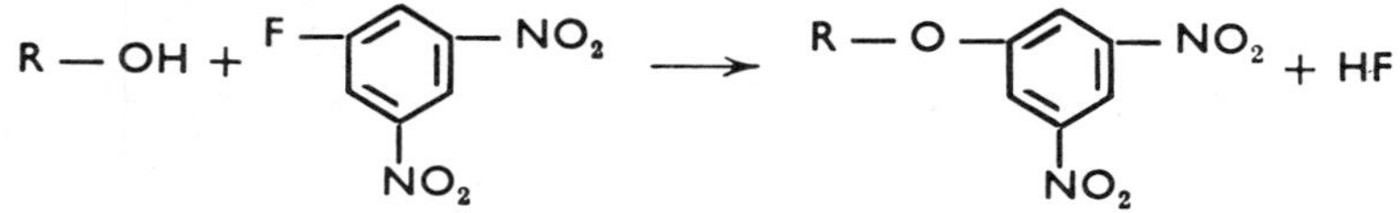

Benzoyl p. nitrochloride

$$R-OH + ClOC-\text{(p-nitrophenyl)} \longrightarrow ROOC-\text{(p-nitrophenyl)} + HCl$$

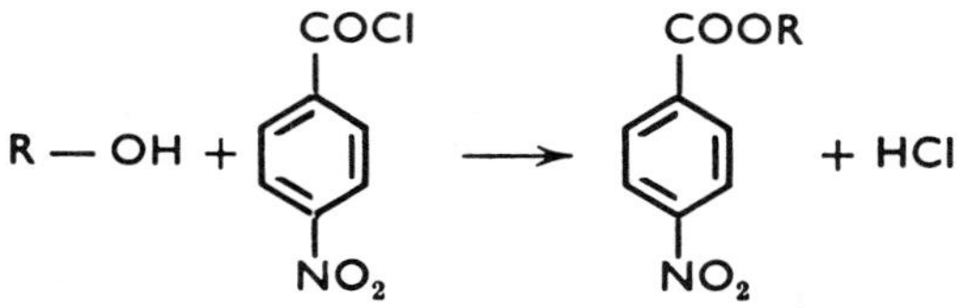

Boehm et al. (73,105)

Dimethyl sulfate

$$R-OH + SO_2(OCH_3)_2 \longrightarrow R-OCH_3 + SO_2(OH)(OCH_3)$$

Donnet (110)
Boehm (56)

Hexamethyldisilazane

$$2R-OH + HN[Si(CH_3)_3]_2 \rightarrow 2ROSi(CH_3)_3 + NH_3$$

Friedman et al. (112)
Hill (113)
Given (99)

Table 13 Quinonic Functions

Reactions of determination	Reference
Stannous fluoborate: quinone $+ \text{Sn}^{++} \longrightarrow$ hydroquinone $+ \text{Sn}^{++++}$	Voet and Teter (108)
Ferric sulfate (hydroquinonic function): hydroquinone $+ 2\,\text{Fe}^{+++} \longrightarrow$ quinone $+ 2\,\text{Fe}^{++}$	Voet and Teter (108)
Lithium aluminum hydride: quinone $\xrightarrow{\ \text{AlLiH}_4\ }$ hydroquinone (OH)	Test and Hansen (109) Rivin (78)
Oxidation of Ti^{3+}: quinone $+ 2\,\text{Ti}^{+++} \longrightarrow$ hydroquinone $+ 2\,\text{Ti}^{++++}$	De Bruin and Plas (72)

Hydroxylamine

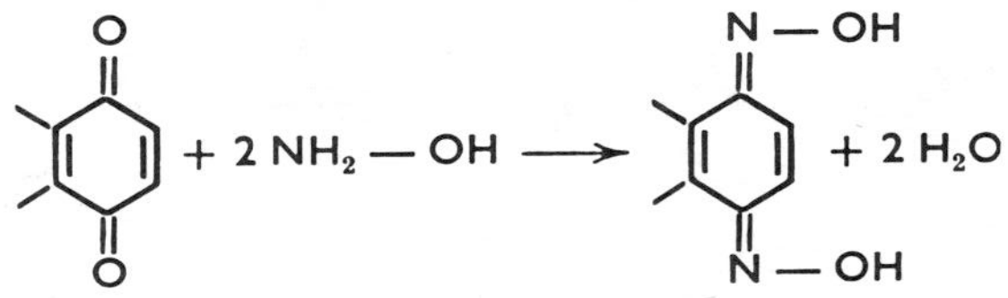

$$+ 2\,NH_2{-}OH \longrightarrow + 2\,H_2O$$

De Bruin and Plas (72)

Diazomethane

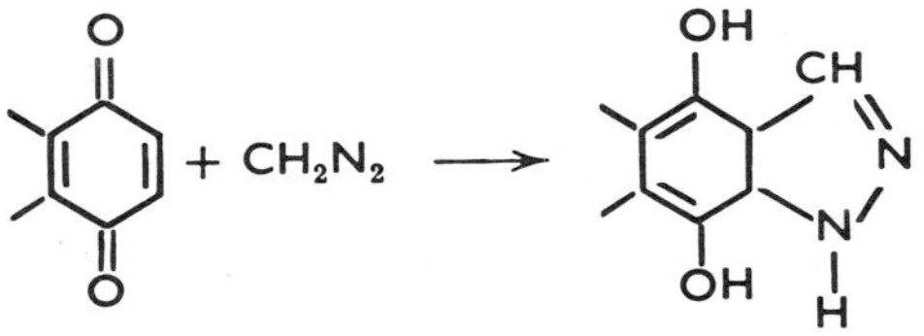

$$+ CH_2N_2 \longrightarrow$$

Studebaker et al. (69)
Papirer and Guyon (94)

Catalytic hydrogenation

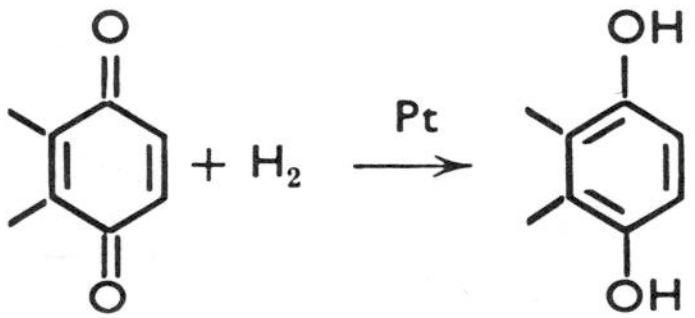

$$+ H_2 \xrightarrow{\ Pt\ }$$

Studebaker et al. (69)

Sodium borohydride

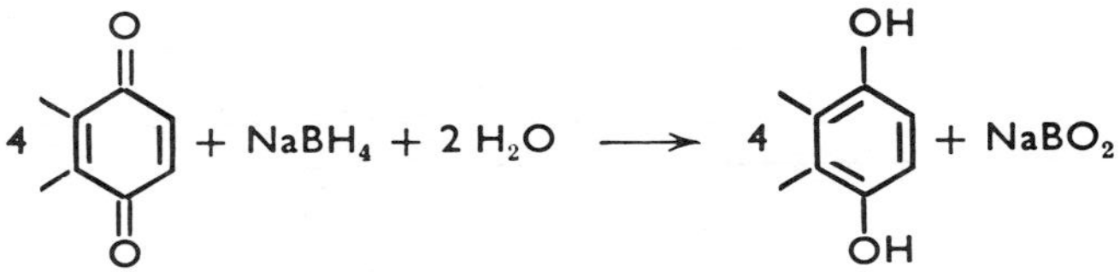

$$4 \quad + NaBH_4 + 2\,H_2O \longrightarrow 4 \quad + NaBO_2$$

Studebaker et al. (69)
Donnet at el. (107)
Matsumura and
Takahashi (96)

Table 13 Continued

Reactions of determination		Reference
Condensation of the isooctane		Hallum and Drushel (102)
Condenstaion of the isobutyric nitrite		Donnet and Henrich (106)
Condensation of the aniline		Donnet and Hendrich (106)

Table 14   Lactone Functions

Reactions of determination	Reference
Ammonia	De Bruin and Plas (72)

$$\begin{array}{c} O \\ \| \\ -C \\ \| \\ -O \end{array} + NH_3 \xrightarrow[\text{méthanol}]{\text{milieu}} \begin{array}{c} O \\ \| \\ -C-NH_2 \\ \\ -OH \end{array}$$

Bromohydric acid	De Bruin and Plas (72)

$$\begin{array}{c} O \\ \| \\ -C \\ \| \\ -O \end{array} + HBr \longrightarrow \begin{array}{c} O \\ \| \\ -C-OH \\ \\ -Br \end{array}$$

Hydride of aluminum and lithium	Rivin (78)

$$\begin{array}{c} O \\ \| \\ -C \\ \| \\ -O \end{array} \xrightarrow{LiAlH_4} \begin{array}{c} O \\ \| \\ -C-OH \\ \\ -OH \end{array}$$

2.3.7   Infrared Spectroscopy

Infrared spectroscopy in its various forms is an important and force-
ful technique which can give useful information about structures,
such as the surface functional groups, and can provide basic spectra
of the activated carbon for comparison with spectra of the same car-
bon-containing adsorbed species. Spectral studies can also provide
useful information regarding the molecular forces involved in the
adsorption processes. Carbon being black in color has a tendency
to absorb most of the radiation, at least in the visible region. Even
its thin sections are opaque. But it can transmit some radiation in
the IR region when examined using extremely thin sections, so that
IR spectra originally could be obtained only on coals, because they
can be cut into very thin sections. Later preparation of halide pellets in which finely divided carbonaceous samples were uniformly distributed was resorted to. But this requires communication of the
carbon material into a very finely ground state. Carbon blacks presented no problem, since they can be obtained in a very fine state
of subdivision, but active carbons, being hard, are difficult to grind.
Friedel and Hofer (114) devised a technique by which active carbons
and chars could be converted into a finely ground state for making
halide pellets for IR spectral studies.

Table 15 Surface Group Analysis by Reaction with Diazomethane

| Carbon sample | Sodium hydroxide neutralization capacity (mEq/100 g) | $-OCH_3$ introduced | | | Sodium hydroxide neutralized minus total $-OCH_3$ n-lactone groups (mEq/100 g) |
		Total	Hydrolyzable = COOH or lactone groups (mEq/100 g)	Non-hydrolyzable = phenolic groups (mEq/100 g)	
Sugar charcoal	81	82	50	32	—
Sugar charcoal heat treated in nitrogen	69	72	34	38	—
Sugar charcoal activated in CO_2	71	72	42	30	—
Eponite	62	107	46	61	—
Sugar charcoal heated in air at					
400°	59	11	05	05	48
600°	14	2	1	1	12
800°	13	8	6	6	5
Sugar charcoal oxidized in air after eva-cuation at 800°					
400°	32	15	10	5	17
600°	22	15	11	4	17
700°	18	15	12	3	3
Carbolac-1	183	146	90	56	37
Mogul-A	118	54	33	21	64
ELF-O	58	24	14	10	34
Spheron-6	37	14	9	6	23
Spheron-9	46	10	5	5	36

Source: Garten and Weiss (55), Studebaker et al. (69), Studebaker and Rinehart (97).

Table 16 Surface Group Analysis by Specific Organic Reactions

Carbon sample	Oxygen as phenol group by Grignard reagent (%)	Oxygen as phenol group by neutraliza-tion of barium hydroxide (%)	Oxygen as quinone group by reduction with sodium borohydride (%)	Oxygen ether group by difference (%)
Pelletex	0.02	0.08	0.16	0.35
Stirling-V	0.03	0.07	0.22	0.23
Kosmos-40	0.08	0.14	0.15	0
Statex-B	0.06	0.17	0.16	0.44
Philblack-A	0.08	0.16	0.24	0.24
Philblack-O	0.11	0.14	0.41	1.16
Philblack-E	0.19	0.33	0.66	1.15
Spheron-9	0.20	0.60	0.92	2.63
Spheron-6	0.18	0.45	0.67	1.82
Spheron-C	0.39	0.76	0.89	1.31
ELF	0.54	0.78	1.10	2.34
Mogul-A	0.52	1.98	1.71	2.84
Mogul	0.71	2.52	2.03	3.18
Carbolac-2	0.96	4.25	2.62	1.84

Source: Studebaker and Rinehart (97).

The technique of IR spectroscopy using halide pellets also presented certain difficulties because of the exposure of the material to atmospheric gases and vapors, which tend to vitiate the results. However, the development of elaborate techniques for obtaining carbonaceous films (85) and preparation of charcoals by carbonization under vacuum conditions (64) broadened the scope of application of IR spectroscopy to the study of carbons and the chemical structure on their surfaces. Furthermore, the sensitivity of IR measurements has been greatly enhanced by using Fourier Transform (115,116), photoacoustic (117,118), and the photothermal beam deflection (26) IR spectroscopy.

Hallum and Drushel (102) first reported IR transmission spectra of a carbon black with a high oxygen content using a Nujol mull of the black. A band at 1600 cm^{-1} (6.3 μm) was attributed to either condensed aromatic ring systems or to hydrogen-bonded conjugated carbonyl groups. The band shifted to normal carbonyl wavelengths 5.7−5.9 μm (1750−1200 cm^{-1}) on treatment of the carbon black with diazomethane. Another band at 1250 cm^{-1} was also noticed after methylation. These bands were attributed to aromatic hydroxyl groups hydrogen-bonded to conjugated carbonyl groups. Garten and Weiss (55) obtained two bands at 1705 and 1600 cm^{-1} in the transmission IR spectra of sugar charcoals. The 1600 cm^{-1} band showed a slight shift on treatment of the carbons with sodium hydroxide, which was attributed to the formation of carboxylate ion from f-lactone-type groups. The disappearance of the 1705 cm^{-1} band on treatment with sodium hydroxide, which they attributed to n-lactones, remains a matter of dispute.

Friedel and Hofer (114) and Friedel and Carlson (119) obtained IR adsorption spectra of Pittsburgh CAL—activated carbon after using an intense grinding technique for preparing the sample. The spectra showed definite bands at 1735, 1590, and 1215 cm^{-1} (Fig. 14), which were attributed to a carbonyl, aromatic structures, or conjugated chelated carbonyl and C—O structures respectively. Donnet's (120) absorption spectra of a carbon black after methylation and subsequent hydrolysis clearly indicated the modification of carboxylic and hydroxyl groups on methylation. But the subsequent hydrolysis of the methylated carbon black did not show a complete reversal of the spectra. The difference in the two spectra was attributed to the existence of keto-enol equilibrium, which exists on carbon black surfaces to be in favor of the enol form.

Studebaker and Rinehart (97) examined 20 samples of commercial-grade carbon black by IR spectroscopy using a potassium bromide pellet technique. They used integrated intensities rather than absorbance at characteristic wavelengths, as the band positions and band shapes for the functional groups are subjected to large shifts due to environmental effects. A comparison of the integrated IR intensities and the functional group analysis by direct chemical reactions indicated that a considerable chemical reactivity of carbon blacks was due to quinones and lactones. Absorption typical of phenols and ethers was observed in the 1205—1195 cm^{-1} and 1275—1265 cm^{-1} regions.

Ishizaki and Marti (121) examined an activated carbon (Filtrasorb) before as well as after neutralization with 0.25 N solutions of sodium hydroxide and hydrochloric acid by direct transmission IR spectroscopy. The spectra of the original Filtrasorb and after neutralization (Fig. 15) showed absorption bands at the 1760—1710 cm^{-1}, 1670—1520 cm^{-1}, 1480—1340 cm^{-1}, 1300—1230 cm^{-1}, and 1180—1000 cm^{-1} regions. Comparison of the spectra for original and sodium hydroxide—neutralized samples showed the acidic nature of the surface structures. The spectrum of Filtrasorb 200 after neutralization with hydrochloric acid (Fig. 15, spectrum 3) showed a shift in the 1760-1710 cm^{-1} absorption band corresponding to carboxylic tautomeric structures. The spectra of the carbon in the 1670—1500 cm^{-1} region presented considerable overlapping of different absorption bands and was attributed to quinonic and carboxylate structures. These chemical structures were supported by the spectra of the carbon after

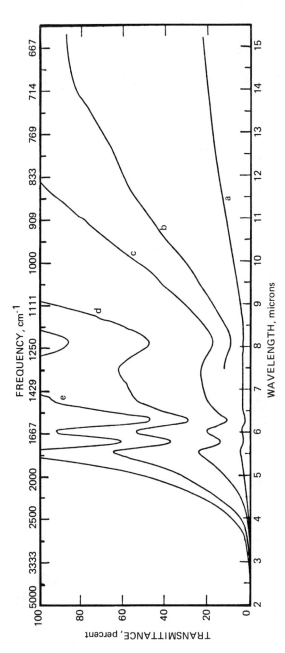

Figure 14 Infrared spectra of an activated carbon (KBr pellet method). *b*, *c*, *d*, and *e* are scale expansion spectra. [From Friedel and Carlson (119).]

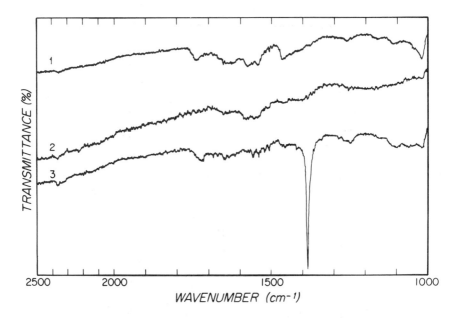

Figure 15 Direct transmission IR spectra of different carbons.
(1) Filtrasorb 200; (2) Filtrasorb 200 neutralized with NaOH; (3)
Filtrasorb 200 neutralized with HCl. [From Ishizaki and Marti (121).]

treatment with alkali and acid. The absorption in this region increased after neutralization with alkali due to the additional carboxylate
structures formed by reaction of alkali with original lactone structures. The absorption decreased after treatment with the acid with
the appearance of a peak in the $1380-1385$ cm^{-1} region. This absorption was attributed to the formation of carbonate structures (65) due
to chemisorbed carbon dioxide (evolved by the action of acid on the
carboxylate or carboxylic acid structure) on semiquinone structures.
In the $1480-1340$ cm^{-1} region, which is characteristic of $-O-H$
bending vibrations, Filtrasorb shows a peak at 1465 cm^{-1}, which
suffers a strong reduction after neutralization with alkali, was assigned to phenolic groups. The absorption in the $1300-1280$ cm^{-1} region,
which is associated with $-C-O$ stretching vibration, was attributed
to lactonic and phenolic structures since it disappeared completely
in the $1180-1000$ cm^{-1} region, which was reduced on treatment with
alkali but remained unchanged on treatment with the acid, probably
arose from phenolic structures.

Papirer et al. (122) examined surface groups on a carbon black before and after methylation followed by hydrolysis using direct transmission with scale expansion and internal reflection IR spectroscopy. At 1250, 1600, and 1740 cm^{-1} (Fig. 16) bands were observed whose intensities decreased when the carbon was heated at 1000°C in hydrogen or argon. Another band at 3200−3400 cm^{-1} was attributed to deformation vibrations of the hydroxyl group of carboxylic structures. There was no dramatic difference in the IR spectra after methylation either with diazomethane or with methanol (Fig. 17) except that the intensity ratio of the peaks at 1740 and 1600 cm^{-1} was decreased. This was attributed to the elimination of ketonic ar aldehyde structures and to the formation of nonchelated carbonyl groups. The spectra of the methylated carbon after hydrolysis (Fig. 18) showed a new band at 1105 cm^{-1} which persisted even when the hydrolyzed product was remethylated to get back the original structures on the carbon black. The authors suggested that methylation brings about irreversible transformation of the carbon surface structures.

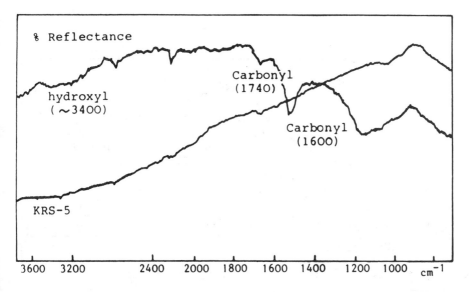

Figure 16 Internal reflectance IR spectra of Carbolac 1. [From Papirer et al. (122).]

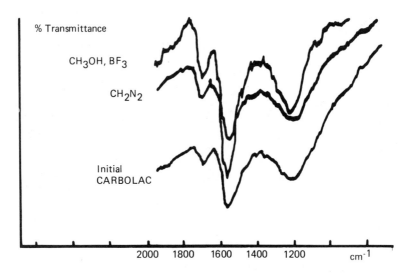

Figure 17 Internal reflectance IR spectra of Carbolac 1 after methylation. [From Papirer et al. (122).]

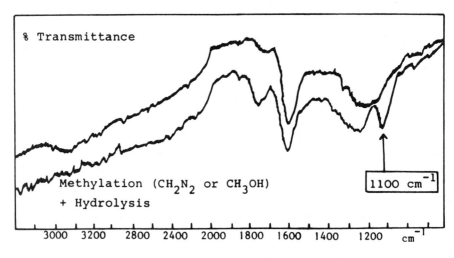

Figure 18 Internal reflectance IR spectra of Carbolac 1 after methylation and subsequent hydrolysis. [From Papirer et al. (122).]

Mattson et al. (123) used internal reflectance spectroscopy for characterizing surface oxygen structures on activated carbons because the technique is easy and does not exihibit the scattering and large energy losses common to transmission IR spectroscopy. The spectra can also be obtained directly on the sample rather than using a mulling agent or a supporting matrix. These workers examined the surface structures developed on laboratory-activated sugar carbons as a function of activation temperature and atmosphere. Two series of sugar carbons activated in 1% oxygen mixture and 5% oxygen mixtures with nitrogen at temperatures between 300 and 700°C were subjected to IRS examination. The IRS spectra of carbons within each series showed the same general features, although the relative intensities of absorption for the same functional group showed some dependence on the temperature of activation.

The IRS spectra show a pair of bands at $1710-1750$ cm^{-1} and $1750-1770$ cm^{-1} (Fig. 19) when the carbons were activated between 300 and 700°C (123). The two bands merged extensively in the samples activated at 500 and 700°C. These two pairs of absorption bands were attributed to a pair of carboxylic acid groups which were created upon oxidation of the edges of the aromatic rings. The other major feature of the spectra in Figure 19 is the pair of absorption bands at $1590-1625$ cm^{-1} and $1510-1560$ cm^{-1}. The variation in the positions and the relative intensities of these two bands in these regions with the temperature of activation indicated the presence of quinone carbonyl groups.

The IRS spectra of sugar carbons activated with a mixture of oxygen, carbon dioxide, and nitrogen did not show the well-defined pair of carboxylic acids. But the adsorption bands were certainly present throughout and showed the general features found for the oxygen and nitrogen-activated carbons.

Zawadzki (65) prepared carbon films by the carbonization of polyfurfuryl alcohol and cellulose and studied the surface oxygen structures produced as a result of oxidation in oxygen using IR spectroscopy in vacuum. Oxidation of the carbon films at 200°C produced absorption bands at 1600 and 1260 cm^{-1} and a C=O stretching vibration band at 1720 cm^{-1}. The oxidation at 300°C increased the intensity of the bands at 1720 and 1260 cm^{-1}. The degassing of the oxidized film between 400 and 600°C caused a decrease in the C=O stretching vibration (1720 cm^{-1} band) and C—O—C vibrations 1260 cm^{-1} band (Fig. 20). The 1720 cm^{-1} band, which is attributed to C=O stretching vibration, showed the presence of lactonic structures. The other acidic structures such as carboxylic and phenolic either were not formed at all or were formed in small smounts and had less thermal stability. The 1590 cm^{-1} together with overlapping absorption bands in the range $1500-1100$ cm^{-1} were attributed to thermally stable carbonate structures. The chemisorption of oxygen at room

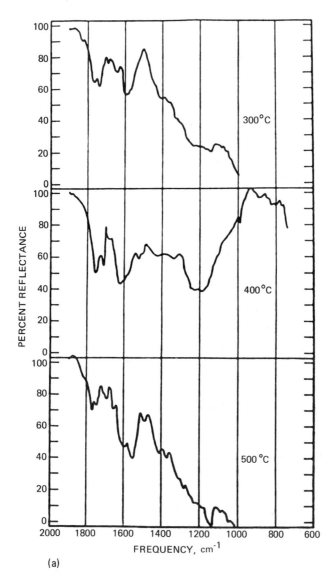

(a)

Figure 19 (a) IRS spectra of sugar charcoals activated (in 1% O_2 and 99% N_2) at 300, 400, and 500°C. (b) IRS spectra of sugar charcoals activated (in 1% O_2 and 99% N_2) at 600° and 700°C. [From Mattson et al. (123).]

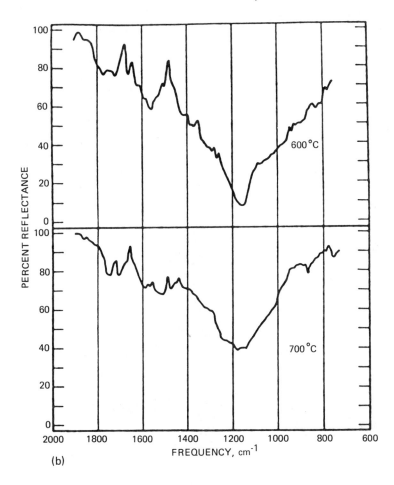

(b)

Figure 19   Continued.

temperature increased the intensity of this band due to the formation of complexes with the condensed aromatic ring system.   These
oxygen structures were present in basic carbons.   The absorption
bands in the range 1500–1100 cm^{-1} were assigned chromene structures.

The oxygen structures formed on oxidation with nitric acid (27)
were more stable and showed their absorption maxima at 3600, 1730,
1620, and 1260 cm^{-1} (Fig. 21).   The intense and broad bands of
O—H stretching vibration (3600 cm^{-1}) and the bands at 1730 (C=O
stretching) and 1260 cm^{-1} (C=O stretching) were attributed to

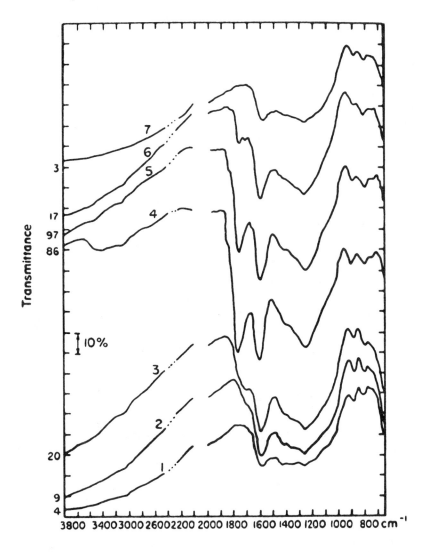

Figure 20 IR spectra of carbonic films after different treatments.
(1) Film desorbed at 600°C; (2) film oxidized with oxygen at room
temperature; (3) film oxidized with oxygen at 200°C for 30 min and
then outgassed at room temperature; (4) film oxidized with oxygen
at 300°C for 30 min and then outgassed at room temperature; (5)
film desorbed for 1 hr at 400°C; (6) film desorbed for 1 hr at 500
°C; (7) film desorbed for 1 hr at 600°C. [From Zawadzki (65).]

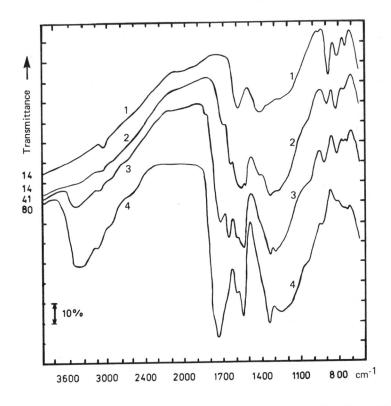

Figure 21 IR spectra of carbonic films after oxidation with nitric acid. (1) Cellulose film carbonized at 400°C and desorbed at 600°C for 1 hr; (2) film oxidized with HNO$_3$ for 1 hr at room temperature; (3) film oxidized with HNO$_3$ for 1 hr at 80°C; (4) film oxidized with HNO$_3$ for 1 hr at 100°C. [From Zawadzki (27).]

carboxyl structures where the bands at 1180 cm^{-1} (C—O stretching vibration) and 1340 cm^{-1} (O—H deformation vibration) was assigned to phenolic structures. Heat treatment at high temperature produced a doublet of bands at 1840 and 1770 cm^{-1} with another band at 730 cm^{-1}. These bands were attributed to the formation of cyclic anhydrides produced by the dehydration of two neighboring carboxylic structures.

The acidic surface structures on the carbon film oxidized with HNO$_3$ were further confirmed by IR studies of the film after neutralization with NaHCO$_3$, Na$_2$CO$_3$, and NaOH solutions and by adsorption of NH$_3$ from the gaseous phase (28). Neutralization with NaHCO$_3$

solution decreased intensities of the bands attributed to C=O stretch-ing in carboxyl groups (1730 cm^{-1}) and O—H stretching vibration (3450 cm^{-1}) (Fig. 22, curve 4). The formation of the sodium salt from carboxylic structures increased absorption at 1600 and at 1380 cm^{-1}, which was due to $CO\bar{O}$ ions. The reaction of the same film with Na_2CO_3 solution caused a decrease in the intensity of the C=O stretching vibration (curve 5) in carboxyl groups which are not neu-tralized by $NaHCO_3$ solution. The intensity of the 1720 cm^{-1} band also decreased after ion exchange with sodium hydroxide solution

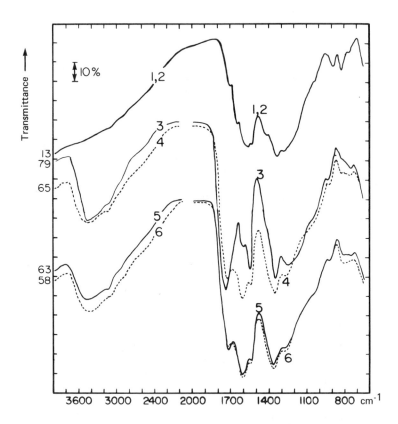

Figure 22   IR spectra of carbon films oxidized with HNO_3 and then neutralized with $NaHCO_3$, Na_2CO_3, and NaOH solutions. (1) Carbon film oxidized with HNO_3 (63%) at room temperature; (2) oxidized film neutralized with 0.05M NaOH solution; (3) film oxidized with HNO_3 at 80°C (1 hr) and at 100°C (1 hr); (4) film neutralized with 0.05M $NaHCO_3$ solution; (5) film neutralized with 0.025M Na_2CO_3 solution; (6) film neutralized with 0.05M NaOH solution. [From Zawadzki (28).]

(curve 6). The presence of the 1720 cm^{-1} band, although of low intensity, after ion exchange with NaOH solution was attributed to carboxyl structures existing in the smallest micropores, which are inaccessible to the base or to C=O groups which were not neutralized by NaOH solution. The intense IR band of OH stretching vibration after neutralization with NaOH solution was suggested to be due to structural phenolic groups, which were not quantitatively neutralized by sodium hydroxide solution. Treatment of the ion-exchanged carbon film with HCl and subsequent desorption at 200°C produced three bands at 1840, 1780, and 1740 cm^{-1} (instead of one at 1730 cm^{-1}) in the region of C=O stretching. There was also a decrease in the intensity of the 3400 cm^{-1} band. The formation of a doublet at 1840 and 1780 cm^{-1} and formation of bands at 900 and 730 cm^{-1} were attributed to formation of cyclic anhydrides by the degradation of neighboring carboxyl structures. The 1740 cm^{-1} band was assigned to laconic structures formed by dehydration of carboxyl and phenolic structures suitably located in space.

Adsorption of ammonia occurred only after the film had been oxidized with nitric acid and caused a strong absorption rise in the region of N–H stretching vibration (Fig. 23). The three bands of the oxidized film at 1840, 1780, and 1740 cm^{-1} disappeared after ammonia adsorption forming a single band at 1720 cm^{-1}. The disappearance of the doublet at 1840 and 1780 cm^{-1} was attributed to the reaction of ammonia with the cyclic anhydrides.

The disappearance of all the three (1840, 1780, and 1740 cm^{-1}) was attributed to the reaction of ammonia with lactonic structures leading to the formation of amide structures and ammonium salts of carboxylic acids. Thus according to these workers mainly the Bronsted-type acidic structures exist on the carbon film oxidized with nitric acid.

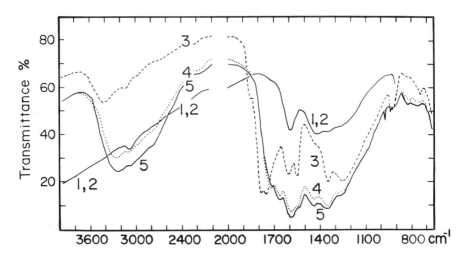

Figure 23 IR spectra of oxidized carbon films after treatment with ammonia. (1) Carbon film carbonized at 400° and desorbed at 600 °C; (2) film after desorption of Nlrs at 1.5 kPa; (3) carbon film after oxidation with HNO_3 at 100°C and desorption at 200°C; (4) film after adsorption of; (5) film after adsorption of. [From Zawadzki (28).]

2.3.8   Fourier Transform Infrared Spectroscopy

The development of computerized FTIR spectroscopy offers several advantages over conventional dispersive methods. The use of an interferometer in place of systems of grating and slits results in a higher energy throughout to the detector. Coupled with internally calibrated computer systems to add a large number of interferograms, this produces markedly superior spectra which can give more precise information concerning the oxidation of carbons and the formation of surface oxygen structures. Furthermore, FTIR can allow measurements of lower concentrations of surface functional groups.

O'Reilly and Mosher (116) examined a few carbon black samples using their dispersions in KBr by FTIR spectroscopy. Two bands at 1720 and 1600 cm^{-1} and a broad envelope at 1400 cm^{-1} were observed. The band at 1720 cm^{-1} was assigned to COOH structures and the broad envelope at 1400−1000 cm^{-1} to C−O stretching vibration of the COOH group. The breadth of the band at 1600 cm^{-1} was not conclusive for any definite assignment. A band observed at 3430 cm^{-1}, which disappeared on dehydration, was attributed to adsorbed water. These workers used an alternative technique involving Lorentzian line shapes to establish semiqualitatively the breadths of the individual curves. A typical comparison of a composite spectrum consisting of the sum of Lorentzian curve shapes with measured spectrum of a carbon black is shown in Fig. 24. The widths of the carboxylic acid band (1720 and 1600 cm^{-1}) are 22 and 43 cm^{-1} respectively, which are comparable with 20−30 cm^{-1} observed for C=O band in solid amorphous polymers. However, two peaks are required in the 1400−1200 cm^{-1} region to reproduce the experimental curve which has a band at 1400−1450 cm^{-1} corresponding to C−O stretching frequency of carboxylic acid and a band at 1120−1190 cm^{-1} probably due to coupled C−O stretching frequency and O−H bending modes of COOH and possibly C−O stretching modes of ethers. The HWHHs (half width at high height) for these curves (200 and 130 cm^{-1} are

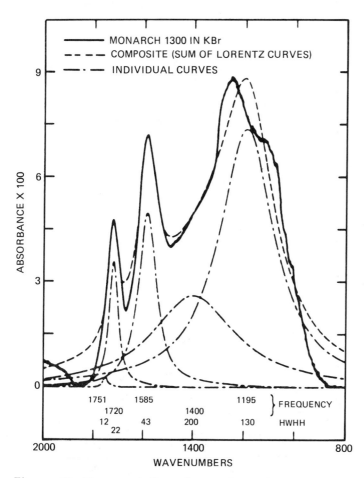

Figure 24 Representation of experimental absorbance as a sum of Lorentzian bands with band maxima and half width at half height. [From O'Reilly and Mosher (116).]

too broad for any conclusion. These workers (116) were of the view that this curve-analyzing procedure was of a limited application and a more detailed line shape analysis including a combination of Lorentzian and Gaussian curve shape elements was required.

Starsinic, Taylor, and Walker (115) carried out a systematic study of the oxidation of a Saran carbon using a potassium bromide pellet by FTIR spectroscopy. The FTIR spectra of the original and oxidized Saran (Fig. 25) show that the oxidation enhances absorption of bands at 1720, 1585, and 1250 cm^{-1} similar to the oxidation of coal (124). The 1720 cm^{-1} band, which is characteristic of the carbonyl group, was attributed predominantly to carboxylic acid structures, the 1250 cm^{-1} being partly associated with C—O stretching and OH bending modes of this functional group.

To assign the 1585 cm^{-1} band these workers (115) carried out FTIR studies after acetylation of the carbon sample to determine reactive hydroxyl groups (125) and successive base-acid washing to differentiate between COOH functions and other carbonyl structures. Acetylation of the 50.4% burnoff sample did not produce any intense band near 1770 cm^{-1}, expected if phenolic hydroxyl groups had been present (125), thus ruling out their presence in any appreciable

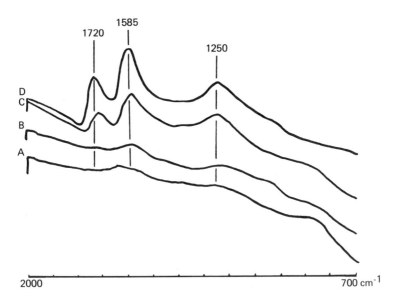

Figure 25 FTIR spectra of Saran charcoals (A) as received; (B) after 16.5% burnoff; (C) after 50.4% burnoff; and (D) after 89.1% burnoff. [From Starsinic et al. (115).]

amounts in the carbon. The 1585 cm^{-1} band cannot, therefore, be assigned to hydrogen-bonded conjugated carbonyl. A comparison of the FTIR spectra of 89.1% burnoff carbon sample after keeping in contact with 0.1 N NaOH solution for about two weeks with the original sample (Fig. 26) showed a complete disappearance of the 1720 cm^{-1} absorption band, leaving a weak band near 1690 cm^{-1} while another band at approximately 1570 cm^{-1} was now superimposed over the 1585 cm^{-1} band, showing that the band at 1720 cm^{-1} was clearly due to COOH structures. There was no evidence for the existence of lactones, which absorb in the region 1750—1790 cm^{-1}, as suggested by several workers (37,55).

A portion of the sodium hydroxide—treated carbon sample (89.1% burnoff) was placed in contact with 1N HCl for 24 hr, so that the carboxylate ions produced by the reaction of the COOH group with NaOH again exchanged cation with HCl. As a result of this treatment the 1570 cm^{-1} band was shifted back to 1720 cm^{-1} but the 1585 cm^{-1} band remained unchanged (Fig. 26). Since this band (1585 cm^{-1}) did not shift on acid treatment, these workers assigned it to a C—C stretching mode. The intensity of this band was found to be dependent on

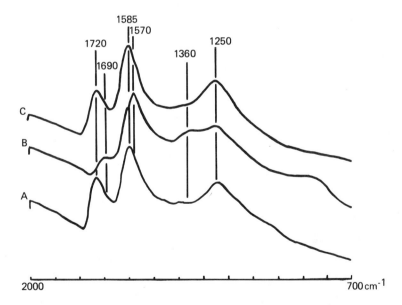

Figure 26 FTIR spectra of (A) 89.1% burnoff Saran charcoals; (B) sample after contacting with 0.1N NaOH for two weeks; and (C) sample after soaking with NaOH followed by soaking with 1N HCl for 24 hr. [From Starsinic et al. (115).]

the amounts of oxygen surface structures (Fig. 27) on carbons and finally leveled off at higher values. This was also cited as an evidence for an aromatic ring stretching mode of this band.

The thermal desoprtion of oxidized Saran charcoals evolved CO and CO_2 in the ratio 4:1, which could not be expected on the basis of the presence of carboxylic structures alone. This was explained by postulating the existence of ether groups as evidenced by the appearance of asymmetric and symmetric $C \stackrel{\textstyle O}{\underset{\textstyle O}{\raisebox{0pt}{$<$}}}$ stretching modes near 1570 and 1360 cm^{-1} respectively.

Van Driel (58) examined an active charcoal by FTIR spectroscopy using the potassium bromide pellet technique before and after oxidation. The difference spectra obtained by substracting the unoxidized from the oxidized sample spectra (Fig. 28) showed that the peaks at 1735 and 1585 cm^{-1} had been enhanced, whereas a peak at 1463 cm^{-1} completely vanished on oxidation. The 1735 cm^{-1} peak was attributed to f-lactone structures.

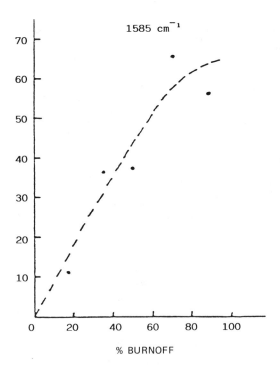

Figure 27 Relationship between percentage of burnoff and the intensity of the 1595 cm^{-1} band for Saran charcoal. [From Starsinic et al. (115).]

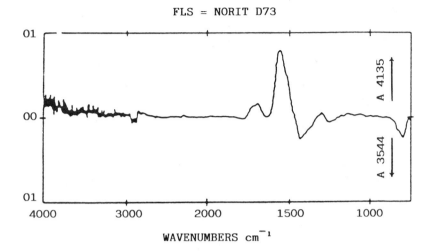

FLS = NORIT D73

WAVENUMBERS cm⁻¹

Figure 28 Difference between FTIR spectra of oxidized and unoxidized sample of an activated carbon. [From Van Driel (58).]

2.3.8.1 *Fourier Transform Infrared Photothermal Beam Deflection Spectroscopy (IR-PDS)*

Morterra, Low, and co-workers (64,126—128) used IR-PDS spectroscopy to study oxidation of carbons prepared by charring cellulose filter paper (Whatmann No. 1, ash content 0.06%) or fiber (Purity 99.6%) under well-defined and controlled conditions leading to the formation of acidic and basic carbons. The oxidation was found to depend on the nature and structure of the char and on the temperature at which it was carbonized. These workers studied the oxidation of chars carbonized between 300 and 880°C (64). Infrared spectra revealed that the oxidation of low-temperature chars started around 200°C—a temperature at which a large number of residual functional groups appeared on the char surface and the IR spectra were therefore quite complex. The high-temperature chars prepared above 600°C had no spectroscopically evident residual surface structures before oxidation and should therefore be more useful for identification of surface oxygen structures formed as a result of oxidation.

The desorption spectra of 560° char oxidized at 400°C (Fig. 29) showed absorptions at 1850, 1790, 1260, 910, and 740 cm⁻¹. These absorptions were attributed to five numbered cyclic conjugated anhydride structures (129) which have characteristic group frequencies at 1860—1850 cm⁻¹ (in phase C—O stretch, weak), 1780—1760 cm⁻¹ (out

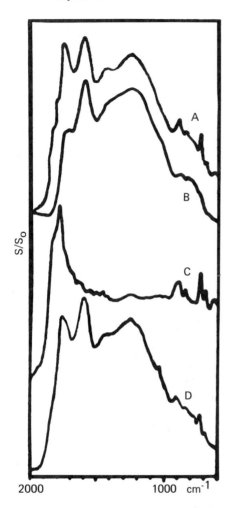

Figure 29 Oxidation—reoxidation IR-PDS spectra of cellulose carbon carbonized at 560°C. (a) Sample oxidized at 400°C; (B) oxidized sample degassed at 560°C; (C) Ratio of A and B; (D) sample B reoxidized at 400°C. [From Morterra et al. (64).]

of phase C—O stretch, strong), and at 1300—1180 cm^{-1} and 955—895 cm^{-1}. These cyclic anhydrides, which evolve CO_2 and are acidic in nature, were also formed on 730° char but were not detectably formed on 880° char. Dauben and Epstein (129) were of the view that these structures were formed by the oxidation of the edge carbon atoms of a polyaromatic network which were gradually annealed or built into a bigger and bigger polyaromatic network as the temperature was increased. The fact that smaller amounts of these structures were formed at 560° char upon reoxidation was cited as an evidence (Fig. 29). The broad and more complex absorptions in the spectra of 360° char at 1780—1720 cm^{-1} and below 1000 cm^{-1} were attributed to carboxylic groups, which were not likely to form on high-temperature chars containing little residual hydrogen, especially after oxidation has taken place.

The spectra of 730° char (Fig. 30) was almost featureless before oxidation at temperatures lower than 300°C. Some features started developing after oxidation at 310°C and showed considerable changes after oxidation at 420°C (64). The sample did not show any burnoff until reaching a temperature higher than 420°C. The spectra of 730° char showed a weaker shoulder at 1850 cm^{-1}, a sharp band at 1770 cm^{-1} with no resolved side components. In addition there was a symmetrical and dominant band at 1600 cm^{-1}, a broad band between 1500 and 950 cm^{-1}, and a sharp and clearly resolved band at 770 cm^{-1} with a few minor components on the high-frequency side. The spectra of 880° char were almost similar to those of 730° char but some of the poorly defined features after oxidation at 420° became well defined after oxidation at 500°C with no further changes occurring at higher temperatures to the burnoff stage (> 650°C). The spectra showed the 1760 and 1600 cm^{-1} bands, the broad absorption at 1500—1000 cm^{-1}, and only one well-resolved band near 740 cm^{-1}, which was more evident than with the 730° char. The 1760 cm^{-1} band (64) was attributed to lactonelike structures most likely of a six-numbered ring unsaturated type (130). These structures sometimes show a doublet 15—30 cm^{-1} apart due to Fermi resonance as in the case of 730° char. The 1600 cm^{-1} band (64) was attributed to a C=C mode of the polyaromatic framework made IR active by asymmetry introduced by the surface oxide species. The experiments carried out by using ^{18}O indicated that the bulk of the 1600 cm^{-1} absorption cannot be ascribed to some ionic carbonatelike or carboxylatelike species, as suggested by Zawadzki.

The spectroscopic examination of basic carbons showed no evidence for the formation of any oxide structures or any of the cellulose chars.

The oxidation pattern of NOC chars (126) (chars prepared by the pyrolysis of NO_2-oxidized cellulose) was almost similar to the oxidation behavior of pure cellulose chars. The low-temperature of

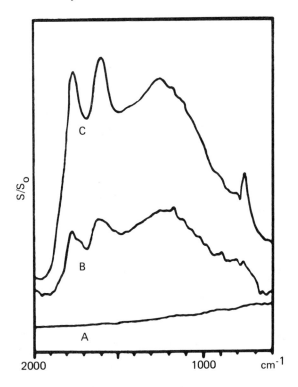

Figure 30 IR-PDS spectra of cellulose carbon prepared at 730°C.
Carbon oxidized (A) at temperatures below 300°C; (B) at 330°C;
and (C) at 420°C. [From Morterra et al. (64).]

NOC 320 and NOC 540 showed mainly a strong absorption at 1850 cm^{-1}
and slightly less strong absorption at about 1800 cm^{-1}. These absorptions were due to anhydride structures. When the NOC 540 char was
oxidized between 400 and 500°C, a strong and complex band system
with apparent maxima at $\cong$ 1770 and $\cong$ 1720 cm^{-1} was observed which
was assigned to lactonelike structures. The differential spectra clearly indicated that while these bands were being formed, the anhydridelike structures formed were progressively being decomposed.

The oxidation behavior of NOC 700° char (126) was also similar
to that of the corresponding cellulose char (Fig. 31). The oxidized
layer developed only on oxidation above 450°C and did not contain
any spectroscopically detectable anhydride structures absorbing above
1800 cm^{-1}. There was formation of lactonelike species absorbing at
1770 cm^{-1} with weak shoulders on the low wave number side. The

amounts of these structures increased up to an oxidation temperature of 500°C and declined at higher oxidation temperatures. On oxidation above 600°C no carbonyl species absorbing above 1700 cm^{-1} were left and there was an appreciable burnoff of the sample (126).

The oxidation behavior of the cellulose chars prepared after carbonizing cellulose impregnated with 5% sodium chloride or potassium bicarbonate (designated NCC and KHC chars) (127,128) was different from that of pure cellulose in NOC chars. The oxidation of NCC chars (127) below 650°C showed a different ratio of the oxidation products and the oxidized layer was more ionic than with cellulose chars. However, when sodium chloride was evaporated at 650°C, the oxidation behavior resembled that of the cellulose char. The presence of

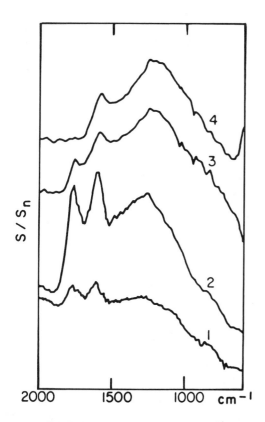

Figure 31 IR-PDS spectra of oxidized cellulose (NOC-700°) carbons carbonized at 700°C. The ratios are NOC-700° oxidized (1) at 450°/ NOC-700°; (2) at 500°/NOC-700°; (3) at 570°/NOC-700°; (4) at 660°/ NOC-700°. [From Morterra and Low (126).]

sodium chloride also changed the oxidation mechanism. The acidic structures which evolved CO_2 on heating were formed to much lower extents (Fig. 32). NCC 360° char on oxidation showed no acidic species. The 1690 cm^{-1} band was due to ketonic or aldehydic groups already present in the char. Oxidation of NCC 550° char showed bands near 1770 and 1720 cm^{-1} corresponding to lactone structures and virtually no band near 1860 and 1750 cm^{-1} corresponding to anhydridelike structures. The spectrum of NCC 650° char, however, showed an abundance of lactonelike structures (127).

The burnoff in the case of KHC chars (128) occurred at much lower temperatures and the gasification temperature range was narrow compared to pure cellulose chars. The spectrum of KHC 590° char showed a continuum and was otherwise featureless. After exposure to oxygen at 350°C some relatively weak spectral features appeared which changed and became more intense at 420°C. At 460°C there was quick burnoff leaving a white crust (Fig. 33). The differential spectra showed relatively weak absorptions around 1200 cm^{-1} and a weak band at 1600 cm^{-1}, which disappeared almost completely on oxidation above 420°C. There was no observable formation of anhydride or lactonelike structures (128).

2.3.9 X-Ray Photoelectron Spectroscopy

X-ray photoelectron spectroscopy (XPS) or ESCA is likely to be of enormous importance in the science and technology of carbons and graphites. Essentially the technique measures the kinetic energy of electrons emitted from atoms under the influence of irradiation with X-rays. The kinetic energy of the emitted electron is related to the binding energy of the electrons, which in turn is dependent upon the chemical environment of the atom from which the electron has been emitted. Thus by measuring the kinetic energy of the emitted electron, the magnitude of the binding energy of an electron can be directly obtained. The variation of binding energy of the electron with the environment of the atom gives rise to different peaks in the ESCA spectrum with the varying intensity for the same atom depending on the nature of the atoms to which the atom is attached. Furthermore, the penetration depth from which the photoelectron emerges is seldom more than 10–15 nm. i.e., about 10–20 atomic layers from the surface, which makes XPS (ESCA) ideal for surface chemical analysis as well as for the study of adsorbed species.

A major part of the work on ESCA studies on carbons has been carried out using carbon fibers and graphites because of their easy manipulation and importance, although a few investigations have been reported on carbon blacks, chars, and activated carbons. However, it is believed that as the technique becomes easily available to investigators at different centers of carbon research, more and more work will be done on surface groups on activated carbons using this technique.

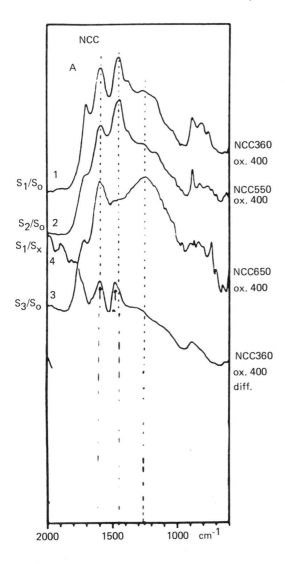

Figure 32 IR-PDS spectra of NaCl-impregnated cellulose carbons. The number indicates the temperature of carbonization and the temperature of oxidation. [From Low and Morterra (127).]

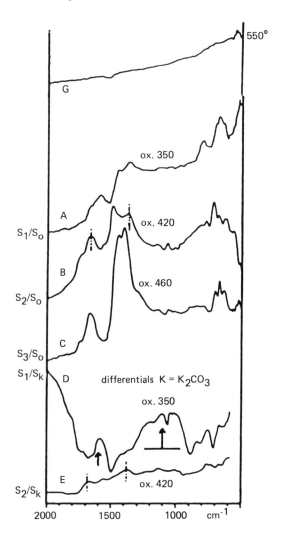

Figure 33 IR-PDS spectra of KHCO$_3$-impregnated cellulose carbons. The number indicates the temperature of carbonization and the temperature of oxidation. [From Morterra and Low (128).]

Thomas et al. (131) carried out surface analysis of type I and type II PAN-based carbon fibers before and after heating and oxidation using ESCA. The nitrogen 1s, oxygen 1s, and carbon 1s spectra of the oxidized fiber (Fig. 34) showed a shoulder on the higher energy side of the carbon 1s peak, indicating two distinct kinds of carbon atoms (deconvolution techniques have now revealed at least four kinds of carbon responsible for this profile of the carbon 1s electron). The spectrum changes completely when the carbon fiber was heated at 1300 K in nitrogen (Fig. 35). The nitrogen and oxygen peaks disappeared almost completely and the carbon peak sharpened considerably. On subsequent oxidation of the heated carbon fiber, a significant oxygen 1s peak reappeared. It was observed that about five oxygen atoms per 100 carbon atoms were present in the penetration depth of 10–15 nm on the surface of the oxidized 1300 K carbon fiber.

Donnet, Papirer, and colleagues (122) applied ESCA to the study of surface groups on a carbon black before and after methylation followed by hydrolysis of the methylated product. The samples were spread on a double-sided adhesive tape and the spectra for $C_{(1s)}$ and $O_{(1s)}$ were recorded at the rate of 0.05 V/sec. The $C_{(1s)}$ spectra for all the three samples showed an asymmetric peak at about 284 eV tailing markedly to the high binding energy side (Fig. 36). The asymmetry suggested the presence of carbon–hydrogen bands (normally around 285 eV) and carbon–oxygen bands (at about 288.5 eV). The

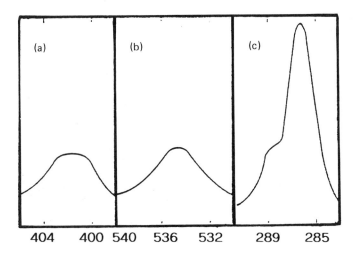

Figure 34 High-energy photoelectron spectra of oxidized Courtelle type II carbon fiber of (a) nitrogen 1s; (b) oxygen 1s; (c) carbon 1s. [From Thomas et al. (131).]

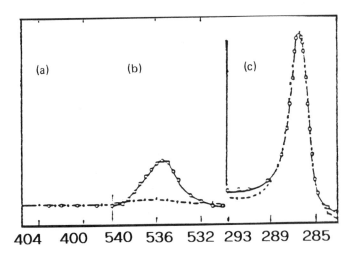

Figure 35 High-energy photoelectron spectra of oxidized Courtelle type II carbon fiber after heat treatment at 1300 K of (a) nitrogen 1s; (b) oxygen 1s; (c) carbon 1s. [From Thomas et al. (131).]

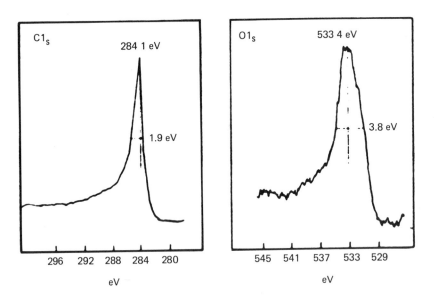

Figure 36 XPS narrow scan photoelectron spectra of Carbolac 1. [From Donnet (120).]

oxygen 1s peak also showed asymmetry, suggesting the presence of two or more forms of bonded oxygen. The deconvoluted $O_{(1s)}$ spectrum (Fig. 37) showed constant positions of the four individual peaks, although their intensities varied. The $O_{(1s)}$ peaks in the original and methylated carbon black were almost similar but for the methylated and the hydrolyzed sample indicated a relatively larger concentration of the higher binding energy form of oxygen. But these workers could not draw any definite conclusions regarding the nature of the surface oxygen structures on the basis of ESCA studies alone.

Ishitani and Takahagi (132,133) analyzed surface functional groups of carbon (CF) and graphite (GF) fibers, introduced by surface oxidation, by ESCA, and digital difference spectrum techniques. The differential spectra for the samples before and after oxidation eliminated the influence of symmetry. A comparison of the XPS spectra (Fig. 38) showed asymmetric $C_{(1s)}$ peaks with tails on the higher binding energy sides both for control CF and GF fibers. But the spectra for control CF was broader than the control GF, probably due to more disorganized lattice structure in control CF fiber. The $O_{(1s)}/C_{(1s)}$ ratio, which was taken as a measure of the degree of oxidation, increased with increase in the extent of oxidation in each carbon fiber and was more in the case of CF compared to GF. The oxygen content of the GF fiber was only about 30% of that of CF fiber. Both the carbon fibers, however, reached the same level of oxygen content on thermal treatment due to decomposition of the oxygen surface structures.

The moderate oxidation of CF resulted in an additional peak at 288−289 eV and a shoulder at 286 eV in the $C_{(1s)}$ spectrum, but the shoulder disappeared and the peak became more noticeable on extensive oxidation. The oxidation of GF fiber added two components at 288−289 eV and at 286 eV. On heat treatment, however, although the CF spectra returned to control, those of GF did not completely return to control, the component at 286 eV still remaining.

The difference spectra (133) obtained by subtracting the $C_{(1s)}$ spectra of control fibers from those of oxidized fibers using the appropriate weight factors showed three components with different chemical shifts, which were assigned to hydroxyl groups (> C−OH, 286 eV, carbonyl groups (> C = O, 287 eV), and carboxyl groups (− C≤$^{OH}_{O}$, 288.6 eV) (Fig. 39). The curve resolving $C_{(1s)}$ difference spectra of CF gave the concentrations of the three functions as carboxyl 17%, carbonyl 10%, and phenolic 73% on moderate oxidation and carboxyl 54%, carbonyl 22%, and phenolic 24% on extensive oxidation. The difference spectrum of GF had two components. The higher binding energy component, which disappeared on heat treatment (288−289 eV), was assigned to carboxyl groups and the one with the lower binding energy, which stayed after thermal treatment (286 eV), was attributed to disordering of the graphite crystal lattice caused by surface oxidation.

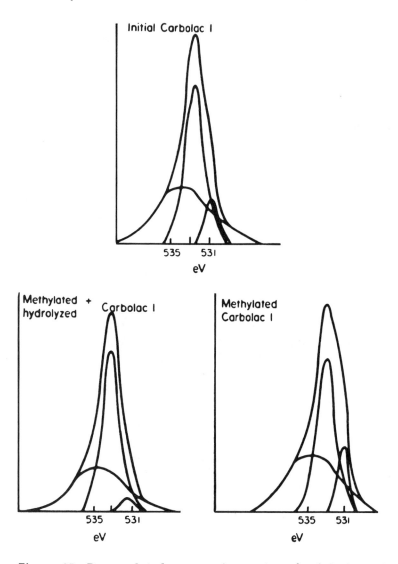

Figure 37 Deconvoluted oxygen 1s spectra of original, methylated, and hydrolyzed Carbolac 1 samples. [From Donnet (120).]

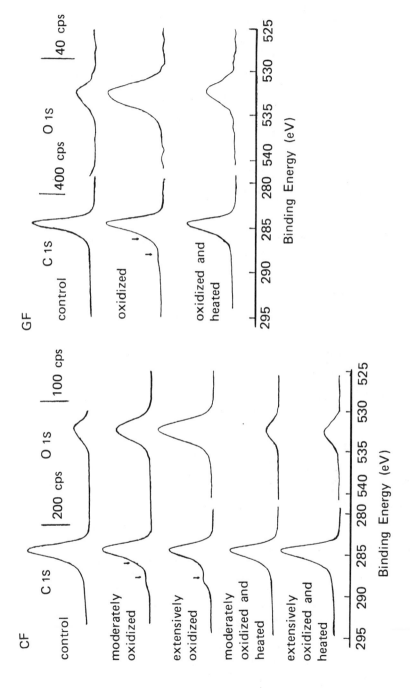

Figure 38 XPS spectra of type II (CF) and type I (GF) carbon fibers after different treatments. [From Ishitani (132).]

C 1s difference spectra

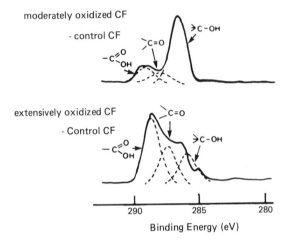

C 1s difference spectrum

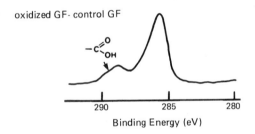

Figure 39 Carbon 1s digital difference spectra of type II (CF) and type I (GF) carbon fiber before and after oxidation. [From Takahagi and Ishitani (133).]

Proctor and Sherwood (134) oxidized carbon fibers electrochemically using sulfuric acid and ammonium carbonate as the electrolytes and analyzed the oxide layer using XPS. The buildup of the oxide layer and the relative proportion of the difference in surface oxygen species varied with the applied potential. The $O_{(1s)}$ spectrum was described by two main component signals corresponding to C=O and C—O type structures, the former dominating at lower potentials and the latter dominating at higher potentials. Assignment of curve-fitted peaks in the $C_{(1s)}$ region was not straightforward and could not be correlated with the $O_{(1s)}$ spectrum. Oxidation in ammonium carbonate solution showed the formation of carbon—nitrogen structures.

Vaneica (135), while studying the oxidation of a carbon foil by angular distribution XPS, observed an oxygen doublet with binding energies of 531.9 and 533.6 eV and with an intensity ratio of 1.5:1 in the untreated foil. The doublet disappeared on heating the foil at 531°C in ultra-high-purity nitrogen and formed ether-type functional groups when bombarded with 500 eV argon ions. The oxidation of the heat-treated foil with Standenmaier method (mixture of H_2SO_4 and HNO_3) produced an intense $O_{(1s)}$ peak at 531.8 eV coupled with a much weaker peak at 533.7 eV (Fig. 40). The $C_{(1s)}$ spectrum showed three peaks at 283.6, 285.1, and 286.4 eV. The 531.7 eV peak in the $O_{(1s)}$ spectrum and the 286.4 eV peak in the $C_{(1s)}$ spectrum were attributed to anhydrous graphite oxide and the 533.7 eV peak in the $O_{(1s)}$ spectrum to chemisorbed water. The two other peaks in the $C_{(1s)}$ spectrum were attributed to massive physical damage of the carbon foil.

Kavan et al. (136) used XPS to investigate the surface structures formed on oxidation in air at room temperature on a carbon obtained by the electrochemical reduction of poly(tetrafluoroethylene) (PTFE).

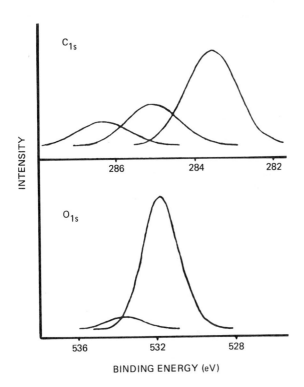

Figure 40 Carbon 1s and oxygen 1s of oxidized carbon foil. [From Veneica (135).]

The carbon was very reactive to oxygen. The oxidation resulted in the formation of surface oxide groups with well-defined chemical shifts (4.3–4.9 eV) of the $C_{(1s)}$ photoemission band, which was attributed to COOH structures. The assignment of the peak 290.2–290.8 eV to carbon in COOH groups, according to these workers, was supported by the fact that the ratio of the concentration of carbon with a higher binding energy to that with a lower binding energy (C_{280}/C_{286}) increased continuously with the degree of oxidation, i.e., with decreasing C/O ratio.

Fischer and Feichtinger (137) examined oxygen surface structures on a carbon oxidized in air at 420°C and subsequently degassed at 800°C in vacuum using ESCA. The oxidized sample was found to contain appreciably larger amounts of oxygen ($\cong$ 1.7 times more) compared to the outgassed sample (Fig. 41). The combined oxygen in the oxidized sample was present in at least two chemical states with bonding energies of 532.2 and 533 eV while the oxygen in the degassed sample seemed to have mainly a binding energy of 532.2 eV. The binding energy of 532.2 eV was attributed to adsorbed oxygen and for 533 eV to chemical C—O structures.

Grint and Perry (138) combined ESCA with TGA and IR spectroscopy to characterize the nature of surface structures formed on oxidation of some bituminous coals. The carbon 1s peak [$C_{(1s)}$] showed considerable broadening (from 2.2 eV to 2.7 eV at FWHM) on the higher binding energy side of the hydrocarbon peak at 285 eV (Fig. 42) due to the formation of carbon—oxygen surface structures by oxidation (138). The groups postulated by these workers were phenolic, ether, and alkoxy structures corresponding to 286.2 eV; ketones and aldehydes at 288 eV; and carboxyl groups including esters at 289.2 eV. The oxygen 1s peak was more complex. The theoretical synthesis of the $C_{(1s)}$ peak using components from these groups (Table 17) showed (138) that the increase in the oxidation temperature from 25 to 385°C increased the formation of all three types of groups but the major part of the chemisorbed oxygen was being incorporated in the form of singly bonded C—O structures. The carboxylate and carbonyl peaks showed a sudden rise at 225°C just below the temperature when oxidative degradation commenced (138).

2.4 BASIC CARBONS

It has been known for a long time that charcoals have a basic character. But the structure of the oxides responsible for this basic character has not yet been elucidated. The basic carbons are those carbons which give basic reactions and adsorb acids and little alkali. These carbons are prepared by contacting outgassed carbons, cooled to room temperature in the absence of oxygen, with oxygen below

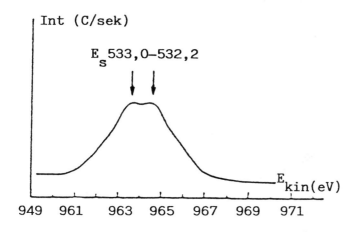

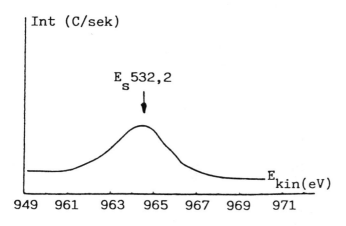

Figure 41 Oxygen 1s spectra of oxidized and degassed carbon. Narrow-scan X-ray photoelectron spectrum of O (1s) emission in two samples. [From Fischer and Feichtinger (137).]

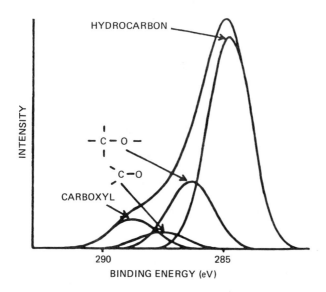

Figure 42   Carbon 1s peak synthesis for coal heated to 325°C.
[From Grint and Perry (138).]

Table 17   Components of $C_{(1s)}$ Peak After TGA Experiment

	Carbon components		
Sample temperature (°C)	Phenolic alkoxy ether groups	Ketones and aldehyde groups	Carboxylic and ester groups
25	12.5	3.1	2.9
50	13.2	3.1	2.9
100	13.6	2.7	2.7
150	15.7	3.3	3.0
225	16.7	6.1	7.6
250	16.7	5.3	8.3
300	17.3	5.2	9.2
325	20.8	4.9	9.2

Source: Grint and Perry (138).

200° C or above 700°C. However, there is no clear demarcation between the temperature within which exposure to oxygen would result in exclusive development of a basic character in carbons.

Bartell and Miller (5) were the first to observe that sugar carbons which were activated at 800−900°C had a basic character and adsorbed acids but little alkali. But it was not until 1929 that Kruyt and de Kadt (66,67) found that a carbon capable of adsorbing acid but little alkali becomes acidic and adsorbs only alkali after heating in air at 400°C. Burstein and Frumkin (50), however, found that a basic charcoal outgassed in high vacuum did not adsorb acid from a deaerated solution unless molecular oxygen was admitted. Frumkin and his group (50,51) proposed an electrochemical theory according to which the adsorption of an electrolyte was determined by the electrical potential at the carbon solution interface and by the capacity of the double layer. But Shilov et al. (52,53) rejected the electrochemical theory, as they claimed that some acid was adsorbed even on completely outgassed charcoals in the absence of oxygen. These workers attributed acid adsorption to surface oxides of a basic nature. These surface oxides were formed when a charcoal, outgassed in vacuum at high temperature, came into contact with oxygen only after cooling.

Steenberg (139) was of the view that acid adsorption resulted largely from physical forces. The protons were held close to the surface by primary adsorption forces and the anions by secondary electrostatic forces forming the outerpart of the diffused double layer. A part of the adsorbed acid could be removed from the carbon surface by nonpolar liquids such as toluene (139). Burstein and Frumkin (50) observed that when acids are adsorbed in the presence of oxygen, hydrogen peroxide is produced, but the amount of hydrogen peroxide produced was not equivalent to the acid adsorbed as expected from their formulation of the reaction. The formation of hydrogen peroxide when moist air and acid reacted with carbon was confirmed by Lamb and Elder (140), Kolthoff (141), and King (142).

Garten and Weiss (55) postulated the existence of chromene groups on basic carbons which were readily oxidized at room temperature in the presence of acid to the corresponding benzopyrylium (carbonium) structures with the adsorption of the anion of the acid and liberation of hydrogen peroxide.

The presence of the carbonium ion, according to Garten and Weiss (55), was responsible for the weakly basic character of these carbons in the presence of oxygen. Frumkin's objection to the oxide theory was met by admitting that some acid may be adsorbed physically as well as chemically. Puri (57), however, argued that there was no independent proof for the existence of chromene structures. Puri et al. (80) and Weller and Young (9) observed that unlike adsorption of bases by carbons, which was almost irreversible, adsorption of acids was not only very much less but was also largely reversible. Moreover, carbons which were outgassed at 1200°C in vacuum and cooled out of contact with air were even better acid adsorbents compared to samples oxidized at 800–1000°C and retained bonded oxygen in the form of basic oxides (57). The basic character of carbons therefore cannot be attributed to the existence of chromene or any other oxygen-containing surface structures. Boehm (56) observed that a graphitized carbon black adsorbed appreciable amounts of acids even though it contained no oxygen. Studebaker suggested that the carbons prepared by the various pyrolytic processes were like polycondensed aromatic hydrocarbons, which are known to be Lewis bases (electron donor) and which can preferentially exchange H^+ ions to OH^- ions from water or aqueous solutions. This would leave the surrounding liquid with excess OH^- ions, giving rise to the basic character. The chemically bonded oxygen, if present, would counteract the basic character in the Lewis sense.

Boehm and Voll (143) observed that active carbon and carbon black which have been freed of oxygen surface structures on heat treatment in vacuum at 1000°C or in nitrogen at 1400°C and then cooled to room temperature in an inert atmosphere adsorb mineral acids from dilute aqueous solutions in the presence of molecular oxygen. This has been attributed to the formation of basic surface oxides (structures), which cannot be represented by pyronelike structures

with the oxygen atoms, in general located in two different rings of a graphite layer, the positive charge being stabilized by resonance. The adsorption of the acid was accompanied by the uptake of oxygen in stoichiometric amounts, one equivalent of acid being adsorbed for each gram atom of oxygen uptake. Both the cation and the anion of the acid were adsorbed. The adsorbed acid could be completely desorbed on treatment with 0.05 N NaOH or with 0.1 N C_2H_5ONa. The product resulting from the description of the chemisorbed acid with 0.05 N NaOH was called the "hydroxide" of the basic structure. The adsorption of the acids was chemical at lower concentration and obeyed Langmuir isotherm equation but was followed by physical adsorption at higher concentrations (Fig. 43). These workers also calculated the dissociation constants of the protonated basic surface oxide (Table 18).

The formation of the basic oxides was also accompanied by the formation of hydrogen peroxide, its content being only 10% of the basic oxide in the case of sugar charcoal (Fig. 44), indicating that it was formed in a side reaction. The rate of acid adsorption was independent of the hydrogen ion concentration but was approximately proportional to the square root of the oxygen pressure above 150 torr. Pyrolysis of the carbons containing basic surface oxides indicated that each basic surface structure contained two differently bonded oxygen atoms, one of which was decomposed into CO and CO_2 at 900 °C and the other at 1200°C.

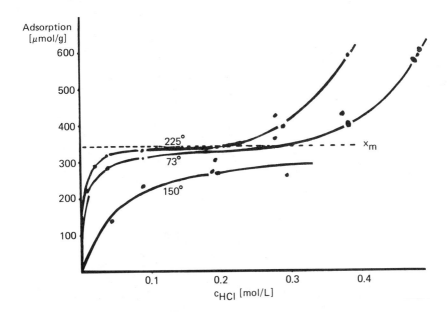

Figure 43 HCl adsorption for sugarcoal ZK5 from aqueous HCl at different temperatures. [From Boehm and Voll (143).]

Table 18 Dissociation Constants of Basic Oxides at Varying Temperatures

Temperature (°C)	K (mol/l)
25	2.42×10^{-3}
50	5.40×10^{-3}
75	1.11×10^{-2}
100	1.97×10^{-2}
125	3.38×10^{-2}
150	5.40×10^{-2}

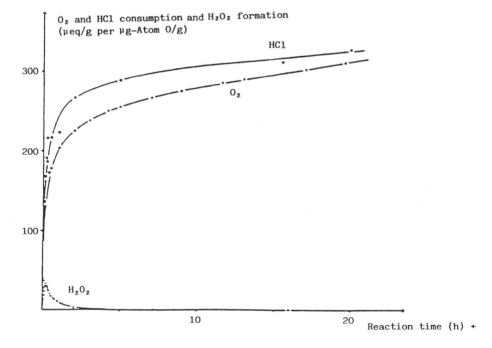

Figure 44 Uptake of oxygen and HCl and the formation of H_2O_2 during the formation of basic oxides on the surface of sugar charcoal ZK-5 at room temperature. [From Boehm and Voll (143).]

These workers (143−146) also determined the active hydrogen on carbon surfaces covered with basic surface oxides in the hydroxide form by determining exchange with D_2O. The weight of the sample decreased very slowly by outgassing at 120° C and no active hydrogen was found on the carbon surface when the weight of the sample became constant after three days. Attempts to determine polar groups on basic carbons by adsorption of H_2O, HCl, BF_3, and $AlCl_3$ from gaseous and from chloroform solutions indicated no specific adsorption in any case. The water adsorption isotherms were not the type II as usually expected on a carbon surface covered with OH groups (Fig. 45). This was attributed to the complete loss of protons chemisorbed in the basic surface oxide on prolonged outgassing at 120° C.

When the basic surface oxides were reacted with aqueous ammonia or ammonium salts, one oxygen atom of the surface groups was exchanged for nitrogen. The same quantity of nitrogen was fixed when

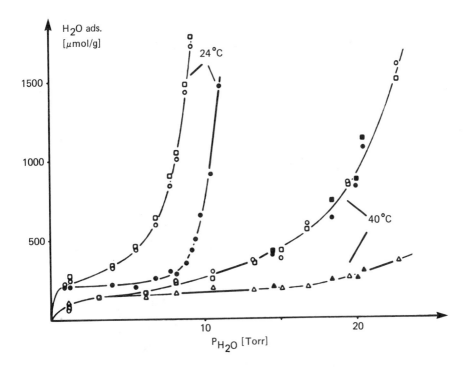

Figure 45 Water vapor adsorption on sugar coal. At 24°C: •, ZK5; ○, ZK5 (freshly prepared with basic surface oxides); □, ZK5-HOH at 40°C. At 40°C. At 40°C: △, ZK5; ○, ZK5-HOH; □, ZK5-HOH at 40°C. (▲ • ■, desorption). [From Voll and Boehm (146).]

basic oxides were formed in the presence of ammonia or ammonium salts. This nitrogen-containing surface group contained one equivalent exchangeable hydrogen and chemisorbed one equivalent of the acid. The amount of acid adsorbed was the same as that adsorbed by the hydroxide form of the basic surface oxide before reaction with ammonia. The basic surface oxides before or after reaction with ammonia could be methylated with methyl iodide or with methanol and sulfuric acid. The amount of the methoxyl group formed was equivalent to the amount of the acid adsorbed by the original basic surface oxide. Methylation and reaction with ammonia gave the same product irrespective of which was done first, indicating that the two reactions occur at different sites. The methylated product did not adsorb any acid but there was an equivalent exchange of sulfate. However, when the methylated product was treated with ammonia, the acid could be adsorbed (Table 19). All these reactions were shown to be possible on the basis of the pyronelike structure (Fig. 46), the two oxygen atoms being located in two different rings of a graphitic layer (Fig. 47).

Table 19 Analysis Values of Reactions of Basic Surface Oxides

Reaction	Zk5-HOH (μeq/g)	CK3-HOH (μeq/g)
Deuterium exchange	30	5
HCl adsorption	130	35
Acetylation with CH_3COCl	130	—
After handling with NH_3		
N content (without pyridine)	130	50
HCl adsorption	130	—
Deuterium exchange	150	—
Methylation with CH_3l	30	—
After methylation with CH_3OH/H_2SO_4		
Methoxyl content	130	32
H^+ adsorption from HCl	0	4
Cl^- adsorption from HCl	120	33
After methylation and subsequent reaction with NH_3		
N content (without pyridine)	130	—
HCl adsorption	140	25

Source: Voll and Boehm (146).

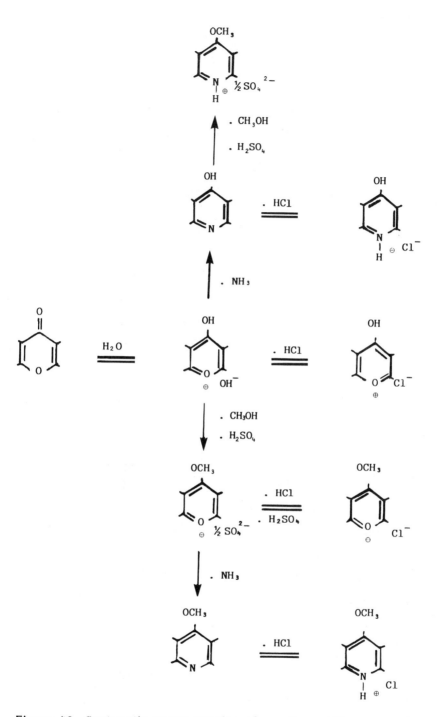

Figure 46 Systematic representation of reactions with γ pyrons as models. The two oxygenations are in fact bonded toward imprecisely fixed sites of the nucleus (hexagonal nucleus). [From Voll and Boehm (146).]

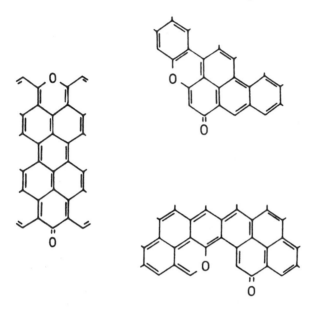

Figure 47 Possible structures of pyrones at the surface.

Zawadzki (65) carried out IR spectral studies of the oxidation of carbon films in the temperature range and under conditions when basic carbons are produced. He observed adsorption at 1590 cm^{-1} with overlapping bands in the 1500–1100 cm^{-1} range, which he attributed to carboxylic carbonate structures. Chemisorption of oxygen at room temperatures was found to enhance the intensity of this band (1590 cm^{-1}) through the formation of complexes with a system of condensed aromatic ring systems. The adsorption bands occurring at low wavenumbers (1500–1100 cm^{-1}) were attributed to chromene structures, as suggested by Garten and Weiss (55).

Morterra et al. (64) prepared basic cellulose chars under carefully controlled conditions both by degassing of the carbon at high temperatures followed by room temperature exposure to oxygen and by high-temperature oxidation. Examination of basic carbons using IR photothermal beam deflection spectroscopy revealed no evidence for the formation of any oxide species or any of the basic cellulose chars. These workers, however, did not rule out the possibility of chemisorption of oxygen but suggested that whatever oxygen was chemisorbed did not produce any structures that could be detected in the spectra under their experimental conditions. Such structures may be either IR inactive due to structure and symmetry considerations or IR active but present in relatively very small amounts so that the discrete weak bands could not be detected in the presence of the strong background or they were strongly coupled to the surface and made IR modes extremely broad. Thus these workers are emphatically of the view that the basic properties of carbons cannot be assigned to well-defined oxygen structures (64). Their data seem to confirm the views of Puri (57) that "the basic character of these carbons cannot be attributed to the existence of chromene or any other oxygen containing surface groups and that the influence of combined oxygen on basic reactions of carbons, at least, is doubtful." These workers are of the view that the adsorption of acids on carbons was due to physical forces (139) or to Lewis coordination of proton or electron donor centers (44). However, they suggest the need for more work to make things clear.

REFERENCES

1. Smith, R. A., *Proc. Roy. Soc. (London)*, *12:*425 (1863).
2. Lowery, H. H. and Hullet, C. A., *J. Am. Chem. Soc.*, *42:*1409 (1920).
3. Rhead, T. F. E. and Wheeler, R. V., *J. Chem. Soc.*, *101:*846 (1912).
4. Lambert, J. D., *Trans. Faraday Soc.*, *32:*249 (1931).
5. Bartell, F. E. and Miller, E. J., *J. Phys. Chem.*, *28:*992 (1924)
6. Garner, W. E. and McKie, *J. Chem. Soc.*, *1927:*2451.
7. Ward, A. F. C. H. and Rideal, E. K., *J. Chem. Soc.*, *1927:*3117.
8. Marshall, M. J. and Bramston, C., *Can. J. Res.*, *B15:*75 (1937)
9. Weller, S. W. and Young, T. F., *J. Am. Chem. Soc.*, *71:*4155 (1948).
10. Puri, B. R., Anand, S. C., and Sandle, N. K., *Indian J. Chem.*, *4:*310 (1966).
11. Dietz, V. R. and McFarlane, E. F., in *Proc. 5th Conf. Carbon*, Vol. II, Pergamon Press, New York, 1961, p. 219.
12. Allardice, D. J., *Carbon*, *4:*255 (1966).
13. Kiselev, A. V., Kovaleva, N. D., Polyakova, M. M., and Tesner, D. A., *Kolloid Zh.*, *24:*195 (1962).
14. Donnet, J. B. and Papirer, E., *Bull. Soc. Chim. France*.
15. Donnet, J. B. and Papirer, E., *Rev. Gen. Caoutchouc Plastiques*, *42:*889 (1965).
16. Smith, R. N., *Quart. Rev.*, *13:*287 (1959).
17. Walker, P. L., Jr., Austin, L. G., and Tietjen, J. J., in *Chemistry and Physics of Carbon*, Vol. I, P. L. Walker, Jr., Ed., Marcel Dekker, New York, 1966, p. 327.
18. Hart, P. J., Vastala, F. J., and Walker, P. L., Jr., *Carbon*, *5:*363 (1967).
19. Barrer, R. M., *J. Chem. Soc.*, *1936:*1261.
20. Boehm, H. P. and Sappok, R. J., *Proc. Symp. Carbon*, Tokoyo VIII-II (1964).

21. Puri, B. R. and Sharma, S. K., *Chem. Ind. London*, *1966*:160.
22. Chang, H. W. and Rhee, S. K., *Carbon*, *16*:17 (1978).
23. Walker, P. L., Jr., Cariaso, O. C., and Ismail, I. M. K., *Carbon*, *18*:375 (1980).
24. Gonzales-Vilchez, P., Linarez-Solano, A., Lopez-Gonzales, J. de D., and Rodriguez-Reinoso, F., *Carbon*, *17*:441 (1979).
25. Joseph, D. and Oberlin, A., *Carbon*, *21*:565 (1983).
26. Low, M. J. D and Morterra, C., *Carbon*, *21*:275 (1983).
27. Zawadzki, J., *Carbon*, *18*:281 (1980).
28. Zawadzki, J., *Carbon*, *19*:19 (1981).
29. Puri, B. R., Myer, Y. P., and Sharma, L. R., *J. Indian Chem. Soc.*, *33*:781 (1956).
30. Anderson, R. B. and Emmett, P. H., *J. Phys. Chem.*, *56*:753 (1952).
31. Anderson, R. B. and Emmett, P. H., *J. Phys. Chem.*, *51*:1308 (1947).
32. Puri, B. R., Myer, Y. P., and Sharma, L. R., *Chem. Ind. London, B. I. F. Rev. R*, *1956*:30.
33. Puri, B. R. and Bansal, R. C., *Carbon*, *1*:451 (1964).
34. Caltharp, M. T. and Hackerman, N., *J. Phys. Chem.*, *72*:1171 (1968).
35. Matsumoto, S. and Setaka, N., *Carbon*, *17*:303 (1979).
36. Trembley, G., Vastola, F. J., and Walker, P. L., Jr., *Carbon*, *16*:35 (1978).
37. Barton, S. S. and Harrison, B. H., *Carbon*, *13*:283 (1975).
38. Bansal, R. C., Dhami, T. L., and Parkash, S., *Carbon*, *15*:157 (1977).
39. Dollimore, J., Freedman, C. M., Harrison, B. H., and Quinn, D. F., *Carbon*, *8*:587 (1970).
40. Bansal, R. C., Vastola, F. J., and Walker, P. L., Jr., *Carbon*, *8*:443 (1970).
41. Barton, S. S., Gillespie, D. J., Harrison, B. H., and Kemp, W., *Carbon*, *16*:363 (1978).
42. Barton, S. S., Boulton, G. L., and Harrison, B. H., *Carbon*, *10*:395 (1972).
43. Barton, S. S., Gillespie, D. J., and Harrison, B. H., *Carbon*, *11*:649 (1973).
44. Studebaker, M. L., *Rubber Chem. Technol.*, *30*:1400 (1957).
45. Redmond, J. P. and Walker, P. L., Jr., *J. Phys. Chem.*, *64*:1093 (1960).
46. Stacey, W. O., Imperial, C. R., and Walker, P. L., Jr., *Carbon*, *4*:343 (1966).
47. Puri, B. R. and Bansal, R. C., *Chem. Ind. London*, *1963*:574.
48. Puri, B. R., Malhotra, S. L., and Bansal, R. C., *J. Indian Chem. Soc.*, *40*:179 (1963).
49. Puri, B. R., Tulsi, S. S., and Bansal, R. C., *Indian J. Chem.*, *4*:7 (1966).

50. Burstein, R. and Frumkin, A., *Z. Phys. Chem.*, *A141:*219 (1929).
51. Frumkin, A., *Kolloid Z.*, *51:*123 (1930).
52. Shilov, N., *Kolloid Z.*, *52:*107 (1930).
53. Shilov, N., Shatunovska, H., and Larowskaja, D., *Z. Phys. Chem.*, *A150:*421 (1930).
54. Garten, V. A. and Weiss, D. E., *Aust. J. Chem.*, *10:*309 (1957).
55. Garten, V. A. and Weiss, D. E., *Rev. Pure Appl. Chem.*, *7:*69 (1957).
56. Boehm, H. P., *Advances in Catalysis*, Vol. 16, Academic Press, New York, 1966, p. 179.
57. Puri, B. R., in *Chemistry and Physics of Carbon*, Vol. 6, P.L. Walker, Jr., Ed., Marcel Dekker, New York, 1970, p. 191.
58. Van Driel, J., in *Activated Carbon: A Fascinating Material*, A. Capelle and F. de Vooys, Eds., Norit N. V., Netherlands, 1983, p. 40.
59. Lang, F. M. and Magnier, P., *Carbon*, *2:*7 (1964).
60. Bonnetain, L., Duval, X., and Letort, M., in *Proc. 4th Conf. Carbon*, Pergamon Press, Oxford, 1962, p. 107.
61. Bonnetain, L., *J. Chem. Phys.*, *58:*34 (1961).
62. Tucker, B. G. and Mulcahy, M. F. R., *Trans. Faraday Soc.*, *65:*247 (1969).
63. Matsumoto, S., Kanda, H., Suto, Y., and Setaka, N., *Carbon*, *15:*299 (1977).
64. Morterra, C., Low, M. J. D., and Severdia, A. G., *Carbon*, *22:*5 (1984).
65. Zawadzki, J., *Carbon*, *16:*491 (1978).
66. Kruyt, H. R. and de Kadt, G. S., *Kolloid Z.*, *47:*44 (1929).
67. Kruyt, H. R. and de Kadt, G. S., *Kolloid Chem.*, *32:*249 (1931).
68. Schweitzer, C. W. and Goodrich, W. C., *Rubber Age*, *55:*469 (1944).
69. Studebaker, M. L., Huffman, E. W. D., Wolfe, A. C., and Nabors, L. G., *Ind. Eng. Chem.*, *48:*162 (1956).
70. Hoffman, U. and Ohlerich, G., *Angew. Chem.*, *62:*16 (1950).
71. Villars, D. S., *J. Am. Chem. Soc.*, *69:*214 (1947).
72. De Bruin, W. J. and Plas, V., *Rev. Gen. Caoutchouc*, *41:*453 (1964).
73. Boehm, H. P., Diehl, E., and Heck, W., *Rev. Gen. Caoutchouc*, *41:*461 (1964).
74. Boehm, H. P., Diehl, E., Heck, W., and Sappok, R. J., *Angew. Chem.*, *3:*669 (1964).
75. Donnet, J. B., Goldreich, L., Ferry, D., and Huber, F., *Compt. Rendu.*, *252:*1146 (1961).
76. Donnet, J. R. and Henrich, G., *Compt. Rendu*, *248:*3702 (1959).
77. Rivin, D., in *Fourth Rubber Technol. Conf. London*, 1962, p. 1.
78. Rivin, D., *Rubber Chem. Technol.* *36:*729 (1963).
79. Puri, B. R., Meyer, Y. P., and Sharma, L. R., *Res. Bull. Panjals University Chandigarh*, *88:*53 (1956).

80. Puri, B. R., Singh, D. D., Nath, J., and Sharma, L. R., *Ind. Eng. Chem.*, *50*:1071 (1958).
81. Puri, B. R., Mahajan, O. P., and Singh, D. D., *J. Indian Chem. Soc.*, *38*:135 (1961).
82. Puri, B. R., Sharma, L. R., and Singh, D. D., *J. Indian Chem Soc.*, *35*:457 (1958).
83. Puri, B. R. and Bansal, R. C., *Carbon*, *1*:457 (1964).
84. Puri, B. R. and Mahajan, O. P., *J. Indian Chem. Soc.*, *41*:586 (1964).
85. Puri, B. R., Talwar, C., and Sandle, N. K., *J. Indian Chem. Soc.*, *41*:581 (1964).
86. Bansal, R. C., Bhatia, N., and Dhami, T. L., *Carbon*, *16*:65 (1978).
87. Puri, B. R., Singh, G., and Sharma, L. R., *J. Indian Chem. Soc.*, *34*:357 (1957).
88. Studebaker, M. L., in *Proc. 5th Conf. Carbon*, Vol. II, Pergamon Press, New York, 1963, p. 189.
89. Epstein, B. D., Dalle-Molle, E., and Mattson, J. S., *Carbon*, *9*:609 (1971).
90. Kinoshita, K. and Bett, J. A. S., *Carbon*, *12*:525 (1974).
91. Kinoshita, K. and Bett, J. A. S., *Carbon*, *11*:403 (1973).
92. Matsumura, Y., Hagiwara, S., and Takahashi, H., *Carbon*, *14*:247 (1976).
93. Matsumura, Y., Hagiwara, S., and Takahashi, H., *Carbon*, *14*:163 (1976).
94. Papirer, E. and Guyon, E., *Carbon*, *16*:127 (1978).
95. Donnet, J. B., Papirer, E., and Couderc, P., *Bull. Soc. Chim. France*, *3*:929 (1968).
96. Matsumura, Y. and Takahashi, H., *Carbon*, *17*:109 (1979).
97. Studebaker, M. L. and Rinehart, R. W., Sr., *Rubber Chem. Technol.*, *45*:106 (1972).
98. Suzuki, S. and Miyazaki, K., *Nippon Kagaku Zasshi*, *88*:299 (1967).
98a. Jones, J. F. and Kaye, R. C., *J. Electroanal. Interface Electrochem.*, *20*:213 (1969).
99. Given, P. H. and Hill, L. W., *Carbon*, *7*:649 (1969).
100. Vaughan, V. A. and Swithenbank, J. J., *Analyst.*, *90*:594 (1965).
101. Greenhow, E. J. and Spencer, L. E., *Analyst.*, *98*:90 (1973).
102. Hallum, J. V. and Drushel, H. V., *J. Phys. Chem.*, *62*:110 (1958).
103. Mattson, S. and Mark, H. B., Jr., *Activated Carbon Surface Chemistry and Adsorption*, Marcel Dekker, New York, 1971.
104. Donnet, J. B. and Marguier, P., *C. R. Acad. Sci. Paris*, *239*:1038 (1954); *242*:771 (1956).
105. Boehm, H. P., Diehl, E., and Heck, W., *Symp. Carbon Tokoyo Preprints*, *8*:10 (1964).

106. Donnet, J. B. and Henrich, G., *Bull. Soc. Chim. France*, 1609 (1960).
107. Donnet, J. B., Lahaye, J., and Schultz, J., *Bull. Soc. Chim. France*, 5:1769 (1966).
108. Voet, A. and Teter, A. C., *Am. Ink Maker*, 38:44 (1960).
109. Test, R. E. and Hansen, R. S., M. S. thesis, Iowa State University, Ames, Iowa, 1961, p. 56.
110. Donnet, J. B., Fr. Patent 1,164,786, Oct. 1958.
111. Donnet, J. B., Heuber, F., Reitzer, C., Oddoux, J., and Reiss, G., *Bull. Soc. Chim. France*, 1727 (1962).
112. Friedman, S., Kaufman, M. L., Steiner, W. A., and Wender, I., *Fuel*, 40:33 (1961).
113. Hill, L. W., Ph. D. dissertation, Pennsylvania State University, 1968.
114. Friedel, R. A. and Hofer, L. J. E., *J. Phys. Chem.*, 74:2921 (1970).
115. Starsinic, M., Taylor, R. L., and Walker, P. L., Jr., *Carbon*, 21:69 (1983).
116. O'Reilly, J. M. and Mosher, R. A., *Carbon*, 21:47 (1983).
117. Rosencwaig, A., *Photoacoustics and Photoacoustic Spectroscopy*, Wiley, New York, 1980.
118. Pao, Y. H., *Optoacoustic Spectroscopy and Detection*, Academic Press, New York, 1977.
119. Friedel, R. A. and Carlson, G. L., *Fuel*, 51:194 (1972).
120. Donnet, J. B., *Carbon*, 20:266 (1982).
121. Ishizaki, C. and Marti, I., *Carbon*, 19:409 (1981).
122. Papirer, E., Guyon, E., and Perol, N., *Carbon*, 16:133 (1978).
123. Mattson, J. S., Lee, L., Mark, H. B. Jr., and Webber, W. J. Jr., *J. Colloid Interface Sci.*, 33:284 (1970).
124. Painter, P. C., Snyder, R. W., Pearson, D. E., and Kwong, J., *Fuel*, 59:282 (1980).
125. Blom, L., Edelhausen, L., and Vaukrevelen, D. W., *Fuel*, 36:135 (1957).
126. Morterra, C. and Low, M. J. D., *Carbon*, 23:301 (1985).
127. Low, M. J. D. and Morterra, C., *Carbon*, 23:311 (1985).
128. Morterra, C. and Low, M. J. D., *Carbon*, 23:335 (1985).
129. Dauben, W. G. and Epstein, W. W., *J. Org. Chem.*, 24:1595 (1959).
130. Jones, R. N., Angell, C. L., Ito, T., and Smith, R. J. D., *Can. J. Chem.*, 37:2007 (1959).
131. Thomas, J. M., Evans, E. L., Barber, M., and Swift, P., in *3rd Conf. Ind. Carbon Graphite London*, Soc. Chem. Ind., 1971, p. 411.
132. Ishitani, A., *Carbon*, 19:269 (1981).
133. Takahagi, T. and Ishitani, A., *Carbon*, 22:43 (1984).
134. Proctor, A. and Sherwood, P. M. A., *Carbon*, 21:53 (1983).

135. Young, V., *Carbon*, *20:*35 (1982).
136. Kavan, L., Bostl, Z., Dausek, F. P., and Jansta, J., *Carbon*, *22:*77 (1984).
137. Fischer, F. G. and Feichtinger, in 15th Bienn. Conf. on Carbon Ext. Abstracts, 1981, p. 472.
138. Grint, A. and Perry, D. L., in 15th Bienn. Conf. on Carbon Ext. Abstracts, 1981, p. 462.
139. Steenberg, B., *Adsorption and Exchange of Ions on Activated Charcoals*, Almquist and Wiksell, Uppsala, 1944.
140. Lamb, A. B. and Elder, L. W., *J. Am. Chem. Soc.*, *53:*137 (1931).
141. Kolthoff, I. M., *J. Am. Chem. Soc.*, *54:*4473 (1932).
142. King, A., *J. Chem. Soc.*, *1932:*842.
143. Boehm, H. P. and Voll, M., *Carbon*, *8:*227 (1970).
144. Voll, M. and Boehm, H. P., *Carbon*, *8:*741 (1970).
145. Voll, M. and Boehm, H. P., *Carbon*, *9:*473 (1971).
146. Voll, M. and Boehm, H. P., *Carbon*, *9:*481 (1971).

3

Porous Structure of Active Carbons

3.1 POROSITY AND SURFACE AREA IN ACTIVE CARBONS

Active carbons are characterized by their strong adsorption capacity, which can be as high as $0.6-0.8$ cm^3/g, and which occurs mostly in cavities of molecular dimensions called micropores. The corresponding adsorption process is described by Dubinin's theory for the filling of micropores (1−12), discussed in detail in Section 3.3.

Depending on their preparation, active carbons also contain larger pores, known as mesopores and macropores in the calssification proposed by Dubinin (13) and now adopted by the International Union of Pure and Applied Chemistry (14). The definition of the different types of pores is based on their width w, which represents the distance between the walls of a slit-shaped pore or the radius of a cylindrical pore. One distinguishes micropores, for which w does not exceed 2 nm; macropores, for which w exceeds about 50 nm; and mesopores, of intermediate size ($2.0-50$ nm).

This classification is not entirely arbitrary, as it takes into account differences in the behavior of molecules adsorbed in micropores and in mesopores. It appears that for pore widths exceeding $1.5-2.0$ nm, the adsorbate condenses in a liquidlike state and a meniscus is formed. As a consequence, a hysteresis loop appears on desorption and its interpretation can lead to the distribution of the mesopores in the solid (15). This hysteresis stops at a relative pressure p/p_0 of approximately 0.4. In the micropores, on the other hand, adsorption should be reversible, in principle, but low-pressure hysteresis has been observed for active carbons by Everett and colleagues (16) and more recently by Linares-Solano et al. (17). This phenomenon is ascribed to the trapping of the adsorbate, following the inelastic distortion of certain micropores. The reverse has been observed by Stoeckli et al. (18), who showed that the accessibility of the micropore system could increase at low pressures, after a number of adsorption—desorption cycles.

The limit between mesopores and macropores at 50 nm is more artificial and it corresponds to the practical limit of the method for the pore size determination based on the analysis of the hysteresis loop.

At the present time, one may postulate for typical active carbons the average structure shown in Figure 1, in agreement with direct observations from Transmission electron microscopy (TEM0 (19,20). It is a limiting case of the model proposed by Oberlin et al. (21), who investigated typical, but not activated, carbonaceous materials.

As suggested by early adsorption experiments with flat and globular molecules, and on the basis of direct observations, it may be assumed that micropores are mostly slit-shaped spaces between twisted aromatic sheets. Figure 2 is a micrograph of carbon U, an active carbon described in Section 3.3. A typical slit-shaped micropore of 0.5 nm in width can be seen at position M. It is located on the edge of a mesopore of approximately 15 nm in diameter and perpendicular to the plane of the micrograph.

Figure 1 Schematic representation of the microstructure of active carbons.

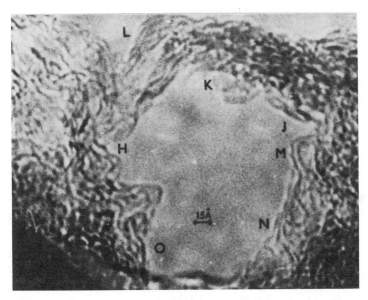

Figure 2 TEM micrograph of active carbon U-02, with a typical slit-shaped micropore M (with 0.5 nm), at the edge of a cylindrical meso-pore. (Courtesy of Dr. J. Fryer, Glasgow.)

The different pores have walls and one may distinguish between two types of surface associated with them. The first is the micro-porous (or internal) surface S_{mi}, represented by the walls of the slits, with an area of several hundred square meters per gram of solid. It is related to the volume W and the accessible width L by the simple geometrical relation

$$S_{mi}(m^2/g) = \frac{2 \cdot 10^3 W \ (cm^3/g)}{L \ (nm)} \qquad (1)$$

Since L is small, the area associated with the volume of micropores is much larger than in the case of mesopores or macropores (2). The second surface is the nonmicroporous or external surface S_e, which includes the walls of the mesopores and macropores, as well as the edges and the aromatic sheets facing the outside. The latter contribution seems to be small for most active carbons.

For typical active carbons, S_e varies between 10 and 200 m^2/g of solid, but higher values can be found in special cases. From a physical point of view, the difference between S_{mi} and S_e is reflect-ed by the value of the adsorption energy, which can be twice as high on the walls of a micropore as on the open surface (22,23). This enhancement is due to the overlapping of adsorption forces from the opposite walls of the micropores, an effect which decreases rapidly as the width increases (24). Calculations show that if the width of a slit-shaped pore is twice the diameter of a simple host molecule, the adsorption potenetial is only 1.2 times larger than on the open reference surface (graphite, in the case of active carbons). As a consequence of the higher adsorption energies, micropore filling will start before adsorption on the external surface.

Another essential difference between adsorption in micropores and in the larger pores lies in the adsorption mechanisms: whereas the former corresponds to the filling of a volume, the latter consists of a gradual and multilayer adsorption process, as described by the classical BET theory (15). The two concepts are distinct and the BET approach is not valid for the interpretation of the early stages of adsorption by active carbons. This is clearly illustrated by the unrealistic surface areas often obtained when applying the BET theory, since the equivalent monolayer capacity of the micropore volume corresponds to S_{mi} only if the pores accomodate two layers of adsorbate.

Adsorption of gases and vapors by a typical active carbon results therefore from distinct contributions, (1) from the filling of the micropores examined in detail in Section 3.3, and (2) from the adsorption on S_e (25,26). Formally,

$$V\left(T; \frac{p}{p_0}\right) = W\left(T; \frac{p}{p_0}\right) + S_e t\left(\frac{p}{p_0}\right) \qquad (2)$$

where V is the total volume adsorbed by the solid at temperature T and at relative pressure p/p_0; W is the volume adsorbed in the micropores; and t represents the thickness of the layer adsorbed on the nonmicroporous surface S_e, a function described in the following section.

Equation (2) is illustrated by Fig. 3, which shows the separate contributions to the overall adsorption isotherm of benzene by carbon A-35 at 293 K. The hysteresis loop on desorption results from capillary condensation in the mesopores. The analysis of the overall isotherm, as expressed by Eq. (2), plays an important role in the characterization of active carbons by adsorption techniques, and it is complementary to the approach based on immersion calorimetry (see Sections 3.4 and 3.6). The first term, V (T; p/p_0), provides information on the micropore system, whereas the second term corresponds to adsorption on the walls of the mesopores and macropores. Before examining the different types of porosity, we deal with the nonmicroporous surface of active carbons, which can be determined by different techniques. As illustrated by Figure 3, the process of micropore filling is usually completed when p/p_0 reaches 0.1, i.e., before adsorp-

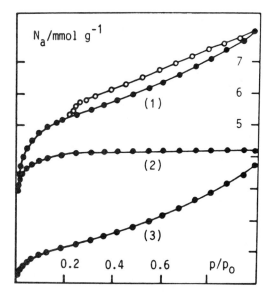

Figure 3 The contributions to the overall adsorption isotherm of benzene by carbon A-35 at 293 K (1). It results from adsorption in the micropores (2) and on the external surface area S_e (3). This solid also contains mesopores, as revealed by the hysteresis loop, which appears on desorption (open circles).

tion becomes significant on the nonmicroporous surface. This means that the micropore system can be saturated at room temperature with *n*-nonane at the appropriate relative pressure, and S_e can be determined from a subsequent nitrogen adsorption isotherm at 77—78 K, by applying the standard BET treatment (15) to it. This technique, proposed by Gregg and Langford (27), has been used and analyzed by different authors (25,28,29). Benzene can also be used at room temperature, instead of *n*-nonane (25). Although the results for S_e are in satisfactory agreement with independent determination, difficulties may arise from the overlap of the micropore filling process and multilayer adsorption, when a broad distribution of micropores is present in the carbon (this corresponds usually to a smooth transition from micropores to mesopores in the case of heterogeneous materials).

A second approach is based on the comparison of the adsorption isotherm with a standard isotherm obtained under similar conditions for a nonporous reference solid (t- and α-plots). As described in detail by Gregg and Sing (15), the method is based on a plot of the amount adsorbed by the active carbon versus the amount adsorbed by the reference at the same relative pressures. The latter may be expressed either in molecules per square meter or in terms of the thickness t of the adsorbed layer (see Section 3.2). Such plots usually display a linear section corresponding to the range of pressures where the adsorption mechanism is the same on both solids (i. e., on the nonmicroporous surface of the active carbon). In principle, the intercept extrapolated from the linear section corresponds to the amount filling the micropores, and the slope is proportional to S_e.

Dubinin and Kadlec (30) suggested an a;ternative approach known as the t/F method, in which both sides of Eq. (2) are divided by the quantity $F = \exp[-(A/\beta E_0)^2]$. which appears in the basic equation of Dubinin and Radushkevich (see Section 3.3). It follows that

$$\frac{V(T; p/p_0)}{F} = W_0 + \frac{S_e t (p/p_0)}{F} \tag{3}$$

and by using suitable function for the thickness t of the adsorbed layer on the carbonaceous surface, a representation of V/F versus t/F should yield a linear plot with a slope equal to S_e and an intercept corresponding to the micropore volume W_0. Figure 4 illustrates such a representation for the adsoprtion of N_2 (78 K) and of benzene (293 K) by carbon AC-11 (30). However, it must be pointed out that linearity may be ill-defined or absent in the comparison plot, as a result of important capillary condensation. This effect gradually reduces the accessible surface area of the sample under investigation as the pressure increases and no linear section is observed. It has also been observed that even if linearity is present, differences in the chemical and physical states of the two solids may lead to incorrect values of the nonporous surface. As suggested by Stoeckli and Kraehenbuehl (25), an alternative approach is the substraction of the micropore contribution from the overall isotherm and the subsequent analysis of the residual isotherm within the framework of the classical BET theory. The main advantage is the absence of a predetermined reference $t(p/p_0)$, as used in the comparison plots, and which is not necessarily compatible with the residual isotherm. This method has also been used recently by Ali and McEnaney (31) and by Martin-Martinez (32).

As seen in Table 1, the different techniques discussed lead to relatively consistent values for the nonmicroporous surface areas S_e of different active carbons (25). These are also in agreement with other determinations described in the following sections (mesoporosity and enthalpies of immersion).

Having examined the problem of the external surface, we next consider the different types of pores found in active carbons.

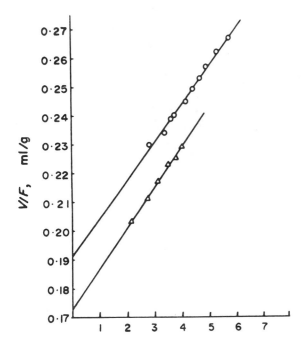

Figure 4 t/F plot of Dubinin and Kadlec for the adsorption of nitrogen (O) (78 K) and of benzene (△) (293 K) on carbon AC-11 (30).

3.2   THE DETERMINATION OF MESOPOROSITY AND MACROPOROSITY

Macroporosity and almost the entire range of mesoporosity can be assessed by mercury porosimetry, a smiple technique based on the penetration of mercury into pores under the effect of pressure.

Mercury is a nonwetting liquid, with a contact angle Θ near 140°, and under a pressure p it fills cylindrical pores of radii $r > r_p$ given by Washburn's equation (15):

$$r_p = -\ 2\left(\frac{\gamma}{p}\right)\cos\ \Theta \tag{4}$$

The values recommended for routine measurements at room temperature are γ = 480 mN/m and Θ = 140°, although the latter may vary from 135 to 150°, depending on the solid (15). For practical purposes Eq. (4) is written in the form

$$r_p\ (nm) = \frac{7300}{p\ (atm)} \tag{5}$$

Table 1 The External Surface Areas S_e Obtained from the Analysis of the Adsorption Isotherms by Applying Different Methods (25)

Carbon	Isotherm	Method	$S_e(m^2/g)$
N-125	C_6H_6 (293 K)	t/F method	175
	N_2 (78 K)	Kelvin (mesopores)	145
	N_2 (78 K)	t-plot	128
	N_2 (78 K)	Preadsorbed C_6H_{20} Decomposition and BET	155
	N_2 (78 K)	Preadsorbed n-C_9H_{20} Decomposition and BET	141
U-02	N_2 (78 K)	t/F method	115
		Kelvin (mesopores)	96
		Preadsorbed n-C_9H_{20}	93
	N_2 (78 K)	Preadsorbed C_6H_6	
		t-plot	115
		t/F method	108
		Decomposition and BET	131
FA	C_6H_6 (293 K)	t-plot	106
		t/F method	111
		Decomposition and BET	118
	N_2 (78 K)	t-plot	109
		t/F method	105
		Decomposition and BET	114

showing that under atmospheric pressure, mercury will fill pores with radii larger than 7300 nm (7.3 μm).

The technical limit of mercury porosimetry is around 3.5 nm (2000 atm) and therefore close to the lower limit of mesoporosity (2.0 nm). A modified version of the Washburn equation [Eq. (4)] also exists for slit-shaped macropores and mesopores.

Figure 5 shows the variation of the cumulative volume V_p occupied by mercury, as the pressure increases from approximately 0.50 to 1900 atm (1.5×10^4 to 3.8 nm) for sample HLY, a typical active carbon of industrial origin. Since the pore sizes extend over several orders of magnitude, they are shown on a logarithmic scale, and the derivation of V_p with respect to r_p or to log (r_p) leads to the corresponding pore size distribution. In the present example, there are two maxima in the distribution, a slight one in the mesopore range and the other near 4000 nm. Such a bimodal distribution is typical for active carbons of vegetable origin, as pointed out by Dubinin (33). The total volume of the pores between 3.8 and 2×10^4 nm is 0.45

cm^3/g of solid, the mesopores of 3.8—50 nm representing only 0.05 cm^3/g. From the variation of V with r_p, it is also possible to calculate the cumulative surface area S_{cum} of the pores filled by mercury. For the range covered in Figure 5, this quantity should be somewhat smaller than S_e, owing to the lack of information on the mesopores between 2.0 and 3.8 nm (large pores, such as macropores, do not contribute significantly to S_e).

As illustrated by Figure 5, mercury porosimetry does not include the full range of mesoporosity, owing to its technical limitations, and it does not cover the domain below 0.38 nm, which often contributes significantly to the overall external surface area in active carbons. However, it appears that in many cases a plot of log (S_{cum}) versus log (r_p) leads to a linear representation in the range of 4—15 nm, from which the total area of the mesopores (r_p > 2.0 nm) can be extrapolated. It also appears that these results are often in good agree-

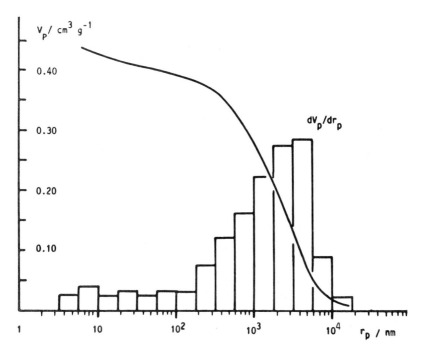

Figure 5 Variation of the macropore and micropore volume V_p of carbon HLY as a function of the pore radius r_p obtained from mercury porosimetry between 0.5 and 1900 atm. The blocks correspond to the pore-size distribution dV_p/dr_p. (Courtesy of Drs. Huber and Lavanchy, Spiez.)

ment with the data obtained from the analysis of the hysteresis loop observed on desorption, following the adsorption of vapors to a value of p/p_0 close to unity (see Fig. 3). This technique provides an important and independent means for the assessment of mesoporosity in active carbons, and the two techniques are in full agreement.

For carbon HLY, the cumulative surface area of the pores with radii larger than 3.8 nm represents 11 m^2/g, whereas for the macropores alone (r > 50 nm) it represents only 0.5 m^2/g. The extrapolation of S_{cum} to 2.0 nm, although inaccurate, suggests a total surface area of 23 m^2/g to be compared with 25 m^2/g obtained from immersion calorimetry. There appears to be a good agreement between the different values.

In the case of another typical active carbon, U-03, one obtains Scum = 22.5 m^2/g from mercury porosimetry with r_p > 3.8 nm, and 43 m^2/g from the extrapolation to 2.0 nm. These values are in good agreement with the corresponding data from the analysis of the hysteresis loop of benzene, 18 and 50 m^2/g for the same domains. In the present case, immersion calorimetry leads to S_e = 84 m^2/g, a larger external surface area. Since carbon U-03 is heterogeneous, the difference of 30–40 m^2/g may be ascribed to the porosity immediately below 2.0 nm and corresponding to supermicropores. (In calorimetry, such pores are seen partly as an open surface, owing to their lower surface energy, a point discussed in Section 3.4.)

Rootare and Penzlow (15,34) reported good agreement between the surface areas of different solids, as obtained from mercury porosimetry and the analysis of the desorption branch of the nitrogen isotherm, but the present examples show that for typical active carbons nonnegligible contribution arises from the mesopores in the range of 2–4 nm, and possibly from supermicropores, if present. This is not surprising, since a continuous transition may exist between micropores and mesopores (the distinction between the two types of porosity is based rather on sorptive than on structural considerations).

The hysteresis loop results from the progressive condensation of the adsorbate in the pore, in the course of adsorption, with the formation of a meniscus. The vapor pressure above the meniscus decreases with the pore size, owing to the small radius of curvature, and it follows that the smallest pores are desorbed last when the pressure in the vapor phase is reduced.

The fundamental relation (15) is the Kelvin equation:

$$\ln\left(\frac{p_d}{p_0}\right) = \frac{-2\gamma V_m}{RTr_K} \tag{6}$$

where p_d is the pressure at which a cylindrical pore of Kelvin radius r_K is desorbed; γ, V_m, and p_0 are respectively the surface tension, the molar volume, and the saturation pressure of the liquid adsorbate at temperature T. It is also assumed that the angle of contact is zero, which implies perfect wetting. It can be shown that in the case of slit-shaped mesopores, r_K represents the Kelvin width L_K of the slits.

Since adsorption occurs naturally on the walls of the mesopores, the quantities r_K and L_K do not correspond to the true pore radius r_p and width L_p. The thickness t of the layer left on the walls after desorption has to be taken into account, and it follows that the real dimensions of mesopores are

$$r_p = r_K + t\left(\frac{p}{p_0}\right) \tag{7}$$

and

$$L_p = L_K + 2t\left(\frac{p}{p_0}\right) \tag{8}$$

This situation is illustrated schematically in Figure 6, for a cylindrical pore. When the desorption pressure $p = p_d$ is reached, only the central core, a cylinder of diameter $2r_K$ (or a slice of width L_K) will effectively be desorbed and a multilayer of thicknes $t(p/p_0)$ will remain on the walls of the pore. Basically, the thickness function corresponds to the natural adsorption isotherm on the corresponding solid, $N_a = N_a (T; p/p_0)$, divided by the monolayer capacity N_{am} and multiplied by a quantity σ representing the average thickness of one layer,

$$t\left(\frac{p}{p_0}\right) = \sigma \cdot N_a\left(T;\frac{p}{p_0}\right) N_{am} \tag{9}$$

On the basis of adsorption experiments carried out on a variety of open surfaces, the following simple equation has been suggested for nitrogen at 78 K (15):

$$t\left(\frac{p}{p_0}\right) = 0.35\left[\frac{5}{\ln\ (p_0/p)}\right]^{1/3} \tag{10}$$

This expression can be used with a good approximation in Eq. (2) and in the comparison plots discussed in the previous section. A similar relation, $t(nm) = 0.45[\ln\ (p_0/p)]^{-2/3}$ (35), has been used successfully for the adsorption of benzene at 293 K (26).

For nitrogen desorption near 78 K, a traditional source of information on mesoporosity, the Kelvin equation [Eq. (6)] becomes:

$$r_K\ (nm) = \frac{0.940}{\ln\ (p_0/p)} \tag{11}$$

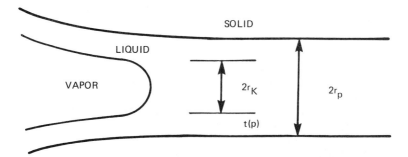

Figure 6 Relation between the Kelvin radius r_K and the true radius r_p of a cylindrical mesopore. The present example corresponds to the desorption of the central core, leaving an adsorbed layer thickness $t = t(p/p_0)$.

if one uses the surface tension (8.72 mN/m) and the molar volume (34.68 cm^3/mol) of the pure liquid. In the case of benzene at 293 K, one obtains a similar equation, with the coefficient 2.107 instead of 0.940.

Equation (11) also illustrates the practical limits of the technique based on the desorption of vapors as p/p_0 approaches unity: for $p/p_0 = 0.95$ and 0.97, the Kelvin radii correspond to 18 and 31 nm and in view of the logarithmic nature of the Kelvin equation, small errors in the pressure lead to increasingly large uncertainties on r_K. Taking into account the corresponding values of t, it appears that the upper limit for a reliable determination of the real pore radius r_p lies near 50 nm, which is the conventional limit of mesoporosity, as defined in Section 3.1. Equations (10) and (11) and their analogues for benzene at room temperature also show that the lower limit of the technique, set by the end of the hysteresis loop near $p/p_0 = 0.3-0.4$, corresponds to effective pore dimensions r_p and L_p of approximately 1.5 to 2.0 nm. This justifies the lower limit for mesoporosity, set by definition at 2.0 nm.

The combination of Eqs. (7), (10), and (11) leads to the real values of the pore radii r_p, given in Table 2 with the corresponding thickness t the remaining layer.

The pore size distributions and the corresponding surface areas of the mesopores can be calculated from the hysteresis loop by the computational techniques described in detail by Gregg and Sing (15).

The determination of mesoporosity based on the Kelvin equation is still a matter of discussion in the literature (15,36,37). The main issues are the variation of the surface tension γ with the curvature of the liquid meniscus, and the exact form of the thickness function

Table 2 Real and Kelvin Pore Radii for the Desorption N_2 near 78 K[a]

p/p_0	$t(p/p_0)$	r_K	r_p
0.90	1.28	8.91	10.19
0.80	0.99	4.21	5.20
0.70	0.85	2.63	3.48
0.60	0.76	1.83	2.59
0.50	0.68	1.35	2.03
0.45	0.65	1.18	1.83
0.40	0.62	1.02	1.64

[a] The model is based on Eqs. (10) and (11).

$t(p/p_0)$ in narrow pores. As shown by Dubinin and Kataeva (37), there may be differences in the pore size distribution, depending on the models, and alternative expressions are proposed for t and r_p, as functions of p/p_0, for the adsorption of nitrogen (78 K) and benzene (293 K) on carbons. However, as shown in Table 1, the approach based on the simple Eqs. (10) and (11) for nitrogen at 78 K, or the equivalent equations for benzene at 293 K, leads to surface areas for the mesopores which are usually in good agreement with the total external (or nonmicroporous) surface S_e of the carbons, obtained from independent techniques.

Although the total volume of the mesopores may be as high as 0.4 to 0.5 cm^3/g in certain active carbons (38,39), their surface area rarely exceeds 200 m^2/g (39). Since adsorption in the mesopores depends primarily on the surface area and not on the volume, the latter is not a fundamental characteristic of the active carbon, as opposed to its micropore volume. This aspect is discussed in Section 3.3.

3.3 PHYSICAL ADSORPTION BY MICROPOROUS CARBONS

Physical adsorption of gases and vapors by microporous solids in general, and by active carbons in particular, is described by Dubinin's theory, developed in successive stages since 1947 (1–12, 15). In its present formulation, the theory of micropore volume filling is expressed by the equation of Dubinin and Astakhov (3,4):

$$W = W_0 \exp [-(A/\beta E_0)^n] \tag{12}$$

where W represents the volume filled at temperature T and relative pressure p/p_0; W_0 is the total volume of the micropores, the quantity $A = RT \ln (p_0/p)$ and n, E_0, and β are specific parameters of the system under investigation.

The DA (Dubinin–Astakhov) equation [Eq. (12)] applies to the adsorption of a variety of organic and nonspecific vapors at relative pressures $p/p_0 < 0.05-0.1$, where the influence of the nonmicroporous surface area S_e is negligable.

For typical active carbons, exponent n is equal to 2, which corresponds to the original empirical equation postulated by Dubinin and Radushkevich in 1947 (1) and known in the literature as the DR equation,

$$W = W_0 \exp \left[-B \left(\frac{T}{\beta} \right)^2 \log^2 \left(\frac{p_0}{p} \right) \right]$$ (13)

In the modern formulation, Eq. (13) becomes

$$W = W_0 \exp \left[-\left(\frac{A}{\beta E_0} \right)^2 \right]$$ (14)

Parameter B, which has dimensions of K^{-2}, is called the structural constant of the given carbon and is related to the characteristic energy E_0 of Eq. (14) by

$$E_0 \text{ (kJ/mol)} = \frac{0.01914}{\sqrt{B}}$$ (15)

The DR and DA equations are based on the observation that a plot of W versus $(RT)^2 \log^2 (p_0/p)$ or A^n leads to a unique curve for a given adsorbate. Moreover, if one introduces a specific and empirical factor β known as the affinity coefficient of the adsorbate, the vapors overlap, benzene being the reference vapor, and by convention $\beta(C_6H_6) = 1$. As a result, adsorption on a given active carbon is described by a unique curve known as its characteristic curve, which illustrates the applicability of Dubinin's equations over a relatively large range of pressures and temperatures. Examples are seen in Figures 7 and 8, in which logarithmic plots are used.

The Dubinin–Radushkevich equation has also been used to described the adsorption of organic compounds (40) and of iodine (41) from aqueous solution by active carbons. For such applications, pressures are replaced by concentrations and for the species adsorbed preferentially from water, Eq. (13) becomes

$$W = W^* \exp \left\{ - k \left[RT \ln \left(\frac{c_0}{c} \right) \right]^2 \right\}$$ (16)

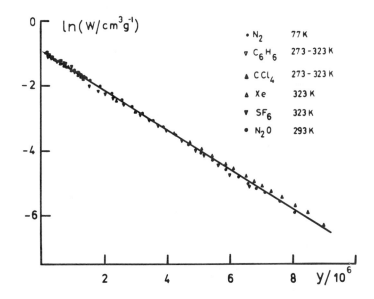

Figure 7 Adsorption of various simple vapors by polymer-based carbon T (42) showing total accessibility to all adsorbates. The variable $y = (T/\beta)^2 \log^2(p_0/p)$.

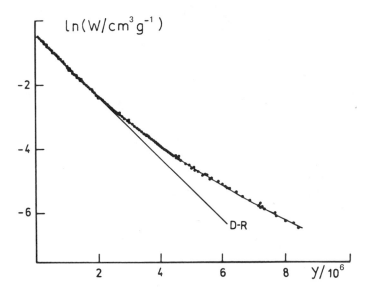

Figure 8 Adsorption of simple vapors by sample F-02, a heterogeneous carbon with $n = 1.5$ in Eq. (12). The experimental conditions are the same as for the carbon in Figure 7.

where W* represents a limiting volume, usually close to W_0; c is the equilibrium concentration; and c_0 represents either the saturation concentration in the aqueous solution or a parameter close to it. Constant k depends on the system under investigation.

The general description of multicomponent adsorption by active carbons, however, requires more elaborate theories. This becomes especially true for concentrated solutions and when all components show a strong affinity for the carbon (this is not case with water). These theories are often based on the Freundlich equation, or modifications of it, which can be derived from the DA equation [Eq. (12)] with n = 1.

In the case of gases and vapors, the DR and DA equations have an extended range of validity and they can be used to predict the static adsorption of a variety of adsorbates. For a given carbon the main requirements are that the degree of accessibility of the micropore system be the same for the different molecules and that specific interactions be absent (water, for example, is an important adsorptive which does not follow Dubinin's theory and, as shown in Section 3.5, its adsorption isotherm requires a different description).

The range of applicability of the classical DR equation [Eq. (13)] is illustrated in Figure 7, which shows the overlap of the data of six different vapors by carbon T (42). The logarithmic representation of Eq. (13), $\ln W$ versus $y = (T/\beta)^2 \log^2 (p_0/p)$, leads to a unique straight line with a slope $B = 0.98 \times 10^{-6} K^{-2}$. This also implies that the same micropore volume is seen by the different molecules.

Table 3 lists the affinity coefficients of a number of adsorbates obtained experimentally by the Soviet school (4) and by other sources (26,43). As pointed out by Dubinin (4), the affinity coefficient β can be calculated from Sugden's parachors and more recently it has also been shown (44) that for simple adsorbates β is proportional to the minimum of the adsorption energy of the corresponding adsorbate on a graphitic surface. This is an indication of the physical meaning of β, but at the present time the best estimate is still provided by adsorption experiments on a carbon showing equal accessibility to the given adsorbate and to the reference vapor.

The state of the adsorbate in the micropores is still open to discussion (4), but to a first and certainly good approximation one may postulate that it corresponds to the condensed adsorbate at the same pressure and temperature. It follows that for typical adsorbates,

$$W = a \cdot V_m(1) \ (T;p) \tag{17}$$

a being the amount adsorbed on the solid and $V_m(1)$ the molar volume of the liquid. This hypothesis is confirmed by the comparison of vapor and liquid adsorption (26) by carbons of known densities in helium (usually $2.0-2.1 \ g/cm^3$). Although pycnometry indicates small differences between the liquid and adsorbed states, in particular for the expansion coefficient α, Eq. (17) can still be used as a very good approximation for the standard adsorbates listed in Table 3. Nikolaev and Dubinin (45) also suggested approximations to predict the molar volumes of gases adsorbed near or above their critical point, as well as a modified version of the Dubinin–Radushkevich equation for $T > T_c$, but this situation falls outside the scope of this book.

The DA equation [Eq. (12)] was originally postulated on an empirical basis and it was shown later, on an experimental basis, that exponent n reflects the degree of homogeneity of the micropore system under investigation. For active carbons, n varies practically from 1.5 to 3 and, as pointed out by Dubinin (4) and by Finger and Bülow (46), the case n = 3 applies to the adsorption by Saran-based carbons, which have a relatively narrow system of micropores of 0.4– 0.5 nm in width. More recently it was shown experimentally that n = 3 also applies to molecular sieve carbons with relatively large but homogeneous micropores near 0.7 nm (43). On the other hand, strongly activated and heterogeneous carbons may lead to a value of n as low as 1.5 (42).

The relation between n and the homogeneity of the micropore system has been evident from recent investigations on the theoretical basis of the Dubinin–Astakhov equation (47–49). It can be shown that this equation results formally from an integral transform involving a local Langmuir isotherm $\Theta_L(T; p; \varepsilon)$ and an adsorption energy distribution $\chi(\varepsilon)$. To a good first approximation, the latter is given by the function

$$\chi(\varepsilon) = n \frac{(\varepsilon - \varepsilon_0)^{n-1}}{(\beta E_0)^n} \exp \left\{ - \left[\frac{(\varepsilon - \varepsilon_0)}{\beta E_0} \right]^n \right\} \tag{18}$$

where ε is the potential adsorption energy of the adsorbate in the micropore and ε_0 represents its lowest value (in practice on the open surface, as the pore width increases).

Mathematical modeling (49) shows that the energy distribution becomes sharper as n increases. This implies that the micropore distribution must become more homogeneous as n increases from 1.5 to 3, the typical range observed for active carbons. Another consequence, confirmed by recent experiments (43,50), is the fact that for n = 2 the micropore system presents some degree of heterogeneity and is not as homogeneous as assumed earlier (42). Figure 9 illustrates the case of carbon CEP-18, resulting from activation to 18% burnoff in CO_2 at 850°C, and for which $E_0 = 33$ kJ/mol. Although the DR plot is linear, about 70% of the pores are found within the range of 0.4–0.6 nm. For typical active carbons, the classical DR approximation with n = 2 holds with a variable degree of approxima-

Table 3 Experimental Affinity Coefficients β and Molar Volumes in the Liquid State at 293 K (77 K for N_2) of Typical Adsorptions Used in the Study of Microporous Carbons

Adsorbate	β	V_m (cm^3/mol)
Benzene	1.00	88.91
Nitrogen	0.34	34.67
Methylene chloride	0.66	64.02
Cyclohexane	1.04	108.10
Carbon tetrachloride	1.05	96.50
Chlorobenzene	1.19	101.70
n-Hexane	1.29	130.52
n-Heptane	1.62	146.56
n-Hexadecane	4.05	292.57
2,5-Norbonadiene	1.62	101.63
α-Pinene	1.70	158.75
Perchlorocyclopentadiene	1.91	159.30

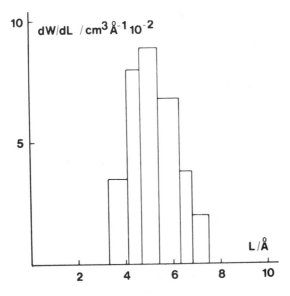

Figure 9 The micropore distribution of carbon CEP-18, for which n = 2, E = 33 kJ/mol, and W_0 = 0.251 cm^3/g (43).

tion, as the burnoff increases and the characteristic energy E_0 decreases from 35−30 kJ/mol to approximately 20 kJ/mol. As illustrated by Figure 8, deviations from linearity begin to appear at low degrees of micropore filling and their detection depends essentially on the range covered by the adsorption experiment.

Figure 8 represents a limiting case, with a relatively important departure from linearity. In such a case, the best fit to the Dubinin−Astakhov equation [Eq. (12)] leads to a value of n smaller than 2, a clear indication of relatively strong heterogeneity in the micropore distribution. It must be emphasized that this type of assessment requires accurate adsorption data over a wide range of temperatures and/or relative pressures. Typically deviations appear in the region of $y > 4−5 \times 10^6$, which corresponds to low degrees of micropore filling, whereas the classical DR approximation is still valid to a good approximation in the range covered by normal experimental conditions (for benzene at 298 K, for example, y is smaller than 2×10^6). It is equally difficult to obtain accurate values of n > 2 for homogeneous carbons, and therefore the DR equation remains the standard approximation for a variety of applications as well as leading to average values and underlining the importance of techniques based on molecular sieve experiments, as described in Section 3.4.

The physical meaning of the structural constant B, or the characteristic energy E_0, was recognized at an early stage, since each varies with the degree of activation and the increasing accessibility of the micropore system to larger molecules. The comparison (5) of simultaneous data from adsorption experiments (2) and from small-angle scattering of X-rays (51), for a series of carbons, leads to the empirical relation

$$E \cdot R_g = 14.8 \pm 0.6 \text{ kJ} \cdot \text{nm/mol} \tag{19}$$

between the characteristic energy of the DR equation and the gyration radius R_g of the micropores. The latter is an average dimension (7,52) of the pore, close but not equal to its half width $x = L/2$. However, in the case of disk-shaped micropores one may postulate, to a first approximation, that R_g and L are proportional. It follows from Eq. (19) that a relation of the type

$$L = \frac{K}{E_0} \tag{20}$$

should hold, in agreement with the observation that with increasing burnoff E_0 decreases and the micropore system becomes accessible to larger molecules (43).

As mentioned earlier, for slit-shaped and open micropores the surface of the walls S_{mi} can be calculated to good approximation from the volume W_0 and the accessible width L by using Eq. (1). This surface is not to be confused with the BET surface area S_{BET}, obtained from the adsorption isotherm, which is found to be close to the monolayer equivalent of the micropore volume W_0. (In fact, the two areas, S_{mi} and S_{BET}, are equal only if the pore accommodates two layers of adsorbate.) Experimentally, the true surface area of the micropores can be obtained from the adsorption of helium, the small-angle scattering of X-rays (52,53), the adsorption of water under certain circumstances (54), or from the distribution curve of the microporosity, resulting from experiments with molecular probes of known dimensions (43), as illustrated by Figure 9.

Let us consider, as a general example, the case of carbon CS, with a relatively homogeneous micropore system of average pore width L close to 0.65 nm. This value is suggested by molecular sieve experiments, showing that CCl_4 is partly excluded from the structure, whereas smaller molecules lead to similar micropore volumes. The parameters obtained from the DR equation are E_0 = 24.7 kJ/mol and W_0 = 0.35 cm^3/g. The adsorption of helium, over a range of temperatures and in the region of Henry's law, leads to an area of 1020 m^2/g. This should correspond to the true surface of the micropore walls, the nonmicroporous surface being negligable for this carbon. Following Dubinin's assumption (54), the amount of water adsorbed at relative pressure p/p_0 = 0.6 should be close to the monolayer capacity on the micropore walls, and in the present case the water adsorption isotherm at 295 K leads to 1033 m^2/g. The two values are in excellent agreement and by using Eq. (1) one obtains an average micropore width L = 0.68 nm from W_0 and S_{mi}. This value is close to the experimental value and illustrates self-consistency between the different parameters. For carbon CEP-18 shown in Figure 9, another typical example, the average pore width of 0.50 nm leads to S_{mi} = 1000 m/g, in good agreement with the cumulative area of 972 m/g derived from the micropore distribution. An extensive study, carried out along these lines, confirmed the validity of Eq. (20), but with a variable parameter K. Table 4 lists the values of K and of the average pore widths L in the range of 0.40−1.8 nm, as a function of the characteristic energy E_0 of the DR equation. The experimental error on L is of the order of 5−10% and it should be kept in mind that the model implied by Eq. (1) is oversimplified, since it corresponds to homogeneous and open micropores. However, the data of Table 4 are useful for a first assessment of microporous carbons on the basis of the DR equation. As illustrated by Figure 9 and described in Section 3.4, a finer description can be obtained for micropore systems in the range of 0.35−0.80 nm by using calorimetry and liquids of different molecular sizes.

Table 4 Characteristic Energies E_0 Derived from the DR Equation and Corresponding Micropore Widths L, Obtained from Different Adsorption and Immersion Experiments[a]

E_0 (kJ/mol)	37	35	30	25	20	17
L (nm)	0.45	0.50	0.60	0.70	13	1.8
K^a (nm·kJ/mol)	16.6	17.5	18.0	17.5	26	30

[a] The proportionality constant K is calculated from Eq. (20).

On the basis of earlier data, Dubinin (5,7) suggested a relation similar to Eq. (20) for $x = L/2$, the half width of the micropore,

$$x = \frac{k_0}{E_0} \tag{21}$$

where k_0 is given by

$$k_0 = 13.03 - 1.53 \times 10^{-5} E_0^{3.5} \quad \text{kJ·nm/mol} \tag{22}$$

It appears that the values of K, given in Table 4, are in good agreement with the corresponding values of $2k_0$ derived from E_0 by using Eq. (22), except in the region of $E_0 = 25$ kJ/mol. For this value, corresponding to a molecular sieve with $L = 0.70$ nm, Eqs. (21) and (22) lead to a width $2x$ of 0.94 nm, which is too large.

A good correlation is often found between S_{mi} calculated from Eqs. (1) and (21) and the areas obtained from the water adsorption isotherm at $p/p_0 = 0.6$ (54) but, as shown elsewhere (26,55), the shape of the water isotherm and therefore the amount of water adsorbed at a given value of p/p_0 depends on the number of hydrophilic sites on a given carbon. The comparison with data from independent methods suggests that the most reliable values of S_{mi} are obtained from water adsorption isotherms at $p/p_0 = 0.6$ when the hydrophilic centers represent approximately 5—10% of the true surface area.

As indicated earlier and illustrated by Figure 8, a strong heterogeneity in the micropore distribution is reflected by the deviation from linearity in the classical DR plot and by a value of exponent n smaller than 2. Such micropore systems may be described either by Eq. (12) with a suitable exponent n < 2 or by a combination of DR equations. The latter approach was used first by Isotova and Dubinin (2), who assumed that a nonlinear graph of the type shown in Figure 8 resulted from the superposition of two extreme ranges of microporosity,

$$W_t = W_{01} \exp \left[-B_1 \left(\frac{T}{\beta}\right)^2 \log^2 \left(\frac{P_0}{p}\right) \right] + W_{02} \exp \left[-B_2 \left(\frac{T}{\beta}\right)^2 \log^2 \left(\frac{P_0}{p}\right) \right]$$

(23)

or, in the modern formalism (5),

$$W_t = W_{01} \exp \left[-\left(\frac{A}{\beta E_{01}}\right)^2 \right] + W_{02} \exp \left[-\left(\frac{A}{\beta E_{02}}\right)^2 \right]$$

(24)

In the case of carbon F-02 (Fig. 8), one obtains, for example (5,56),

$$W_{01} = 0.178 \text{ cm}^3/\text{g}, \ E_{01} = 25.5 \text{ kJ/mol } (B_{01} = 0.56 \times 10^{-6} \text{K}^{-2})$$

$$W_{02} = 0.462 \text{ cm}^3/\text{g} \ E_{02} = 17.7 \text{ kJ/mol } (B_{02} = 1.17 \times 10^{-6} \text{K}^{-2})$$

According to Table 4 and Eqs. (21) and (22), the two types of micropore correspond to 0.7 and 1.6 nm respectively, the latter being called supermicropores. Although the existence of two distinct classes of micropore is questionable, Eqs. (23) or (24) provide useful information on the extent of heterogeneity, clearly not reflected in the linear domain of the DR representation. In the present case, the linear section of Figure 8 leads to $W_0 = 0.64 \text{ cm}^3/\text{g}$ and $E_0 = 18.9 \text{ kJ/mol}$, corresponding to an average pore width L of 1.4 nm.

A refinement of Isotova and Dubinin's approach has been suggested by Stoeckli and Huber (42,57) and was developed further in collaboration with Dubinin and colleagues (4,6−12,56,58). It is based on the idea of a continuous distribution of the micropores, as a function of B, and consequently

$$W_t = \int_0^\infty f(b) \exp(-By) \, db$$

(25)

As a first approximation, and in view of the limitations introduced by the experimental errors on the overall isotherm, a normalized gaussian of half width Δ

$$f(B) = \left(\frac{W_0}{\Delta \sqrt{2\pi}}\right) \exp \left[-\left(\frac{B_0 - B}{2\Delta^2}\right)^2 \right]$$

(26)

was postulated for the distribution $dW/dB = f(B)$. This analytical form is also convenient for solving the integral transform (25), which leads to

$$W_t = W_0 \exp(-B_0 y) \exp \left(\frac{y^2 \Delta^2}{2}\right) \times 0.5 \ [1 - \text{erf}(z)]$$

(27)

and where

$$z = \frac{(y - B_0/\Delta^2)\, \Delta}{\sqrt{2}} \tag{28}$$

In Eq. (27), heterogeneity is expressed by Δ, a measure of the spread of B around B_0. Table 5 gives typical values obtained for strongly activated carbons.

Since B is related to E_0, and consequently to L, parameter Δ has a clear physical meaning and it reflects the spread of the approximate micropore distribution dW/dL corresponding to f(B). From Eqs. (15), (21), and (22), one obtains

$$\frac{dW}{dL} = \left(\frac{MW_0L}{2\Delta\sqrt{2\Pi}}\right) \exp\left[\frac{-M^2(L_0^2 - L^2)^2}{32\Delta^2}\right] \tag{29}$$

where L_0 is calculated from B_0 and $M = (0.03828/K)^2$ or $(0.01914/k_0)^2$. As a good approximation, one may use the average value of $k_0 = 12$– 13 kJ·nm/mol for carbons with characteristic energies E_0 near 20 kJ/ mol. In terms of the half width x of the micropores, the parameter introduced and used by Dubinin (56), the normalized distribution (29) is

$$\frac{dW}{dx} = \left(\frac{2MW_0x}{\Delta\sqrt{2\Pi}}\right) \exp\left[\frac{-M^2(x_0^2 - x^2)^2}{2\Delta^2}\right] \tag{30}$$

Figures 10 and 11 show the distributions dW/dL calculated for carbons F-02 and CEP-59. In the case of carbon CEP-59, with a degree of heterogeneity similar to that of FO-2, the distribution is shifted to smaller values of L, since $E_0 = 25$ kJ/mol, and a direct comparison with another technique is possible. As shown in Fig. 11 the lefthand side of the distribution, obtained from molecular sieve experiments, is in good agreement with the curve calculated from the parameters of Eq. (27). This important example is a test and an illustration of the approach based on the generalization of the DR equation.

Table 5 gives the values of the different parameters of Eqs. (12), (13), and (27) obtained from the best fit of adsorption data. It is found, in practice, that the values of E_0 obtained from the linear DR region and from Eq. (27) often are similar. This means that the extra information provided by Δ will be complementary to the average parameters obtained from W_0 and E_0 and it is a direct measure of the extent of heterogeneity, as expressed by Eq. (29) or (30). On the basis of experimental evidence it may be assumed, as a general rule, that any active carbon with a characteristic energy E_0 of less than

Table 5 Comparison of Parameters Obtained for Various Active Carbons (5) by Fitting the Overall Adsorption Isotherms to Eqs. (12), (13), and (27)

Carbon	Eq. (27), W_0 (cm^3/g)	B_0 ($10^{-6}K^2$)	Δ($10^{-6}K^2$)	Eq. (13), E_0 (kJ/mol)	Eq. (12), n
U-02	0.43	0.92	0.21	20.0	1.65
F-02	0.64	1.03	0.29	18.7	1.35
F-6	0.78	1.00	0.36	19.1	1.28
T	0.40	0.61	0	24.5	2.0
CAL	0.44	0.99	0.26	19.2	1.63
CEP-59	0.48	0.66	0.18	25.0	1.67

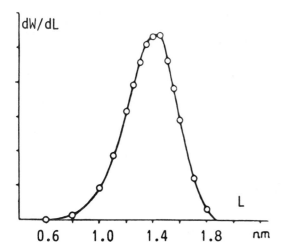

Figure 10   Micropore distribution of carbon F-02, following Eq. (29). See also Figure 8 and Table 5. (Adapted from ref. 11.)

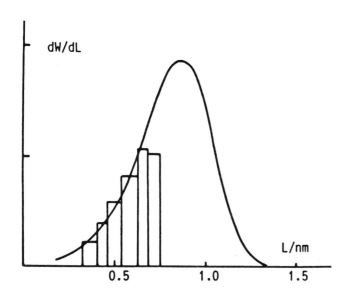

Figure 11   Micropore distribution of carbon CEP-59, calculated from Eq. (29), and the experimental data obtained from molecular sieve experiments in the region of L < 0.75 nm.

20−22 kJ/mol and with a large micropore volume is potentially heterogeneous, and further information requires extended adsorption data
of the type described above. As illustrated by Table 5, an increase
in the degree of heterogeneity corresponds to a decrease in exponent
n of the Dubinin−Astakhov equation [Eq. (12)], and it was shown by
Dubinin and Stoeckli (5) that there exists a linear relation between n
and Δ. This correlation, suggested by mathematical modeling based
on Eqs. (12) and (27), is illustrated in Figure 12. The corresponding empirical relation is

$$n = 2.00 - 1.78 \times 10^{6} \cdot \Delta \tag{31}$$

which means that an approximate micropore distribution can be calculated directly from the DA equation [Eq. (12)].

The possibility of different analytical expressions for f(B) has
been discussed (5,59), and recently Dubinin suggested the use of a
gaussian distribution for dW/dx, rather than for f(B), which leads to
interesting developments (10,11). However, it should be kept in mind
that owing to the experimental error on the overall isotherm, a definite choice of distribution is very difficult, if not impossible. On the
basis of the gaussian normalized on W_0,

$$\frac{dw}{dx} = \left(\frac{W_0}{\delta \sqrt{2\Pi}}\right) \exp\left[\frac{-(x_0 - x)^2}{2\delta^2}\right] \tag{32}$$

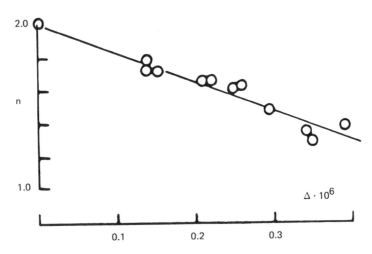

Figure 12 Empirical relation between exponent n of the Dubinin−
Astakhov equation [Eq. (12)] and parameter Δ of the generalized DR
equation [Eq. (27)].

A procedure similar to the foregoing leads to another generalized adsorption equation for the filling of micropores (10−12,58), which contains the information on the pores themselves,

$$W = \frac{W_0}{2\sqrt{1 + 2m\delta^2 A^2}} \exp\left(\frac{-mx_0^2 A^2}{1 + 2m\delta^2 A^2}\right)\left\{1 + \mathrm{erf}\left[\frac{x_0}{\delta\sqrt{2(1 + 2m\delta^2 A^2)}}\right]\right\}$$

(33)

In this equation $A = -RT \ln(p_0/p)$, as in the DA equation [Eq. (12)], and $m = (1/\beta k_0)^2$. For k_0, use of the average value of 12 kJ·nm/mol for typical heterogeneous carbons with characteristic energies E_0 near 20 kJ/mol is suggested.

Equation (33), which relates directly the micropore distribution to the adsorption data, has been called the Dubinin−Stoeckli equation by Dubinin, to distinguish it from the generalized DR equation [Eq. (27)]. For practical reasons, Eqs. (32) and (33) may also be expressed in terms of the accessible pore width $L = 2x$, rather than x.

A comparison of the parameters of Eqs. (27) and (33) is presented in Table 6. As illustrated by Figure 13, the micropore distributions resulting from the two approaches are in good agreement. (This example illustrates again the point made on the choice of different analytical functions, when extending Dubinin's theory to very heterogeneous systems.) A major advantage of Eq. (33) over Eq. (27) is the fact that it directly relates the adsorption isotherm to W_0, x_0, and δ, a set of physically relevant quantities.

As pointed out recently (43), the case where n = 3 in the general equation of Dubinin and Astakhov [Eq. (12)] corresponds to very homogeneous micropores, as found in MSC-5 and MSC-7. It is therefore obvious to extend the treatment outlined above to the DR equation itself and to consider it, formally, as a combination of the type

$$W\left(T;\frac{p}{p_0}\right)_{DR} = \Sigma\, W_{0i} \exp\left[-\left(\frac{A}{\beta E_{0i}}\right)^3\right]$$

(34)

This approach also takes into account the slight heterogeneity observed for carbons such as CEP-18 (Fig. 9), which follow the DR equation over a large range of relative pressures. In this approach, the values of E_{0i} must be calculated from the corresponding pore widths L_i by using an expression of the type

$$E_{0i} = \frac{k_i}{L_i}$$

(35)

Table 6 The Parameters of Eqs. (27) and (33) Obtained from the Adsorption Data for Strongly Activated Carbons (11)[a]

Carbon	Eq. (27)			Eq. (33)		
	W_0 (cm^3/g)	$B_0 \times 10^6 (K^{-2})$	$\Delta \times 10^{-6} (K^{-2})$	W_0 (cm^3/g)	x_0 (nm)	δ (nm)
F-02	0.64	1.03	0.29	0.64	0.62	0.09
AU-I	0.46	0.57	0.03	0.46	0.48	0.05
AU-II	0.647	1.03	0.28	0.65	0.64	0.11
AU-III	0.228	0.46	0.42	0.22	0.50	0.30

[a]Values of L_0 are calculated by using Dubinin's equation [Eq. (22)].

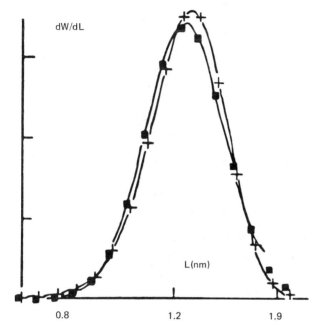

Figure 13 Micropore distribution of carbon AU-II obtained from the best fit of the adsorption data to Eqs. (27) and (33), with their underlying distributions (+ and ■, respectively). (Adapted from ref. 11.)

The data for the adsorption of benzene by MSC-5 and MSC-7 lead to characteristic energies E_0 of 27.6 and 24.0 kJ/mol with n = 3, and consequently k_i varies from 14 to 17 nm·kJ/mol. Some success has already been obtained in applying Eq. (34) to carbons such as CEP-18, but owing to the fact that the general variation of K with E or L is unknown when n = 3, the practical use of Eq. (34) is still limited at the present time. The approach based on Eq. (34) is, nevertheless, of great importance from a theoretical point of view.

3.4 THERMODYNAMIC CONSEQUENCES OF DUBININ'S THEORY. ENTHALPIES OF IMMERSION

Although Dubinin's theory does not include Henry's law (4), the general equation of Dubinin and Astakhov [Eq. (12)] leads to a number of interesting consequences in the framework of classical thermodynamics. By combining Eq. (12) with the definition of the isoteric heat of adsorption q^{st} (15)

$$\left(\frac{\partial \ln p}{\partial T}\right)_a = \frac{q^{st}}{RT^2} \tag{36}$$

it is found (4) that

$$q^{st}(T; \Theta) = \beta E_0 \left\{ \left[\ln\left(\frac{a_0}{a}\right) \right]^{1/n} + \frac{\alpha T}{n} \left[\ln\left(\frac{a_0}{a}\right) \right]^{(1/n)-1} \right\}$$
$$+ \Delta H_{vap}(T; p) \tag{37}$$

where a represents the amount adsorbed at relative pressure p/p_0; a_0 corresponds to the limiting adsorption value; Θ is the degree of micropore filling; and α is the thermal expansion coefficient of the adsorbate. Equation (37) generalizes the earlier expression derived by Bering and Serpinksii (60,61) on the basis of Eq. (13), the DR equation.

From the definition of the net heat of adsorption,

$$q^{net} = q^{st} - \Delta H_{vap} \tag{38}$$

it appears that q^{st} can be expressed in the exact form

$$q^{net}(T; \Theta) = \beta E_0 \left\{ \left[\ln\left(\frac{a_0}{a}\right) \right]^{1/n} + \frac{\alpha T}{n} \left[\ln\left(\frac{a_0}{a}\right) \right]^{(1/n)-1} \right\} \tag{39}$$

A useful consequence of Eq. (39) is found in the field of immersion calorimetry, an important technique described in Section 3.6. It can be shown (15,29,31,62) that for a microporous solid without any external surface, the enthalpy of immersion into a liquid is given by

$$-H_i\Delta(T) = \int_0^1 q^{net}(T; \Theta)\, d\Theta \tag{40}$$

where the minus sign takes into account the fact that ΔH_i is a negative quantity, whereas q^{st} is positive, by convention. This enthalpy change corresponds to the transfer, at temperature T, of 1 mol of adsorbate from the liquid state into the micropores.

As pointed out by Stoeckli (63), Eq. (39) implies a straightforward relation between the enthalpy of immersion into an organic liquid and the parameters obtained from the adsorption of its vapor on the same solid. In spite of the absence of a limiting value for q^{st} as $\Theta \to 0$, Eq. (39) can be integrated and for the case where n = 2 one obtains the molar enthalpy of immersion

$$\Delta H_i \ (J/mol) = - \frac{\beta E_0 \sqrt{\pi} \ (1 + \alpha T)}{2} \tag{41}$$

for the liquid filling the micropores at temperature T. For a carbon with a micropore volume W_0 and a liquid of molar volume V_m, Eq. (41) can be expressed in joules per gram of solid, and

$$\Delta H_i \ (J/g) = - \frac{\beta E_0 W_0 \sqrt{\pi} \ (1 + \alpha T)}{2V_m} \tag{42}$$

The validity of Eq. (42) has been tested by comparing calculated and experimental values of ΔH_i obtained for microporous carbons with negligable external surface areas (26,64).

Equation (42) clearly shows that the enthalpy of immersion corresponds to a process of micropore filling and depends on the pore size, unlike the wetting of open or nonporous surfaces. By combining Eq. (42) with Eqs. (1) and (20) and eliminating E_0, W_0, and L, one obtains

$$\Delta H_i \ (J/g) = - \frac{\beta \cdot K \cdot S_{mi} \sqrt{\pi} \ (1 + \alpha T)}{4 \cdot 10^3 \times V_m} \tag{43}$$

and

$$\frac{\Delta H_i}{S_{mi}} \ (J/m^2) = - \frac{\beta \cdot K \sqrt{\pi} \ (1 + \alpha T)}{4 \cdot 10^3 \times V_m} \tag{44}$$

This shows that the enthalpy of adsorption of microporous carbons into a given liquid is not proportional to the micropore surface S_{mi}, since K (or Dubinin's parameter k_0) depends on the pore width L as shown in Table 4. For typical values of E_0 between 30 and 20 kJ/mol, Eq. (44) leads to specific enthalpies $\Delta H_i/S_{mi}$ of approximately 0.13—0.18 J/m^2 for benzene at room temperature, against 0.115 J/m^2 for the open graphitic surface. It appears that no definite standard value can be proposed for the micropore walls. This is obviously the result of the overlap of the adsorption forces from opposite walls and their variation with the pore width (22—24), and it is not surprising that the specific enthalpies are approximately 1.5 times larger in typical micropores than on the open surface, as observed for the potenetial adsorption energy from the gas phase. In the case of supermicropores, on the other hand, when E_0 is smaller than 17—18 kJ/mol, the specific enthalpy of immersion should be close to that of the geometrical surface of the pores. This was pointed out in Section 3.2, when dealing with sample U-03.

Moreover, since the BET surface area derived from the adsorption isotherm is usually different from the true micropore surface S_{mi}, the quantity $\Delta H_i / S_{BET}$ appears to have little or no physical meaning in the case of microporous carbons. On the other hand, the concept of wetting applies to the external or nonmicroporous surface S_e, often present in active carbons, which can be as high as 200–300 m^2/g in special cases (39). The experimental enthalpy of immersion of active carbons therefore contains a variable contribution from the wetting of S_e, and Eq. (42) becomes (65)

$$-\Delta H_i \ (J/g) \ \exp = \frac{\beta E_0 W_0 \sqrt{\pi} \ (1 + \alpha T)}{2V_m} - h_i S_e \qquad (45)$$

where h_i represents the specific enthalpy of wetting, in joules per square meter. It appears that the values quoted for the wetting of graphitic surfaces by typical liquids (66–68) lead to consistent results in the case of well-characterized active carbons (26,65), and Eq. (45) can be used to estimate S_e if E_0 and W_0 are known. The values of α and h_i used for standard adsorbates in the liquid state are given in Table 7. (As mentioned earlier, α and consequently V_m are not necessarily the same in the adsorbed and the liquid states, but in the absence of reliable information, it was decided to use systematically the date of the liquid state.)

Equations (42) and (45) are useful in the case of carbons showing molecular sieve properties below 0.80 nm: when such effects are present, the volume W(L) actually accessible to a molecule of critical dimension L is inferior to that seen by a small reference molecule such as CH_2Cl_2 or benzene. Consequently, the experimental enthalpy

Table 7 Thermal ExpansionCoefficient α and Specific Enthalpies of Wetting h_i and Critical Diameters L of Standard Liquid Adsorbates Near Room Temperature[a]

Liquid	$\alpha \ (10^{-3}K^{-1})$	$-h_i(J/m^2)$	L (nm)
Benzene	1.24	0.114	0.41
Methylene Chloride	1.34	0.152	0.33
Cyclohexane	0.96	0.101	0.54
Carbon tetrachloride	1.22	0.115	0.63
2,5-Norbornadiene	1.20	0.110	0.66
α-Pinene	1.02	0.110	0.68/0.80
Perchlorocylopentadiene	1.17	0.110	0.75/0.88

[a]See Table 3 for β and molar volumes.

of immersion of the test molecule will be smaller than the value calculated by using E_0 and W_0 of the reference, and parameters β, V_m,
h_i, and α of the given molecular probe. From Eq. (29) it follows that

$$W(L) = - [\Delta H_i (exp) - h_i S_e] \cdot \frac{2V_m}{\beta E_0 \sqrt{\Pi} (1 + \alpha T)} \qquad (46)$$

and a histogram of the type shown in Figures 9 and 11 can be established from experiments with a series of organic liquids. For selected systems, the volumes $W(L)$ obtained calorimetrically have also been
compared with the limiting volumes $W_0(L)$ obtained from the adsorption of the corresponding vapor, and a good agreement has been found.

Table 7 lists effective diameters L. These values are in good
agreement with the data found in the literature (69–72). From the
molecular structure of solid cyclohexane, determined by X-ray crystal-
lography (73), it appears that in the chair configuration this molecule
can fit into slits of approximately 0.54 nm in width. This value is
0.07 nm higher than suggested by Kippling and Wilson (69).

In a similar way, the width of 2.5-norbornadiene, estimated from
the data on a derivative (74), is in the range of 0.60–0.64 nm.
Since the apparent volumes $W(L)$ obtained from norbornadiene and
carbon tetrachloride are similar in the case of carbons with micropore
width L between 0.5 and 0.7 nm, the best-fit value of 0.63 nm for
carbon tetrachloride seems reasonable, compared with 0.60 nm, as
predicted by a molecular model (72). The data for L given in Table
7 results from an overall comparison of experiments with different
carbons (26,43) and is not absolute or definite. Its systematic use,
as described in ref. 43, can nevertheless lead to useful information
when comparing series of carbons.

3.5 THE INTERACTION OF WATER WITH ACTIVE CARBONS

The adsorption of water vapors by microporous carbons does not follow Dubinin's theory, but it can be described by an equation recently proposed by Dubinin and Serpinskii (75,76) and used in a series
of investigations (23,24,46–49,54,55,77–80).

$$\frac{p}{p_0} = \frac{a}{c(a_0 + a)(1 - ka)} \qquad (47)$$

In this equation, a represents the amount of water adsorbed at relative pressure p/p_0 and a_0 is the number of so-called primary centers
(usually expressed in mmol/g of solid); c is the ratio between the
rate constants of adsorption and desorption, and k is a constant whose
magnitude is fixed by the condition that for $p/p_0 = 1$, the total amount

of water adsorbed in the micropores is a_s. Implicity, Eq. (47) is valid for a given type of primary sites and in the case of sites with different energies, parameters a_0 and c become apparent quantities. It should also be noted that Eq. (47) applies to the adsorption branch of the water isotherm and there exists no model for the description of the hysteresis loop associated with the desorption of water from the micropores.

Typical water adsorption isotherms (55) are shown in Figures 14 and 15. As revealed by the latter example, and confirmed by mathematical modeling, for a given carbon the position of the isotherm also depends on the values of variable a_0 and c. This means that water adsorption depends on both the chemistry and the physics of the microporous carbon. It follows, as indicated earlier, that the value of $a_{0.6}$, the amount adsorbed at relative pressure $p/p_0 = 0.60$ and used by Dubinin (54), does not necessarily correspond to the monolayer capacity of water on the micropore walls.

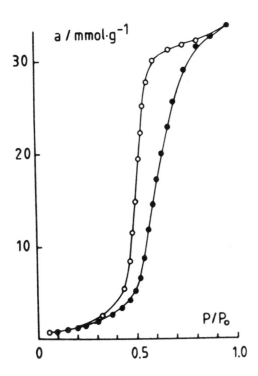

Figure 14 Adsorption (•) and desorption (○) of water by carbon U-02 at 293 K (55).

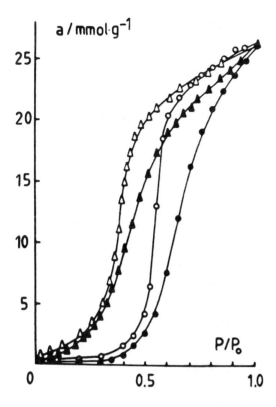

Figure 15 Adsorption of water by samples MSC5-1 (▲ △) and MSC5-2 (● ○) at 293 K, containing, respectively, 26.5 and 4.6% of primary sites (55).

It was shown recently by Stoeckli et al. (55,80) that the number of primary sites a_0 is directly related to the enthalpy of immersion of active carbons into water. For typical industrial carbons, treated in vacuo at 400–600°C and containing a uniform type of hydrophilic site, the relation is

$$\Delta H_i \ (J/g) = -25 \ (J/mmol) \ a_0 \ (mmol/g) - 0.6 \ (J/mmol)$$
$$\cdot(a_s - a_0) \ (mmol/g) \tag{48}$$

These sites are probably of the carbonyl type, since their desorption at high temperatures leads to carbon monoxide, as shown by Barton and Harrison (81).

For carbons resulting from a treatment with hydrogen peroxide and containing acidic sites (80), the values of a_0 obtained from the isotherms are in good agreement with the direct titration, as shown in Table 8. The corresponding enthalpies of immersion are also related to a_0 and it is found that

$$\Delta H_i \ (J/g) = -55 \ (J/mmol) \ a_0 \ (mmol/g) \ - \ 0.6 \ (J/mmol)$$
$$\cdot \ (a_s - a_0) \ (mmol/g) \tag{49}$$

In the case of a carbon with both types of site (or other sites) present simultaneously on its surface, the enthalpy of immersion would be expressed as a sum of contributions. On the other hand, the quantity a_0 obtained from the isotherm would not be equal to the total number of different sites, since the Dubinin–Serpinskii equation [Eq. (47)] applies only to the adsorption by a single type of site.

Stoeckli et al. (81) have shown, on the basis of a thermodynamic development of Eq. (47), that parameter c is related to the enthalpy of immersion through

$$c = c_0 \exp \left(-\frac{\Delta H_{im}}{RT} \right) \tag{50}$$

and it appears experimentally that $c_0 = 1.00 \pm 0.11$.

ΔH_{im} is the molar enthalpy of immersion, expressed in joules per mol of water filling the micropores and obtained by the simple relation

$$\Delta H_{im} \ (J/mol) = \frac{\Delta H_i \ (J/g)}{a_s \ (mol/g)} \tag{51}$$

Equation (50) establishes a direct relation between constant c of the Dubinin–Serpinskii equation [Eq. (47)] and the enthalpy of immersion into water, a quantity which can be measured easily and quickly. Moreover, through Eqs. (48) and (49) or their combination, c is related to the total number of primary centers, whether they are uniform or not. It follows that in the case of a single type of site, the adsorption branch of the water isotherm can be calculated from the corresponding enthalpy of immersion by using Eq. (48) or (49) for a_0 and Eq. (50) for c. The remaining constant, k, is given by the condition that $a = a_s$ as p/p_0 tends to 1 and, following Dubinin and serpinskii (76),

$$k = \frac{c - 1}{c \cdot a_s} \tag{52}$$

Table 8 Parameters of the Water Adsorption Isotherm at Room Temperature (62) and the Enthalpies of Immersion of 13 Active Carbons into Water (80)

Solid	$-\Delta H_i$ (J/g)	a_s (mmol/g)	a_0 (mmol/g)[a]	a_0 (mmol/g)[b]	a_0 (mmol/g)[c]	c
UO2-1	27.9	26.1	0.35	0.50	—	1.79
UO2-2	40.7	26.1	0.95	1.03	—	1.85
UO2-3	70.0	26.3	2.01	2.22	—	2.49
UO2-4	51.2	28.0	0.59[d]	0.63[d]	0.52[d]	1.92
UO2-5	56.2	31.0	0.70[d]	0.69[d]	0.70[d]	2.04
UO3-1	31.2	31.0	0.22[d]	0.23[d]	—	1.63
FO2	47.0	32.6	1.49	1.12	—	1.62
N-125	31.6	31.5	0.49	0.52	—	1.66
MSC5-1	23.2	9.5	0.74	0.72	—	2.35
MSC5-2	14.4	9.5	0.38	0.36	—	2.05
SP	17.1	7.0	0.58	0.53	—	3.10
AU-1	40.4	22.0	1.18	1.61	—	1.98
AU-2	49.4	35.0	0.95	1.16	—	1.80

[a]a_0 values obtained from the isotherms.
[b]a_0 values obtained from the enthalpy of immersion and Eq. (48) or (49).
[c]a_0 values obtained by backtitration with hydrochloric acid after neutralization with sodium hydroxide in excess.
[d]Acidic sites.

In the case of a carbon with different types of sites, following, for example, Eqs. (48) and (49), the enthalpy of immersion into water is given by the linear combination

$$-\Delta H_i = 25a_{01} + 55a_{02} = 0.6 \ (a_s - a_{01} - a_{02}) \tag{53}$$

where a_{01} and a_{02} (in millimolecules per gram) represent the number of sites of each type. The backtitration of the acidic sites leads to a_{02}, and a_{01} can be calculated from Eq. (53). By using a_{01} and a_{02} and the corresponding values of c_1 and c_2 given by Eq. (50), it is possible to calculate the overall water adsorption isotherm resulting from a combination of individual isotherms of the Dubinin—Serpinskii type.

To illustrate the techniques described in this chapter, Table 9 presents the results obtained from the full characterization of a typical industrial carbon, N-125. An extensive characterization was carried out, in order to confirm the results obtained from a minimum set of experiments. The benzene adsorption isotherm at 298 K leads to the fundamental parameters W_0 and E_0 from the low-pressure data, whereas the upper part could be used to assess the external surface area S_e. As shown in Table 1, several methods lead to the average value of 157 ± 20 m^2/g, in good agreement with 144 m^2/g derived from the enthalpy of immersion into liquid benzene at 293 K, by using Eqs. (42) and (45).

With Eqs. (20) and (1), W_0 and E_0 lead to L = 1.6 nm and S_{mi} = 834 m^2/g, and it follows that the total surface area S_t of the carbon (microporous and external) is approximately 1000 m^2/g. (This area is somewhat larger than the value obtained from the adsorption of helium.) As pointed out earlier, the amount of water adsrobed at $p/p_0 = 0.6$ does not agree with S_{mi} owing, probably, to the relatively low concentration of hydrophilic adsorption centers. The values of a_0 obtained either from the Dubinin—Serpinskii equation [Eq. (47)] or from the enthalpy of immersion are in good agreement (0.49 and 0.52 mmol H_2O/g) and represent approximately 3% of a_t, the monolayer equivalent of the total surface area S_t. This is important information for the characterization of the sample, which appears to be hydrophobic in the present case.

The foregoing example suggests that the minimum set of experiments required for a good characterization of a pure active carbon could include the following:

The adsorption isotherm of benzene near 293—298 K, leading to
W_0 and E_0 and implicitly to an average value of L and S_{mi},
The immersion into benzene to yield the nonmicroporous surface
S_e, and the immersion into water to gain information on the
hydrophilic or hydrophobic character of the material.

Table 9 Characterization of a Typical Active Carbon (N-125) by Different Techniques

W_0 (cm^3/g)	E_0 (kJ/mol)	L (nm)	S_{mi} (m^2/g)	S_e (m^2/g)	S_{He} (m^2/g)	S_t (m^2/g)
0.64	16.61	1.6	834	157 ± 20	715	991(±20)

S_{H_2O} (m^2/g)	c	a_0 (mmol H$_2$O/g)	a_t (mmol H$_2$O/g)	a_0/a_t	d_{He}(g/cm^3)
494	1.66	0.49/0.52	15.7	0.031	2.09 ± 0.02

These important techniques are briefly described in Section 3.6.

3.6 ADSORPTION AND IMMERSION TECHNIQUES

The adsorption isotherm of gases and vapors on active carbons can be determined volumetrically or gravimetrically (15). In view of the high sorption capacity of this material, the latter technique is adequate. Adsorption is measured directly by monitoring the weight of the sample (approximately 0.2–0.5 g) contained in a small bucket and suspended at the end of a helicoidal spring made of quartz. The sensitivity of the spring is in the range of 4–8 cm/g and by using a suitable micrometer, with an accuracy of $\pm 5 \times 10^{-3}$ mm, it is possible to determine weight changes within $\pm$ 0.1 mg. In the case of nitrogen adsorbed at 78 K, this corresponds to approximately $\pm$ 0.3 m^2 or 10^{-3} cm^3, which is small when compared with S_e and W_0 of typical active carbons. The micropore volume W_0 can therefore be determined from the adsorption isotherm with a final accuracy of approximately $\pm$ 0.01 cm^3/g.

The adsorbent and the spring are contained in a vacuum system, allowing it to be treated in situ. Prior to adsorption experiments, the carbon is originally outgassed under 10^{-5} torr at least, for 25 hr at 300°C and for 1–2 hr at 400–450°C. Between experiments the time of outgassing varies, depending on the nature of the vapors. The pressure in the gas phase is measured with transducer gauges, in the range of 10^{-4} to 10^3 torr and the sample is kept at a constant temperature by means of an external bath (either liquid nitrogen near 78 K or a thermostatted bath, kept within $\pm$ 0.01 K near room temperature).

Immersion calorimetry is carried out with a calorimeter of the Tian–Calvet type, shown in Figure 16 and designed specifically for the energies evolved in the experiments described in this chapter (2–50 Joules). The solid (usually 0.3 g) is outgassed in its glass bulb (1), originally attached to a vacuum line and sealed off under vacuum. The liquid (2), which is contained in the brass cell (3) of the calorimeter, penetrates into the bulb from the bottom when the tip is broken mechanically from outside. This avoids the formation of a vapor phase in the bulb and the solid is fully wetted by the liquid. The energy released by the process of immersion is led through 180 copper-constantan thermocouples (4) to a thermal buffer (5) consisting of 1 kg of copper powder. This buffer is itself surrounded by a water jacket and an external bath near 298 K, with a stability of $\pm$ 0.01 K. This ensures very good thermal stability in the system. The energy released by the immersion experiment, usually 2–50 J, is proportional to the area under the signal recorded at the extremities of the thermocouples, connected in series and having a total sensitivity of 9.5 mV/K

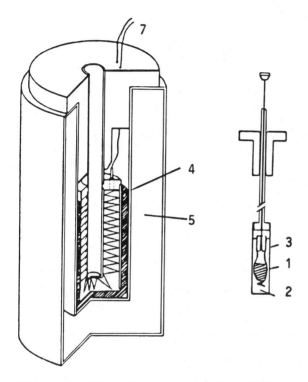

Figure 16 Schematic representation of the immersion calorimeter used for active carbons (25,26).

near room temperature. The equipment is calibrated electrically and with standard systems. The reproducibility of such a calorimeter is ± 1% for typical experiments with active carbons immersed into organic liquids or water, and up to three experiments can be performed in one day.

REFERENCES

1. Dubinin, M. M., Zaverina, E. D., and Radushkevich, L. V., *Zh. Fiz. Khim.*, *1947*:1351.
2. Isotova, T. I. and Dubinin, M. M., *Zh. Fiz. Khim.*, *1965*:2796.
3. Dubinin, M. M. and Astakhov, V. A., *Izv. Akad, Nauk. SSSR* (ser. khim), *1971*:5.
4. Dubinin, M. M., in *Progress in Surface and Membrane Science*, Vol. 9, D. A. Cadenhead, Ed., Academic Press, New York, 1975.

5. Dubinin, M. M. and Stoeckli, H. F., *J. Colloid Interface Sci.*, 75: 34 (1980).

6. Dubinin, M. M., *Carbon, 19:* 321 (1981).

7. Dubinin, M. M., *Carbon, 20:* 195 (1982).

8. Dubinin, M. M., *Carbon, 21:* 359 (1983).

9. Dubinin, M. M., *Izv. Akad. Nauk. SSSR* (ser. khim.), *1983:* 487.

10. Dubinin, M. M., *Carbon, 23:* 373 (1985).

11. Dubinin, M. M., Efremov, S. N., Kataeva, L. I., and Ustinov, E. A., *Izv. Akad. Nauk. SSSR* (ser. khim.), *1985:* 255.

12. Dubinin, M. M. and Polyakov, N. S., *Izv. Akad. Nauk. SSSR* (ser. khim.), *1985:* 1943.

13. Dubinin, M. M., *Zh. Fiz. Khim.,* 34: 959 (1960).

14. *IUPAC Manual of Symbols and Terminology*, Appendix 2, Pt. 1, Colloid and Surface Chemistry, Pure and Appl. Chem. 31, 578 (1972).

15. Gregg, S. J. and Sing, K. S. W., *Adsorption, Surface Area and Porosity*, Academic Press, London, 1982.

16. Bailey, A., Cadenhead, D. H., Davies, D. A., Everett, D. H., and Miles, A. J., *Trans. Faraday Soc.,* 67:231 (1971).

17. Linares-Solano, A., Rodriguez-Reinoso, F., Martin-Martinez, J. M., and Lopez-Gonzales, J. D., *Adsorp. Sci. Technol.,* 1: 317 (1984).

18. Stoeckli, H. F., Perret, A., and Mena, P., *Carbon, 18:* 443 (1980).

19. Fryer, J. R., *Carbon, 19:* 431 (1981).

20. Marsh, H., Crowford, D., O'Grady, T. M., and Wennenberg, A., *Carbon, 20:* 419 (1982).

21. Oberlin, A., Villey, M., and Combaz, A., *Carbon, 18:* 347 (1980); *J. Microscop. Spectrosc. Electron.,* 7: 327 (1982).

22. Stoeckli, H. F. and Perret, A., *Helv. Chim. Acta,* 58: 2318 (1975).

23. Everett, D. H. and Powl, J. C., *J. Chem. Soc. Faraday Trans. I, 1976:* 619.

24. Stoeckli, H. F., *Helv. Chim. Acta,* 57: 2195 (1974).

25. Stoeckli, H. F. and Kraehenbuehl, F., *Carbon, 22:* 297 (1984).

26. Kraehenbuehl, F., Ph. D. dissertation, University of Neuchâtel, 1983.

27. Gregg, S. J. and Langford, J. F., *Trans. Faraday Soc.,* 65: 1394 (1969).

28. Linares-Solano, A., Lopez-Gonzalez, J. D., Martin-Martinez, J. M., and Rodriguez-Reinoso, F., *Adsorp. Sci. Technol.,* 1: 123 (1984).

29. Bohara, J. N. and Sing, K. S. W., *Adsorp. Sci. Technol.,* 2: 89 (1985).

30. Dubinin, M. M. and Kadlec, O., *Carbon, 13:* 263 (1975).

31. Ali, S. and McEnaney, B., *J. Colloid Interface Sci.,* 107: 355 (1985).

32. Martin-Martinez, J. M., Rodriguez-Reinoso, F., Molina-Sabio, M., and McEnaney, B., *Carbon*, *24*:255 (1986).
33. Dubinin, M. M., in *Chemistry and Physics of Carbon*, Vol. 2, P. Walker, Ed., Marcel Dekker, New York, 1966.
34. Rootare, H. M. and Prenzlow, C. F., *J. Phys. Chem.*, *71*:2733 (1967).
35. Li, G., Liu, N., and Wu, C., *Cinhua Xuebao*, *2*:268 (1981); *Chem. Abstracts*, *96*:110665 k (1982).
36. Dubinin, M. M., Kataeva, L. I., and Ulin, V. I., *Izv. Akad. Nauk. SSSR* (ser. khim.), *1977*:510.
37. Dubinin, M. M. and Kataeva, L. I., *Izv. Akad. Nauk. SSSR* (ser. khim.), *1977*:516; *1980*, 238, 498.
38. Dubinin, M. M. and Fedoseev, D. V., *Izv. Akad. Nauk. SSSR* (ser. khim.), *1982*:246.
39. Andreeva, G. A., Polyakov, N. S., Dubbinin, M. M., and Nikolaev, K. M., *Izv. Akad. Nauk. SSSR* (ser. khim.), *1981*: 2193.
40. Koganowskii, A. M. and Levtsenko, T. M., *Zh. Fiz. Khim.*, *46*:1789 (1972); Stabnite, A. M. and Eltekov, Yu., *Zh. Prikladnoi Khim.*, *48*:186 (1975); Jaronec, M. and Derylo, A., *J. Colloid Interface Sci.*, *84*:191 (1981).
41. Meguro, T., Torikai, N., and Watanabe, N., *Carbon*, *23*:137 (1985).
42. Huber, U., Stoeckli, H. F., and Houriet, J. P., *J. Colloid interface Sci.*, *67*:195 (1978).
43. Kraehenbuehl, F., Stoeckli, H. F., Addoun, A., Ehrburger, P., and Donnet, J. B., *Carbon*, *24*:483 (1986).
44. Stoeckli, H. F. and Morel, D., *Chimia*, *34*:502 (1980).
45. Nikolaev, K. M. and Dubinin, M. M., *Izv. Akad. Nauk. SSSR* (ser. khim.), *1958*:1165.
46. Finger, G. and Bülow, M., *Carbon*, *17*:87 (1979).
47. Cerofolini, G. F., *Surface Sci.*, *24*:2391 (1971).
48. Cereofolini, G. F., *Thin Solid Films*, *23*:129 (1974).
49. Stoeckli, H. F., Lavanchy, A., and Kraehenbuehl, F., in *Adsorption at the Gas—Solid and Liquid—Solid Interface*, J. Rouquerol and K. S. W. Sing, Eds., Elsevier, Amsterdam, 1982.
50. Stoeckli, H. F., Kraehenbuehl, F., Lavanchy, A., and Huber, U., *J. Chim. Phys.*, *81*:787 (1984).
51. Dubinin, M. M. and Plavnik, G. M., *Carbon*, *2*:261 (1964); *6*:183 (1968).
52. Guinier, A. and Fournet, B., *Small—Angle Scattering of X-rays*, Wiley, New York, 1955.
    Guinier, A., *Theorie et Technique de la Radiocristallographie*, Dunod, Paris, 1964.
53. Janosi, A. and Stoeckli, H. F., *Carbon*, *17*:465 (1979).

54. Dubinin, M. M., *Carbon*, *18:*355 (1980).
55. Stoeckli, H. F., Kraehenbuehl, F., and Morel, D., *Carbon*, *21:*589 (1983).
56. Dubinin, M. M., *Izv. Akad. Nauk. SSSR* (ser. khim.), *1980:*18.
57. Stoeckli, H. F., *J. Colloid Interface Sci.*, *59:*184 (1977).
58. Dubinin, M. M., Polyakov, N. S., and Ustinov, E. A., *Izv. Akad. Nauk. SSSR* (ser. khim.), *1985:*2680.
59. Rozwadowski, M. and Wojsz, R., *Carbon*, *22:*363 (1984).
60. Bering, B. P. and Serpinskii, V. V., *Dok. Akad. Nauk. SSSR*, *114:*1257 (1957); *148:*1331 (1963).
61. Bering, B. P. and Serpinskii, V. V., *Dok. Akad. Nauk. SSSR*, *148:*1331 (1963).
62. Clint, H. J., *J. Chem. Soc. Faraday Trans. I*, *1973:*1320.
63. Stoeckli, H. F., *Izv. Akad. Nauk. SSSR* (ser. khim.), *1981:* 63 (Festschrift M. M. Dubinin).
64. Stoeckli, H. F. and Kraehenbuehl, F., *Carbon*, *19:*353 (1981).
65. Stoeckli, H. F. and Kraehenbuehl, F., *Carbon*, *22:*297 (1984).
66. Bartell, F. E. and Sugitt, R. M., *J. Phys. Chem.*, *58:*36 (1954).
67. Robert, L., *Bull. Soc. Chim. France*, *1967:*147.
68. Wade, W. H., Deviney, M. L., Brown, W. A., Hnoosh, M. H., and Wallace, D. R., *Rubber Chem. Tech.*, *45:*117 (1972).
69. Kippling, J. J. and Wilson, R. B., *Trans. Faraday Soc.*, *56:* 557 (1960).
70. Adams, L. B., Boucher, E. A., and Everett, D. H., *Carbon*, *8:*761 (1970).
71. Fitzer, E. and Kalka, J., *Carbon*, *10:*173 (1972).
72. Ainscough, A. N., Dollimore, D., and Heal, G. R., *Carbon*, *11:*189 (1973).
73. Kahn, R., Fourme, R., Renaud, D., and Renaud, M., *Acta Cryst. B 29:*131 (1973).
74. Griffin, A. M. and Sheldrick, G., *Acta Cryst. B*, *31:*895 (1975).
75. Dubinin, M. M. and Serpinskii, V. V., *Izv. Akad. Nauk. SSSR* (ser. khim.), *1981:*402, 1151.
76. Dubinin, M. M. and Serpinskii, V. V., *Carbon*, *19:*402 (1981).
77. Andreeva, G. A., Polykov, N. S., Dubinin, M. M., Nikolaev, K. M., and Ustinov, E. A., *Izv. Akad. Nauk. SSSR* (ser. khim), *1981:*2188.
78. Vartapetyan, R. Sh., Voloshtshuk, A. M., Dubinin, M. M., Polyakov, N. S., and Serpinskii, V. V., *Izv. Akad. Nauk. SSSR* (ser. khim.), *1982:*1215.
79. Dubinin, M. M., Nikolaev, K. M., Petukhova, G. A., and Polyakov, N. S., *Izv. Akad. Nauk. SSSR* (ser. khim.), *1984:*743.
80. Kraehenbuehl, F., Quellet, C., Schmitter, B., and Stoeckli, H. F., *J. Chem. Soc. Faraday Trans. I*, *1986:*3439.
81. Barton, S. S. and Harrison, B., *Carbon*, *10:*245 (1972).

4

Characterization of Active Carbons

Carbon surface has a unique character. It has a porous structure
which determines its adsorption capacity, it has a chemical structure which influences its interaction with polar and nonpolar adsorbates, it has active sites in the form of edges, dislocations, and
discontinuities which determine the chemical reactions of its surface
with other heteroatoms. Thus the adsorption behavior of an active
carbon cannot be interpreted on the basis of surface area and pore
size distribution alone. Carbons having equal surface area but prepared by different methods or given different activation treatments
show markedly different adsorption characteristics. The determination of a correct model for adsorption on active carbon adsorbents
with complex chemical structure is, therefore, a complicated problem.
A proper model must take into consideration both the chemical and
the porous structure of the carbon, which includes the nature
and the amounts of the surface chemical structures, surface area,
and pore size distribution as well as the chemical and physical
characteristics of the adsorbate such as its chemical structure,
polarity, and molecular dimensions. It is therefore essential to characterize the surfaces of carbons with respect to number and nature
of the chemical structures on the surface, the polarity of the surface, pore size distribution, micropore volume, and surface area.
This chapter deals with various methods that have been used to
characterize the surfaces of carbons.

4.1 CHARACTERIZATION BY IMMERSION CALORIMETRY

When a solid is brought into contact with a nonreacting liquid, a thermal effect is produced which is called the heat of immersion or heat of wetting. This thermal effect comprises the net integral heat of adsorption evolved in forming an adsorbed layer at saturated vapor pressure plus the heat of wetting of this film. In the case of active carbons which are highly microporous the heat of wetting involves the thermal effects related to the wetting of the external surface (nonporous structure) and the filling of micropore volume. Heat of immersion studies were reported as early as 1802 by Leslie (1), although attempts were usually aimed at measuring surface areas. However, it is now well known that the heat of immersion for the same carbon is different in different liquids, so that it cannot be related only to surface area. It is in fact a measure of the accessibility of the surface to a given liquid. The heat of immersion depends on the chemical structure of the carbon surface as well as on the nature of the immersion liquid and consequently has been rated very high for determination of surface polarity, site heterogeneity, hydrophobicity, and several other surface properties of microporous carbons (2−6). Attempts have been made to relate heats of immersion to thermal pretreatment of carbons (7−9), but they have been complicated by changes in surface area and possible changes in the surface structure. As the immersional energetics are higher for specific interactions between water and surface oxygen groups on carbons compared to nonspecific interactions, immersion calorimetry has been suggested as a valuable tool for characterization of these surface groups as well (10).

Barton and co-workers (11−13) evaluated the average polarity of the surface of graphite covered by oxygen structures by measuring heats of immersion in a series of butyl derivatives of different dipole moments. Heat of immersion of graphite in n-hexane differed only slightly compared to that for a completely oxygen-free sample, indicating that the oxygen complexes, which covered about 30% of the graphite surface, did not markedly effect the immersional energetics in n-hexane. However, the heats of immersion values were appreciably different in butyl derivatives (Table 1). This was attributed to the interaction of the electrostatic field at the surface due to the polar oxygen groups with the dipole moment of the molecules of the immersion liquid. These workers calculated the value of this interaction energy ($E\mu$) by using the equation suggested by Zettlemoyer and Chessick (14):

$$E\mu = - \mu F$$

where F is the electrostatic field and μ is the dipole moment of the immersion liquid. Assuming that the interaction with the oxygen-free

Table 1 Heat of Immersion of Graphite in Various
n-Butyl Derivatives at 27°C

Derivative	Heat of immersion	
	cal/g	erg/cm
Hexane	- 4.47	- 115
n-Butanol	- 5.36	- 138
n-Butyric acid	- 5.82	- 149
n-Butyl chloride	- 5.46	- 140
n-Butyraldehyde	- 6.30	- 162
n-Butyl nitrile	- 6.45	- 166
n-Butylamine	- 11.07	- 282

Source: Barton et al. (11).

surface of graphite could be described purely by dispersion forces, the interaction energy with the dipoles of the n-butyl derivatives was related to the surface covered by oxygen groups. The average electrostatic field at the graphite surface was found to be $1.1 \pm 0.1 \times 10^5$ e.s.u.

The influence of different types of carbon—oxygen surface structures on heats of immersion in polar and nonpolar liquids was examined by evacuating samples of graphite and a carbon black, Spheron-6, at different temperatures up to 1000°C (12,13). The heats of immersion both in water and in methanol increased with evacuation temperatures up to 200°C due to the removal of adsorbed water and exposure of additional surface oxides. At temperatures of evacuation above 200°C, while the heat of immersion in water decreased continuously up to 1000°C, in the case of methanol heats decreased up to 600°C and increased thereafter (Fig. 1). The continuous decrease in the case of water resulted from the removal of surface oxides, which provided hydrophilic centers for hydrogen bonding. The final heat of immersion value, when all the surface oxygen was removed (40 erg/cm^2), agreed fairly well with the value (32 erg/cm^2) obtained by Wade (8) on graphon which was completely free of any combined oxygen and which had a predominantly hydrophobic basal plane surface. The small difference between the two values could be understood in terms of the fact that the graphite used in these studies (12) had about 30% of the surface exposed as edge planes. The small increase in the heat of immersion in the case

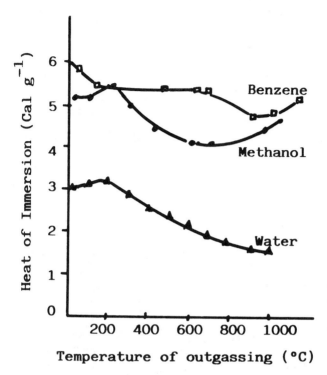

Figure 1 Immersional heats of graphite as a function of outgassing temperature. [From Barton and Harrison (12).]

of methanol at evacuation temperatures between 600 and 1000°C indicated a higher interaction energy of methanol with the bare surface than with the surface covered with oxygen evolved as carbon monoxide. The decrease in the heat of immersion on evacuation in the case of water was found to be related linearly to the total oxygen content of the carbon (Fig. 2) and not to the oxygen evolved as carbon dioxide, as shown by Puri et al. (15,16) for charcoals. However, in the case of methanol, the decrease in the heat of immersion varied linearly with the oxygen desorbed as carbon dioxide on evacuation.

The heat of immersion in the nonpolar benzene remained more or less unaffected by the removal of CO_2-complex in the evacuation temperature range up to 550°C and then decreased at higher temperatures. This decrease was attributed to the decrease in the interaction between π electron clouds of benzene molecule with the partial positive charge on the carbonyl carbon atom which evolved as carbon monoxide at temperatures above 550°C (16).

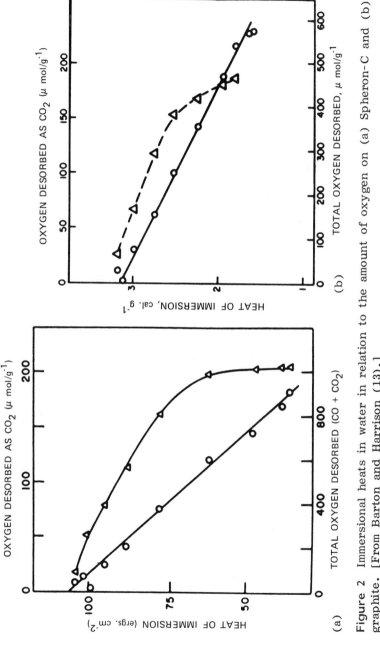

Figure 2 Immersional heats in water in relation to the amount of oxygen on (a) Spheron-C and (b) graphite. [From Barton and Harrison (13).]

Table 2 Surface Area and Heat of Immersion of Casuarina Charcoals

Charcoal	Carbonization temperature (°C)	Surface area (m²/g)			Heat of wetting (cal/g)	
		Carbon dioxide 25°C	Methanol 35°C	Water 30°C	Methanol 35°C	Water 30°C
CI	500	294	286	55	13.50	6.80
CII	650	315	313	49	14.60	6.80
CIII	800	358	354	43	16.60	6.70
CIV	900	382	375	19	17.10	1.60
CV	1000	58	59	16	3.20	1.60
CIV	1200	14	14	8	1.80	1.55

Source: Youssef (17).

Youssef (17) observed that heat of immersion in methanol decreased with temperature of carbonization of wood charcoal between 500 and 900°C and increased thereafter and followed the same trend as surface area (Table 2). However, the heat of immersion per unit area, as calculated from the slope of the linear plot, 192 erg/cm^2, was less than half the value obtained by Bangham and Razouk (18), which was 433 erg/cm^2. Thus the concept of heat of immersion as a measure of surface area has a limited applicability and probably is useful only for comparative data on a series of charcoals obtained from the same material. The heat of immersion in water (Table 2) showed no simple relationship either with surface area or with the number of oxygen functional groups on these charcoals (17).

Maggs and Robins (19) examined a relationship between surface area and heat of immersion for a number of commercial activated carbons and carbon cloths using liquids of varying molecular dimensions. The accessible area as a function of the molecular diameter of the immersion liquid decreased as the size of the molecule increased in all carbons and carbon cloths (Fig. 3). The shape of the plots also indicated a marked difference in the porous structure of the different carbons, the highly activated charcoals showing a more pronounced open structure. Comparison of the accessible areas with BET (N$_2$) areas showed that only a part of the BET area was accessible to even the smallest molecule such as benzene. Thus these workers suggested that the BET monolayer capacity, in fact, represents a pore volume capacity. When these carbon cloths were activated with carbon dioxide for varying periods of time, the areas accessible to all the molecules increased appreciably with the degree of activation (Fig. 4). Almost all of the BET area became accessible to benzene at a burnoff of about 80%.

Atkins et al. (20), while measuring the heats of immersion of a charcoal cloth having predominantly narrow pores and a sample of Amoco-activated carbon having a wide distribution of micropores (1–2 nm) in a number of organic liquids, observed a wide variation in the heat of immersion in different liquids. This indicated that heat of immersion is not a measure of surface area but a measure of the volume of the pores accessible to the molecule of the wetting liquid (Table 3). Since the concept of BET surface area is completely unrealistic to microporous carbons, these workers suggested that the representation of heat of immersion per unit surface area is not completely correct for microporous carbons. According to them it is better to represent heat of immersion in terms of adsorption capacity or in terms of per unit weight of carbon (in joules per gram); the latter units were sufficient to indicate trends in the molecular sieve properties of carbons with the help of a series of molecular probes.

Zettlemoyer et al. (21) determined the heat of immersion of a polymer-based carbon, Carbosieve-s, with narrow micropore size

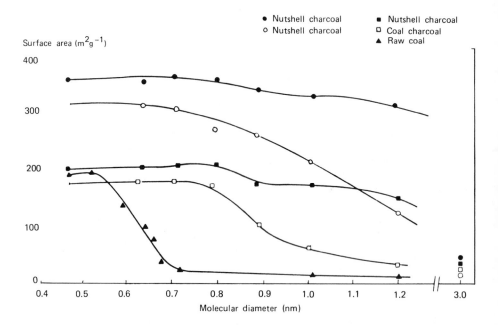

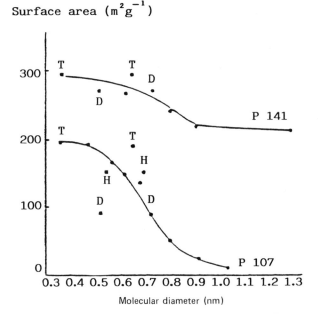

Figure 3 Accessible surface area as a function of molecular dimensions of the wetting liquid for granular charcoals. [From Maggs and Robins (19).]

Table 3 Heats of Immersion of Carbons in Organic Liquids

Immersion liquid	Activated carbon Amoco Px21		Carbon cloth AM/4		Carbon black Vulcan 3G	
	J/g	mJ/m^2	J/g	mJ/M^2	J/g	mJ/m^2
Toluene	271	73	155	123	8.2	115
Mesitylene	300	81	161	128	9.8	138
Isodurene	305	82	154	122	10.7	150
n-Hexane	245	66	94	74	6.1	86
Cyclohexane	190	51	97	7	5.8	82
Neohexane	190	52	72	58	5.3	75

Source: Atkins et al. (20).

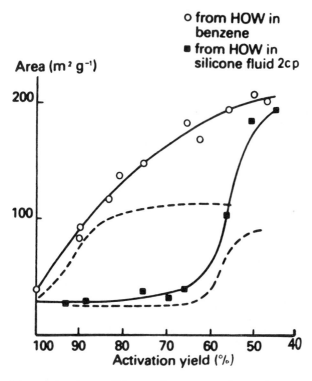

Figure 4 Accessible surface area of a charcoal cloth as a function of activation yield. [From Maggs and Robins (19).]

distribution (around 0.5 nm) in water, methanol, and 2-propanol as a function of precoverage of the surface with the immersion liquid. The immersional heats decreased to a plateau at 20% micropore volume filling followed by a decrease after 80% micropore volume filling in the case of water; two plateaus at 25% and at 80% in the case of methanol; and a sharp decrease and a plateau in the case of 2-propanol (Fig. 5). The differential heats of adsorption calculated from the slopes of the heats of immersion against surface-covered plots showed a rapid fall from 180 kJ/mol at 1 mg/g adsorption to 82 kJ/mol at 2 mg/g adsorption in the case of water. The heat of adsorption approached the heat of liquefaction at about 23% micropore volume filling. In the case of methanol there was also a sharp fall and the adsorbate approached liquidlike properties at about 30% micropore volume filling. The heats of adsorption in the case of 2-propanol, however, showed two maxima (Fig. 6) occurring at 40% and 90% of micropore volume filling. A large fall in the heat of

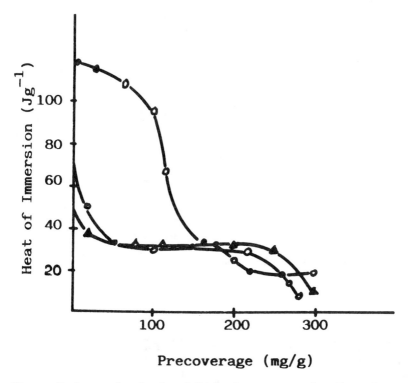

Figure 5 Immersion heats of Carbosieve-s as a function of precoverage. Δ-Δ water; -●-●- methanol; -O-O- 2-propanol. [From Zettlemayer et al. (21).]

adsorption at low coverages was attributed to surface heterogeneity with adsorption occurring at acid group sites, whereas the large difference between methanol and 2-propanol in the initial stages of surface coverages was due to differences in their molecular diameters.

Takahashi and co-workers (22—24) observed that the immersional heats of carbon blacks in ethanol and n-butanol increased linearly with an increase in the active hydrogen content of the carbon black (Fig. 7). The energy contributions of the activated hydrogen-containing functional groups to the heats of immersion in ethanol and n-butanol were 37.7 and 34.2 kJ/mol respectively. These values were approximately 25% of that for water (142 kJ/mol) (23,24). Hagiwara et al. calculated the hydrogen-bonding energy between

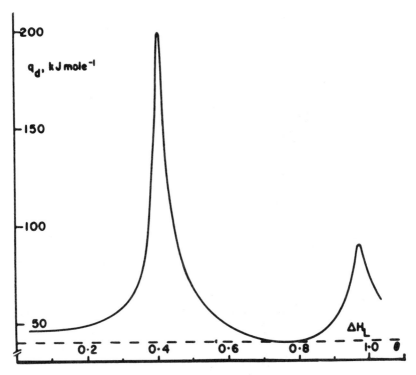

Figure 6 Differential heat of adsorption of 2-propanol on Carbo-sieve at 25°C. [From Zettlemoyer et al. (21).]

alcohol molecules and active hydrogen sites by applying the analytic-al procedure suggested by Zettlemoyer (25) and using the equation

$$\Delta H_i - H_\ell = (q_d + q_u + q_w) + q_h$$

where ΔH_i is the heat of immersion, H_ℓ the surface enthalpy of the liquid, q_d the energy of London dispersion interaction, q_u the electrostatic interaction energy between the electrostatic field of the carbon surface and the dipole of the liquid, q_w the polarization energy of the liquid molecule by the surface field, and q_h the hydrogen-bonding energy.

The electrostatic interaction energy of the field on the carbon black surface with the dipole of the n-butanol, q_u, and the hydro-gen bonding energy of its active hydrogen site with the n-butanol molecule, q_h, were found to be 17.8 and 16.4 kJ/mol respectively for 1 mol of active hydrogen. The sum of these two energy

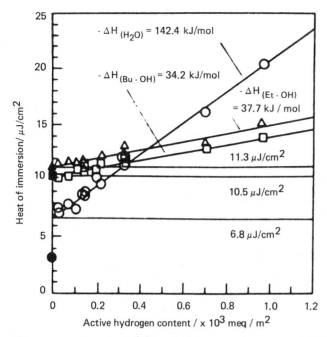

Figure 7 Heats of immersion in relation to active hydrogen content of various carbon blacks. Δ ethanol; □ *n*-butanol; O water. [From Hagiwara et al. (22).]

contributions agreed closely with the energy contribution of the active hydrogen-containing functional groups to the heat of immersion (34.2 kJ/mol). These results showed clearly that heats of immersion 37.7 kJ/mol for ethanol and 34.2 kJ/mol for *n*-butanol estimated from a linear plot between heat of immersion and the active hydrogen content on the carbon surface (Fig. 7) were the sum of the electrostatic interaction energy and of the hydrogen-bonding energy so that the interaction of the active hydrogen with the immersion liquid was electrostatic of the hydrogen-bonding type. The high value 142.2 kJ/mol for water probably included the energy of dissociation hydration reactions.

Stoeckli and Kraehenbuehl (10) tried to establish a link between the enthalpy of immersion ΔH_i and the parameters of Dubinin's theory of volume filling of micropores (TVFM) using enthalpy of immersion values on a large range of microporous carbons. According to these workers the enthalpy of immersion is a function of the characteristic energy βE_0 of the Dubinin—Astakhov equation and can be expressed as

$$- \Delta H_i (T) = \beta E_0 (1 + \alpha T) \frac{\sqrt{\pi}}{2} \tag{1}$$

where α is the thermal expansion coefficient of the adsorbate at temperature T and ΔH_i is in joules per mole of the adsorbate. The enthalpy of immersion in terms of joules per centimeter of micropores can be obtained by dividing the right-hand side of the Eq (1) by molar volume of the pore liquid V_m (in cubic centimeters per mole):

$$- \Delta H_i (J/cm^3) = \beta E_o (1 + \alpha T) \frac{\sqrt{\pi}}{2 V_m} \tag{2}$$

These workers compared the enthalpies of immersion determined experimentally, using an extremely sensitive calorimeter designed in their laboratory, for 10 different microporous carbons in benzene, n-heptane, and water with the values calculated using Eq. (1) and (2) (Table 4). The experimental and calculated values for benzene and n-heptane agreed closely, the proportionality constants being 1.05 for benzene and 1.02 for n-heptane. Similar correlation was observed for several of the carbons when immersed in carbon tetrachloride and toluene. However, these workers pointed out that it was not possible in the case of active carbons to use a standard value of ΔH_i to calculate directly the micropore volume, W_0, from enthalpies of immersion alone because ΔH_i for a given liquid was also a function of E_0 and consequently of the micropore system. Since the adsorption of water vapor did not follow the TVFM (27), these new relations did not apply to immersion into water and thus no trend was observed.

Radeke (28) proposed equations for determining the average slit pore width (x) and the external geometric surface area S_e of the micropores of the activated carbons based on Dubinin's TVFM but using data from immersion calorimetry. According to Radeke the two parameters are related to enthalpy of immersion by

$$S_e = - \frac{h_i}{0.169} \tag{3}$$

and

$$x = - K, (T, E_0) \frac{V_m}{h_i} \tag{4}$$

Table 4 Comparison of Experimental and Calculated Enthalpies of Immersion (ΔH^i in J/cm^3 of Micropores)

Carbon	Heat of immersion in benzene		Heat of immersion in n-heptane		Heat of immersion in water
	Experimental	Calculated	Experimental	Calculated	
1	394	413	—	—	146
2	377	395	355	362	90
3	366	383	346	353	39
4	349	366	321	327	110
5	295	309	293	299	42
6	287	301	267	272	68
7	284	297	—	—	76
8	273	285	276	282	65
9	265	276	256	261	79
10	232	243	242	247	73

Source: Stoeckli and Kraehenbuehl (10).

where h_i is the specific heat of immersion and is related to molar heat of immersion by

$$\Delta H_i = h_i \frac{V_\beta}{V_m}$$

where V_β is the molar volume of benzene and V_m is the micropore volume. These relations are a direct consequence of the Stoeckli equation [Eq. (6)]. These micropore parameters were calculated from heats of immersion values of 11 different activated carbons in benzene and in water using Eq. (3) and (4) and compared with those obtained from adsorption isotherms using the Dubinin−Redushkevich equation. The values for the two parameters agreed well (Table 5) for benzene but differed appreciably for water. The difference in the case of water was attributed to its smaller molecular diameter so that it could penetrate even very small pores. Stoeckli et al. 29−31) recently showed that the external surface S_e of active carbons with good approximation can be calculated from the immersion calorimetry using the equation

$$\Delta H_{i(\text{expt.})} = \Delta H_{i(\text{micropores})} + S_e h_i \tag{5}$$

where h_i is the specific enthalpy of immersion of the open reference surface (carbon black, for example, and $\Delta H_{i(\text{micropores})}$ is given by the equation

$$\Delta H_i \ (\text{J/g}) = - \beta E_0 \ (1 + \alpha T) \frac{\sqrt{\pi} \ W_0}{2 \ V_m} \tag{6}$$

A comparison of the S_e values by this method with those obtained by other independent methods (Table 6) indicates a fair agreement.

Stoeckli et al. (26) also obtained a satisfactory relationship between a_0, the number of primary adsorption centers derived from adsorption isotherms using the Dubinin−Serpenski equation, and the enthalpy of immersion ΔH_i in water by the equation

$$\Delta H_i \ (\text{J/g}) = - \ 25.0 \ (\text{J/mmol H}_2\text{0}) - 0.6 \ (\text{j/mmol H}_2\text{0}) \ (a_s - a_0) \tag{7}$$

where a_s and a_0 are expressed in millimoles of water per gram of the active carbon. In this equation -25.0 kJ/mol represents the enthalpy of interaction of water molecules with the oxide sites and

Table 5 Comparison of Surface Area and Average Slit Pore Width Obtained from Adsorption Data (using the Dubinin–Radushkevich equation) and from Calorimetric Data

Carbon sample	Surface area from adsorption data (m²/g)	Average slit pore width from adsorption data X (nm)	Calculated from calorimetry data		
			Benzene surface area (m²/g)	Average slit pore width (nm)	Water surface area (m²/g)
1	470	0.61	460	0.69	610
2	700	0.78	575	0.74	775
3	850	0.47	610	0.52	935
4	545	0.57	500	0.73	—
5	610	0.59	590	0.79	790

Source: Radeke (28).

Table 6 Comparison of External Surface Areas

Carbon	S_e (m^2/g) (immersion)	S_e (m^2/g) (adsorption)
N-125	144 (C_6H_6)	175 (t/F method, C_6H_6 at 293 K)
E_O = 16.6 kJ/mol	194 (n-C_7H_{16})	145 (Kelvin/mesopores)
W_O = 0.64 cm^3/g	176 (n-$C_{16}H_{34}$)	141 (decomposition of N_2 isotherm after n-C_9H_{20} pre-adsorption)
FA	123 (C_6H_6)	111 (t/F method, C_6H_6 at 293 K0
E_O = 2.0 kJ/mol	112 (n-C_7H_{16})	118 (decomposition of C_6H_6 isotherm)
W_O = 0.29 cm^3/g	106 (n-$C_{16}H_{34}$)	105 (t/F method, N_2 at 78 K)

Source: Stoeckli et al. (31).

−0.6 kJ/mol is the enthalpy change associated with pore filling. The data (Table 7) showed clearly that the agreement between the values of a_0 obtained from the Dubinin−Sperpenski equation and by using Eq. (7) with experimental enthalpies of immersion is fairly good. These workers caution that Eq. (7) is valid only for pure microporous carbons since enthalpies of immersion are very sensitive to mineral impurities whose dehydration or dissolution may affect ΔH_i. These workers also observed a linear decrease in enthalpy of immersion in water with the amount of oxygen desorbed as carbon monoxide at various temperatures and this value corresponded to − 30 kJ/mol of atomic oxygen.

Barton et al. (32) measured the heats of immersion in water and cyclohexane of a BPL-active carbon oxidized to different degrees with nitric acid. These workers agreed with Stoeckli et al. (26) that the major factor influencing the reaction between the carbon surface and water was the presence of hydrophilic oxide sites, but they found no relationship between the enthalpy of immersion and a_0 for the oxidized carbon samples. The Stoeckli relationship was, however, applicable to unoxidized BPL-active carbon. These workers observed a rapid increase in the enthalpy of immersion with increasing a_0 for low values of a_0 (Fig. 8), with the initial rate of increase being about −25.0 kJ/mol. Above a_0 = 6 mmol/g, however, the

Table 7 Comparison of Amounts of Adsorption Centers (a_0) Obtained from Water Adsorption Isotherms and Immersion Calorimetry

Sample	a_0(exp)	a_0(calc)	$-\Delta H_i$ (J/g)
1	0.35	0.50	27.9
2	2.01	2.22	70.0
3	1.49	1.12	47.0
4	0.74	0.72	23.3
5	0.37	0.36	14.4
6	0.15	0.10	14.7
7	0.50	0.56	31.6
8	0.31	0.46	25.4
9 [7]	0.58	0.53	17.1

Source: Stoeckli et al. (26).

enthalpy of immersion assumed a constant value (Fig. 8), which according to Barton et al., indicated that at a_0 = 6 mmol/g an adsorbing water molecule experiences the maximum interaction with the surface and that further increase in a_0 did not enhance the amount of hydrogen bonding-type reactions. These workers are of the view that the total micropore volume does not markedly influence the interaction of the surface with water. Rather it is the concentration of the oxide sites which dictates the magnitude of the interaction enthalpy change. Once the surface oxide concentration has reached a threshold value, however, further increase does not affect the amount of the interaction.

In the case of cyclohexane, the net molar enthalpies of interaction (obtained by dividing enthalpy of immersion by the micropore adsorption amount) were all smaller than the molar enthalpy of vaporization (-33 kJ/mol), and decreasing at first rapidly and then very slowly (Fig. 9) above a = 6 mmol/g as the hydrophilic nature of the surface increased. Thus the basic difference between the interaction of cyclohexane and water with a porous carbon was that in the former case the pore filling itself produced the major enthalpy change, whereas in the latter case the interaction involving a large decrease in enthalpy of immersion was governed by the surface oxide concentration up to a threshold value which probably represented the concentration of the oxide sites permitting maximum hydrogen bonding with water molecules and subsequent pore filling occurring with small enthalpy changes.

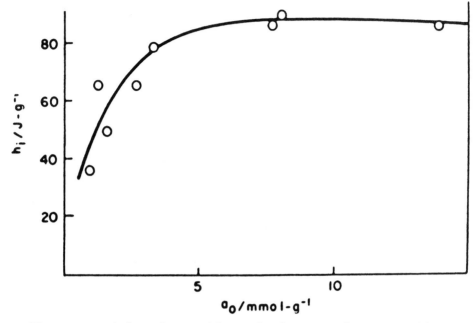

Figure 8 Variation of heat of immersion in water with hydrophilic adsorption centers a_0. [From Barton et al. (32).]

Table 8 Immersional Heats of Carbon U-02 before and after Adsorption—Desorption Cycle of Tertiobutyl-Benzene

	ΔH_i (J/g solid)	
Sample	Benzene	t-butyl benzene
Initial carbon	117.3	114.3
After adsorption—desorption cycle	119.8	122.4

Source: Stoeckli et al. (33).

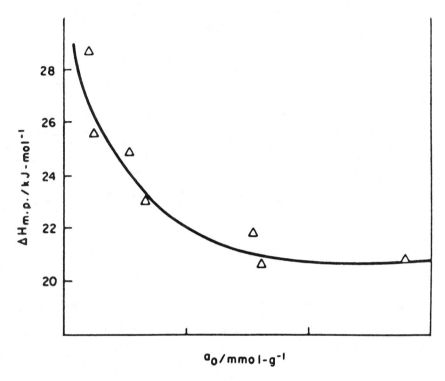

Figure 9 Variation of net molar enthalpy of adsorption ΔH_{mp} δ with hydrophilic adsorption centers a_0. [From Barton et al. (32).]

Stoeckli et al. (33) also observed that the immersion calorimetry can be used to measure porous structural modifications introduced by the adsorption of large-sized adsorbates in microporous carbons. The enthalpies of immersion were found to change appreciably with the adsorption of tertiobutyl benzene (Table 8). These differences in the enthalpies of immersion have been attributed to the work needed for the deformation of the micropore structure in the untreated sample.

4.2 CHARACTERIZATION BY ADSORPTION OF VAPORS

Adsorption of vapors has been carried out for many decades and has been the main source for deriving adsorption models both from a theoretical and an experimental point of view. However, most of these models are based on simple and idealized systems using non-porous solids whose surfaces are considered perfectly homogeneous and good approximations of the geometrical or external surfaces. The adsorption equilibrium has been described by equations based on a two-dimensional analogue of the van der Waals equation which includes dispersive attractive interactions between the adsorbate molecules and the repulsive interactions caused by the finite volume of the molecules. Consequently, the adsorption models, although of immense value in explaining adsorption of nonpolar adsorbents, failed or showed deviations when applied to adsorption of nonhomogeneous or polar adsorbents. In the case of adsorption on polar adsorbents and of polar adsorbates, several factors can contribute. For example, besides the dispersion forces, specific interactions resulting, for instance, from hydrogen bonding play an important role in the adsorption of polar compounds. In addition, part of the polar adsorbate molecule may react with the carbon surface due to the presence of the oxygen functional groups.

4.2.1 Adsorption of Water Vapors

Study of the mechanism of adsorption of water vapors on active carbons and interpretation of the respective adsorption isotherms have been undertaken by many groups of investigators for the last several decades. Nonetheless, several ambiguities remain unresolved. The principal points generally raised and argued concern the role of associated oxygen and capillary condensation in determining the adsorption of water vapor on active carbons. For example, Pierce and co-workers (34,35) and Dubinin et al. (36,37) are of the view that certain active sites on carbons made available by oxygen-containing groups act as primary adsorption centers on which water gets bonded at low relative pressures. These adsorbed water molecules then act as secondary adsorption centers where more water can be adsorbed by hydrogen bonding. With increasing relative pressure two-dimensional islands of the condensed phase appear on the carbon surface; these later merge into each other as the pressure is increased, forming a continuous film of water over a part of the carbon surface. Hysteresis arises from the fact that while adsorption proceeds through the formation of clusters, desorption occurs from the continuous layer in nonporous carbons and from the menisci in the case of porous carbons. The forces in the former process were confined to active sites only, whereas in the latter they extended to the whole layer so that the desorption occurs at higher relative pressures.

Dubinin and Serpinski (36,38) were able to derive an equation to calculate the number of these primary adsorption centers and which could explain the sharp rise in the adsorption isotherm. According to them if a_0 is the number of primary adsorption centers and a (both expressed in millimoles per gram) is the amount adsorbed at relative pressure h $(h = p/p_0)$, then the dynamic equilibrium condition of adsorption can be represented as

$$K_i\ (a_0 + a)\ (1 - ka)h = k_2 a \tag{8}$$

where $(a_0 + a)$ represents the total number of adsorption centers and $(1 - k_a)$ takes into account the decrease in the number of acting adsorption centers with increase in adsorption. The parameter k can be determined from the condition $a = a_s$ when $p/p_0 = 1$ where a_s is the saturation adsorption value. The isotherm equation can also be written as

$$h = \frac{a}{[c\ (a_0 + a)\ (1 - ka)]} \tag{9}$$

where $c = k_1/k_2$. This isotherm equation has three parameters, a_0, c, and k, which can be determined from a single adsorption isotherm as follows.

Equation (9) can be written as

$$\frac{a}{h} = A_1 + A_2 a - A_3 a^2 \tag{10}$$

where $A_1 = ca_0$ $A_2 = c\ (1 - a_0 k)$ and $A_3 = ck$.

The equation can be linearized by introducing a function Z as

$$Z = \frac{\dfrac{a}{h} - \dfrac{a_1}{h_1}}{(a - a_1)} \tag{11}$$

where a and a_1 are adsorptions at two relative pressures h and h_1. Thus from Eqs. (9) and (10) we get

$$Z = (A_2 - A_3 a_1) - A_3 a = Z_0 - A_3 a \tag{12}$$

so that a plot between Z and a is a straight line with slope equal to A_3 and intercept equal to Z_0. Thus A_2 and A_3 can be calculated from the slope and the intercept as

$$A_2 = Z_0 + A_3 a$$

and

$$A_1 = \frac{a_1}{h_1} - A_2 a_1 + A_3 a_1^2$$

Thus if one knows A_1, A_2, and A_3, the three parameters of Eq. (2) can be calculated. This equation was found to be applicable partially over the entire range of relative pressures and could provide approximate information about the number of primary adsorption centers, i.e., carbon–oxygen structures, capable of forming hydrogen bonds with water molecules.

Stoeckli et al. (26) recently calculated the parameters of the Dubinin–Serpinski equation for water adsorption by direct curve fitting using a computer and found that the equation could be more accurately fitted to the adsorption data at relative pressures between 0.05 and 0.1 and the inflection point of the ascending branch of the isotherm. Deviations at lower relative pressures were attributed to Langmuir-type adsorption in the very early stages. At high stages of micropore filling a_0 represented only a small percentage of the amount of water adsorbed and the influence of the primary centers became less important. These workers suggested that the primary centers are important in the type III part of the isotherm and in the actual position of the steep rise. The upper part of the isotherm, as p/p_0 tends to saturation, reflects the limiting filling capacity of the micropore system and was not related to the primary centers of adsorption a_0.

Puri and co-workers (39,40) studied the adsorption of water vapor on charcoals associated with varying amounts of oxygen and observed that it was the oxygen present as CO_2-complex (oxygen evolved as CO_2 on evacuation or acidic oxygen complex) which provided active sites for the sorption of water vapor and that the sorption–desorption isotherms did not meet even at zero relative pressure as long as CO_2-complex was present. This indicated that a certain amount of water was fixed on the CO_2-complex sites by hydrogen bonding and could not be removed by evacuation at the adsorption temperature. This amount of water fixed corresponded roughly to 1 mol of water for each mole of CO_2-complex. The rest of the combined oxygen had little effect on the sorption of water or the magnitude of hysteresis. When the CO_2-complex was enhanced by the oxidation of charcoal or reduced by degassing, the adsorption of water vapor was enhanced or reduced correspondingly. Thus according to these workers capillary condensation does not play any part in the adsorption of water vapor.

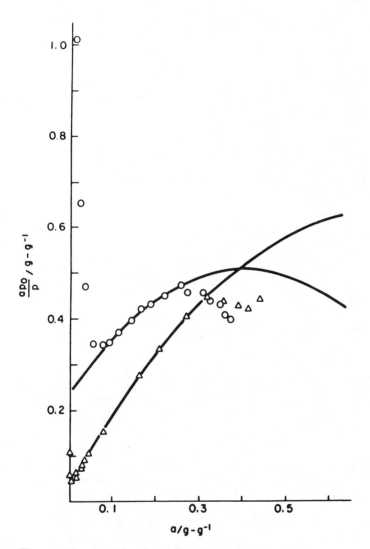

Figure 10   Verification of Dubinin—Serpinski equation for water adsorption of carbons. Triangle, BPL-activated carbon; circle, BPL oxidized with NHO_3. [From Barton et al. (32).]

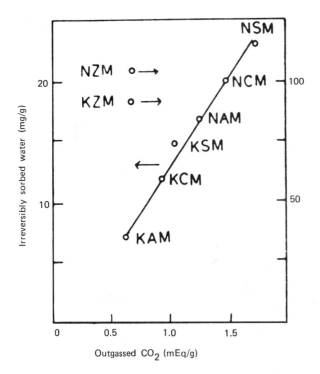

Figure 11 Relationship between amount of water adsorbed irreversibly and amount of CO_2 desorbed on outgassing for different carbons. [From Youssef et al. (42).]

Barton et al. (32) observed that the adsorption of water vapor on a BPL-active carbon before and after oxidation to different degrees with nitric acid was influenced not so much by the pore dimensions as by the presence of hydrophilic centers provided by acidic oxygen structures. BET monolayer values obtained from the low-pressure part of the isotherm showed a linear relationship with acidic surface structures as determined by neutralization with 0.1 M potassium hydroxide solution. The line passed through the origin and the slope of the linear plot indicated that 1.6 mol of water were adsorbed per acid group. These workers also calculated the concentration of oxide sites using the Dubinin—Serpinski equation and found that the equation was applicable only to the type III part of the adsorption isotherm (Fig. 10). Youssef and co-workers (17,41, 42) found that the water vapor adsorption isotherms, on coals carbonized at temperatures between 300 and 800°C and followed by

activation with HNO_3, $KMnO_4$, and $ZnCl_2$ and on wood charcoals
carbonized between 500 and 1200°C, were type II with an open
hysteresis loop, which did not close even at zero relative pressure
except in the case of charcoals carbonized at 800°C and above. In
the case of oxidized charcoals, the irreversibly adsorbed water
showed a linear relationship with the amount of CO_2 obtained on outgassing the samples (Fig. 11). However, this linear relationship
could not be observed in the case of charcoal activated with ZnCl,
followed by oxidation with HNO_3, or $KMnO_4$. This was attributed to
the fixation of oxygen deep into the micropores so that these charcoals retained much more water compared to the other samples. The
surface areas calculated from water adsorption isotherms using the
BET equation did not agree with the surface areas calculated from
CO_2 adsorption, indicating that in the case of adsorption of water
vapors, the monolayer represents the number of reacting groups
rather than the number of adsorbed molecules required to form a
monolayer.

Adsorption of water vapor on a charcoal cloth having predominantly narrow micropores and a sample of Amoco-activated carbon
having a wider distribution of micropores extending up to 1—2 nm
showed type V isotherms with characteristic hysteresis extending to
low relative pressures (20). The very low affinity between water
and these carbons was caused by the absence of polar groups that
could be involved in specific interactions of the hydrogen-bonding
type. The adsorption increased only at relative pressures above 0.5
when the adsorbate interactions were significant. This indicated that
the dispersion forces in the micropores were not sufficient for an
enhanced adsorption of water at low relative pressures. The pronounced hysteresis was attributed to strong adsorbate interactions
rather than to being a reflection of the pore structure. Zettlemoyer
et al. (21) also obtained type V isotherms on Carbosieve-s but without any hysteresis. The adsorption of water was greatly influenced
by the hydrophobicity of the external surface and the micropores.
This indicated that the hydrogen-bonded structure, by which water
is generally adsorbed, was not viable within the constricted environment of the hydrophobic micropore. The low-pressure water adsorption isotherm was found to be related closely to the number of titrable acidic groups, indicating that low-pressure adsorption occurs
on these localized sites. These sites must also exist within the
micropores, since argon adsorption showed very small external surface for this sample. Thus these workers were of the view that
water adsorption involved localized adsorption until a sufficient hy-
drostatic pressure exists to compress the adsorbate into liquidlike
behavior in the micropores. Stoeckli, on the other hand, using the
same Carbosieve sample observed a hysteresis loop in water
adsorption—desorption isotherms.

Some investigators, however, have a different point of view regarding the adsorption of water vapors on carbons. They are of the opinion that capillary condensation plays an important role in the adsorption process. Wiig and Juhola (43–45), for instance, are of the view that the filling of even micropores occurs by capillary condensation. According to them, physical and chemisorption of water become significant only in the case of highly microporous carbons associated with hydrophilic groups on their surface and that this was shown by the failure of the desorption curve to meet the adsorption curve even at zero relative pressure. They even suggested the use of the desorption branch of the water vapor adsorption isotherm to calculate the pore size distribution by using the Kelvin equation. McDermot and Arnell (46), agreed with Pierce, Smith, and Dubinin et al. regarding the mechanism of adsorption of water vapor by active carbons and calculated the pore size distribution from the desorption branch using the Kelvin equation and found the values to be reasonably comparable with those obtained from nitrogen adsorption isotherms.

Kiselev and co-workers (47–49) compared the adsorption isotherms of water on porous and nonporous carbons containing varying amounts of associated oxygen and observed that the adsorption was determined by the degree of oxidation of the carbon surface as well as by the porosity of the carbon and that the effect of capillary condensation became increasingly significant as the hydrophobicity of the surface increased. They suggested that the steep rise of the isotherm in the medium range of relative pressure was caused not by merging of discrete islands of adsorbed water but by filling of pores by capillary condensation, which was also responsible for the occurrence of hysteresis. According to these workers adsorption first occurs in the most constricted portions of the pores where the adsorption potential is likely to be highest. The thickness of the adsorbed layer of water increases with increasing pressure until a continuous film is formed when capillary condensation can take place. The process extends to larger portions of the pores with increased relative pressure. Thus hysteresis was a result of adsorption in the narrow parts of the neck and desorption from the wider parts of the body of the pore. These workers calculated the pore size of the constricted portions of the pores from the desorption isotherms and found the values to be reasonable (47).

Bansal et al. (50) studied the adsorption isotherms of water vapor on several polymer charcoals having different porosities and associated with varying amounts of combined oxygen. The shapes of the isotherms were different on different charcoals, indicating that the adsorption involved different mechanisms. The charcoals associated with similar amounts of combined oxygen but possessing different porosities showed similar adsorptions at lower relative pressures but

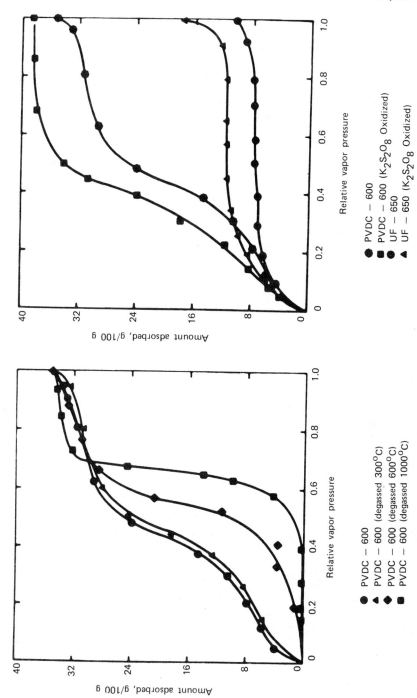

Figure 12 Water vapor sorption isotherms on polyvinyledene chloride (PVDC) and urea formaldehyde resin (UF) carbons before and after outgassing and oxidation. [From Bansal et al. (50).]

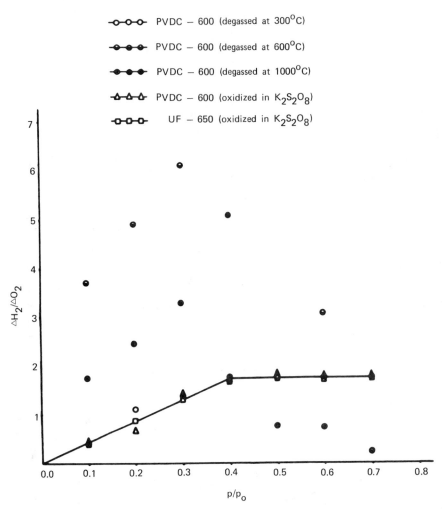

Figure 13 Variation of water adsorption in relation to variation of associated oxygen in different carbons. [From Bansal et al. (50).]

differed appreciably in their adsorption values at higher relative pressures, indicating the influence of associated oxygen and of porosity in the two ranges of relative pressure. These workers examined the influence of associated oxygen on the adsorption of water vapor by a systematic increase or decrease of the oxygen content of these charcoals. The adsorption isotherms (Fig. 12) showed clearly that there was a continuous decrease in the

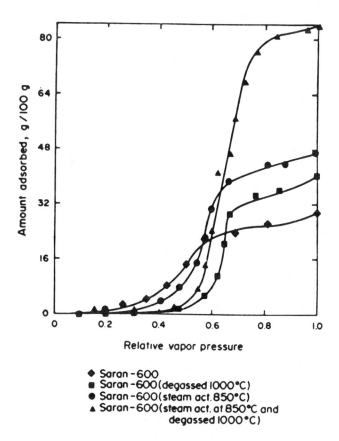

◆ Saran-600
■ Saran-600(degassed 1000°C)
● Saran-600(steam act. 850°C)
▲ Saran-600(steam act. at 850°C and
     degassed 1000°C)

Figure 14   Water vapor sorption isotherms on outgassed and steam-activated Saran carbon. [From Bansal et al. (50).]

adsorption of water vapor as more and more of the associated oxygen was removed from the sample. But this decrease was generally restricted only to adsorption at low relative pressures ($p/p_0 <$ 0.5). In fact there was a slight increase in adsorption at higher relative pressures. It was interesting that a 1000° outgassed charcoal sample, which was almost completely free of any associated oxygen, adsorbed only negligible amounts of water at $p/p_0 < 0.5$. Similarly in the case of PVDC and UF charcoals oxidized with potassium persulfate solution, the increase in the adsorption of water vapor was limited to relative pressures up to 0.5 and terminated thereafter. A plot between the increase or decrease in the amount of water vapor adsorbed on oxidation or degassing per millimole of associated oxygen at different relative pressures (Fig. 13) clearly shows that the increase or decrease in water adsorption at lower relative pressures was directly related to the increase or decrease of associated oxygen.

When Saran charcoal was degassed at 1000° or activated in steam at 850°C, the adsorption of water vapor decreased at lower relative vapor pressures but increased at higher relative pressures (Fig. 14). The decrease at lower relative pressures was attributed to the elimination of combined oxygen, whereas the increase at higher relative pressures was due to widening of some of the micropores as a result of the activation processes. This was further supported by a big increase in the adsorption of water vapor on outgassing the steam-activated Saran charcoal at 1000°C. The widening of the microcapillary pores was evident from the pore size distribution curves.

Thus these workers concluded that the adsorption of water vapor by carbons at low relative pressures (p/p_O < 0.5) is determined by the associated oxygen which is present in the form of oxygen structures attached to individual carbon atoms forming the walls of the micropores. However, as the carbon becomes more hydrophobic, the effect of the porsity tends to predominate and capillary condensation becomes more significant. Further evidence of capillary condensation was provided by adsorption—desorption hysteresis, which persisted even when all the combined oxygen was removed by outgassing at 1000°C (Fig. 15).

4.2.2 Adsorption of Polar Organic Vapors

The best description of adsorption on microporous carbons is given by the Polanyi—Dubinin potential theory (51—53). However, this equation is more accurately followed only when the heterogeneity of adsorption engergy is not large and when state properties of the adsorbate liquid phase and normal liquid phase do not differ considerably (51,54,55). In the case of adsorption of polar adsorbates, the state properties of the adsorbed phase differ from the state properties of the normal liquid phase. Therefore, a precise interpretation of the results requires a modification of the Dubinin—Polanyl theory.

Rozwadowski and co-workers (56,57) studied the adsorption of spectroscopic-grade aliphatic alcohols and amines on active carbons prepared from chemically pure sugar and activated with O at 450°C (sample A), CO_2 at 850°C (sample B), and $ZnCl_2$ (samples D and E). These workers compared the state properties of the adsorbate liquid phase with those of the normal liquid phase and determined the share due to adsorption of dispersive adsorbate—adsorbent interactions.

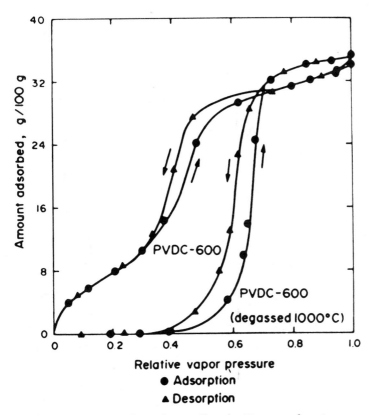

Figure 15 Adsorption—desorption isotherms of water vapor on PVDC carbon before and after outgassing at 1000°C. [From Bansal et al. (50).]

The adsorption of polar adsorbates such as methanol and ethanol was in accordance with the Dubinin—Polanyi theory of volume filling of micropores in the higher ranges of adsorption. However, devia-tions were observed at lower adsorption values, indicating that a different mechanism was involved at low pore fillings (Fig. 16). The differences between the adsorption curves calculated on the basis of the Dubinin—Polanyi equation and the experimental curve were at-tributed to the differences in the degree of association of the polar adsorbates in the liquid and the adsorbed states. The curves calcu-lated after making allowances for the degree of these associative in-teractions were closer to the experimental curve. According to these workers the degree of association in the adsorbed phase will actually depend on the amount of the substance adsorbed so that the p_0

Table 9 Values of Activity Coefficient γ (at 298.2 K) for Alcohol and Amine Adsorbates on Carbon D (I, pure carbon; II, carbon covered with amine)

a (mol/kg)	MeOH I	MeOH II	EtOH I	MeNH$_2$ I	MeNH$_2$ II	Me$_2$NH I	Me$_2$NH II	EtNH$_2$ I	EtNH$_2$ II
0.2	0.36	0.23	0.98	—	—	—	—	—	—
0.3	0.60	0.35	1.17	—	—	—	—	—	—
0.4	0.80	0.46	1.31	—	—	—	—	—	—
0.5	0.97	0.55	1.41	—	—	—	—	—	—
0.6	1.12	0.63	1.49	—	—	—	—	—	—
0.8	1.35	0.77	1.61	—	—	—	—	—	—
1.0	1.50	0.87	1.69	—	—	—	—	—	—
1.25	1.64	0.97	1.75	—	—	—	—	—	—
1.6	1.72	1.05	1.78	—	—	0.13	0.03	—	0.04
1.75	1.77	1.10	1.79	0.05	0.03	0.14	0.04	—	0.05
2.0	1.80	1.14	1.79	0.06	0.04	0.15	0.05	0.20	0.08

2.5	1.80	1.19	1.76	0.08	0.05	0.17	0.07	0.21	0.09
3.0	1.77	1.20	1.71	0.09	0.07	0.19	0.10	0.22	0.11
3.5	1.72	1.20	1.64	0.11	0.10	0.21	0.12	0.23	0.13
4.0	1.65	1.18	1.56	0.13	0.12	0.23	0.16	0.24	0.16
4.5	1.59	1.16	1.48	0.15	0.14	0.25	0.19	0.25	0.19
5.0	1.51	1.13	1.40	0.16	0.18	0.27	0.22	0.26	0.23
5.5	1.44	1.09	1.31	0.18	0.21	0.29	0.25	0.28	0.26
6.0	1.37	1.06	1.22	0.20	0.23	0.31	0.28	0.29	0.29
6.5	1.30	1.02	1.14	0.22	0.26	0.33	0.31	0.31	0.33
7.0	1.24	0.98	1.06	0.24	0.29	0.35	0.34	0.32	0.36
7.5	—	—	0.97	0.26	0.31	0.38	0.36	0.35	0.39
8.0	—	—	0.89	0.28	0.33	0.40	0.37	0.37	0.42
8.5	—	—	—	0.29	0.35	0.43	0.38	0.41	0.45
9.0	—	—	—	0.31	0.36	—	—	0.46	0.47

Source: Rozwadowski and Wojsz (56).

Table 10 Values of Activity Coefficient γ (at 298.2 K) for Alcohol and Amine Adsorbates on Carbon B (I, pure carbon; II, carbon covered with amine)

a (mol/kg)	MeOH I	MeOH II	EtOH I	MeNH$_2$ I	MeNH$_2$ II	Me$_2$NH I	Me$_2$NH II	EtNH$_2$ I	EtNH$_2$ II
0.2	0.21	0.26	—	—	—	—	—	—	—
0.3	0.41	0.38	1.11	—	—	—	—	—	—
0.4	0.62	0.49	1.24	—	—	—	—	—	—
0.5	0.81	0.59	1.33	—	—	—	—	—	—
0.6	0.99	0.68	1.41	—	—	—	—	—	—
0.8	1.29	0.83	1.54	—	—	—	—	—	—
1.0	1.53	0.95	1.63	—	—	—	—	—	—
1.25	1.75	1.07	1.72	—	—	—	—	—	—
1.5	1.91	1.17	1.77	—	—	0.20	0.06	—	0.07
1.75	2.02	1.24	1.81	0.10	0.03	0.21	0.07	—	0.08

2.0	2.09	1.30	1.83	0.11	0.04	0.22	0.08	—	0.09
2.5	2.15	1.38	1.84	0.12	0.06	0.23	0.11	—	0.12
3.0	2.14	1.41	1.81	0.15	0.08	0.25	0.13	—	0.15
3.5	2.09	1.42	1.76	0.17	0.11	0.26	0.16	—	0.19
4.0	2.02	1.41	1.68	0.20	0.15	0.27	0.20	0.32	0.23
4.5	1.92	1.38	1.59	0.21	0.18	0.28	0.23	0.32	0.27
5.0	1.82	1.33	1.48	0.23	0.21	0.29	0.25	0.32	0.30
5.5	1.71	1.28	1.36	0.25	0.25	0.29	0.27	—	0.34
6.0	1.59	1.22	1.23	0.27	0.28	0.29	0.28	—	0.37
6.5	1.48	1.15	1.09	0.29	0.31	—	—	—	0.38
7.0	1.37	1.08	0.94	0.30	0.33	—	—	—	—
7.5	—	—	0.78	0.32	0.35	—	—	—	—
8.0	—	—	—	0.32	0.36	—	—	—	—
8.5	—	—	—	0.32	—	—	—	—	—

Source: Rozwadowski and Wojsz (56).

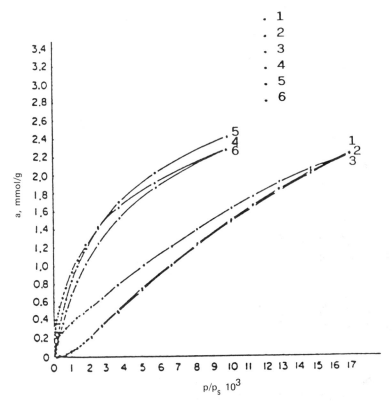

Figure 16 Comparison of theoretical and experimental adsorption isotherm of alcohols on carbon. (1) Experimental for methanol; (2) theoretical for methanol; (3) theoretical on the basis of association for methanol; (4) experimental for ethanol; (5) theoretical for ethanol; (6) theoretical with association for ethanol. [From Rozwadowski and Wojsz (56).]

values used for calculating the differential molar adsorption work A in the Dubinin–Polanyi equation $\theta = (W/W_0 \{exp -(A/\beta E_0)^2\}$ where $A = RT \ln (P_0/p)$ depends not only on temperature but on the amount of the adsorbed substance as well. Thus the fugacity of the saturated vapor at a given stage of adsorption will be the product of saturated vapor pressure P_0 and an activity coefficient ν dependent on the degree of surface filling (adsorption)

$$f^* = \nu P_0$$

Rozwadowski et al (56) calculated the activity coefficients for the adsorption of alcohols and amines before and after covering the surface with amine to neutralize chemisorption forces, as a function of the degree of adsorption. The values of activity coefficients (Tables 9 and 10) differ appreciably from unity, indicating that the state properties of the adsorbed phase are different from those of the bulk liquid phase. The values less than unity (< 1) indicate the existence of strong association of the adsorbate, which was clearly greater in the adsorbed phase than in the bulk phase. This association is specially strong in the case of amines. Alcohols showed activity coefficients less than unity only at low adsorptions.

Thus in the case of adsorption of polar substances, there are strong perturbations of the adsorbed state in the adsorption force field resulting in an increase or decrease of differential molar-free energy compared to the nonperturbed adsorbate in the bulk liquid phase. These workers calculated the perturbation thermodynamic functions such as differential perturbation entropy ΔS^* (which represents the change of differential molar adsorption entropy) and the differential perturbation enthalpy ΔH^* (which represents the change of differential molar adsorption enthalpy) for the adsorption of amines and alcohols on carbons B, D, and E. The plots between these perturbation thermodynamic functions with surface coverage (Fig. 17) showed clearly that at higher adsorption levels, the modified liquid state approximates the normal liquid state. The perturbation enthalpies for amines were generally positive and decreased with increase in adsorption, which indicated that the associative interactions in the adsorbed phase were stronger than in the normal liquid phase. Positive values of differential perturbation entropy ΔS^* showed a lower mobility of amine molecules in the adsorbed phase compared to that in the bulk phase. However, only methanol showed positive ΔH^* values and that too at low adsorptions, ethanol showing negative values at all adsorptions. The positive values of ΔH^* for methanol probably resulted from strong specific interactions (56) between alcohol molecules and adsorption centers existing on the surface of these carbons.

On the basis of these thermodynamic considerations, Rozwadowski and Wojsz (56) were of the view that the adsorption of polar molecules on microporous carbons involves chemisorption or strong specific adsorption at low relative pressures, the energy of these interactions being much larger than that of dispersive interactions so that the activity coefficients have low initial values (Tables 9,10). The positive values of ΔS^* and ΔH^*, in the range of middle and higher pressures, with the activity coefficient values remaining still less than one, in the case of amines indicate the presence of strong and considerably greater association of amine molecules in the adsorbed phase than in the bulk phase. As the amount of adsorption

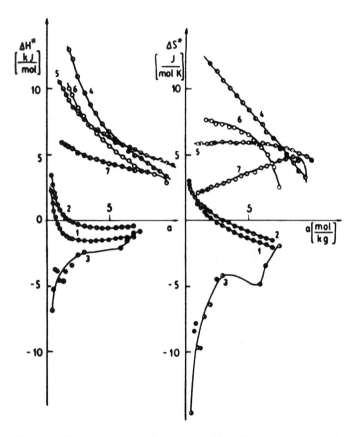

Figure 17 Variation of perturbation thermodynamic functions (ΔS^* differential entropy and ΔH^* differential enthalpy) as a function of amount adsorbed for amines and alcohols. (1) original carbon—methanol system; (2) amine-covered carbon—methanol system; (3) original carbon—ethanol system; (4) amine-covered carbon; (5) original carbon—methylamine system; (6) amine-covered carbon—methylamine system; (7) original carbon—dimethylamine system. [From Rozwadowski and Wojsz (56).]

increases in this range, the activity coefficient values start increasing and the ΔH^* and ΔS^* values tend to approach zero. This shows that the mean degree of association of the amines in the adsorbed phase decreases with increase in adsorption. In general, the adsorbed phase is characterized by a higher degree of association than the bulk liquid phase. It is known that in the case of amines, the mean degree of association in the liquid phase is low (58,59), so that the specific interactions as well as the strongly increased adsorption potential of the adsorbents can cause a considerable increase in the degree of association. Alcohols, on the other hand, have a high degree of association in the bulk liquid phase compared to amines, so that the change in the electron distribution of alcohol molecules, caused by specific interactions and the increased adsorption potential, is of a lower degree. Thus the mean degree of association in the adsorbed phase of alcohols is clearly lower than that in the bulk liquid phase. The difference in the behavior of methanol in the early range of adsorption is due to the strong specific interactions on the most active centers provided by polar oxygen groups (118). The adsorption of polar adsorbates at higher pore fillings is according to the theory of volume filling of micropores.

Rozwadowski et al. (57) separated the share due to adsoprtion of specific and associative interactions for alcohol adsorption using a modified TVFM equation. The results of these calculations (Table 11) showed that the contribution of the specific interactions can be considered as finished after the adsorption of 2.267 mmol/g of methanol and 2.280 mmol/g of ethanol on carbon A. These values were obtained from the points of convergence of experimental and theoretical curves (Fig. 16). It was also clear from the data in Table 11 that the extent of the specific interaction was determined both by the nature of the adsorbate and by the nature of the carbon surface. Thus the shape of the experimental adsorption isotherm will be the result of overlapping of two opposing interactions: adsorbate–adsorbent and adsorbate–adsorbate, which influenced the value of molar adsorption work.

4.2.3 Adsorption of Organic Vapors

The adsorption of nonpolar organic vapors is physical and involves dispersion attractive forces and short-range repulsive interactions. The dispersion forces are characterized by an increase in energy of adsorption in micropores due to the superposition of adsorption potentials of the opposite micropore walls. The adsorption isotherms are satisfactorily explained by the Dubinin theory of volume filling of micropores (TVFM). The characteristic energy, E_0, of the TVFM increases between 8.5 and 35 kJ/mol for benzene, depending on the parameters of the microporous structure, and this increase can explain the abrupt rise in the adsorption isotherm, especially at low relative pressures. The major thrust in the study of adsorption of organic vapors has been to derive pore size distributions and to characterize molecular sieve behavior of microporous carbons. Since detailed description of the adsorption isotherms of organic vapors is not the purpose of this book, we discuss only briefly the important research directed toward organic compounds as molecular probes.

Table 11 Amounts of Adsorption Showing Adsorbate–Adsorbate Associative Interactions and End of Adsorbate–Adsorbent Interactions

Sorbent	Association interaction				Specific interaction			
	Methanol		Ethanol		Methanol		Ethanol	
	p/p_0	a (mmol/g)	p/p_0	a (mmol/g)	p/p_0	a (mmol/g)	p/p_0	a (mmol/g)
A	0.0165	2.175	0.0025	1.405	0.0170	2.267	0.0098	2.280
B	0.0039	0.520	0.0004	0.425	0.0378	3.664	0.0144	3.778
D	0.0027	0.280	0.0002	0.170	0.0424	2.882	0.0146	2.620
E	0.0029	0.250	0.0002	0.170	0.0082	0.484	0.0035	0.653

Source: Rozwadowski et al. (57).

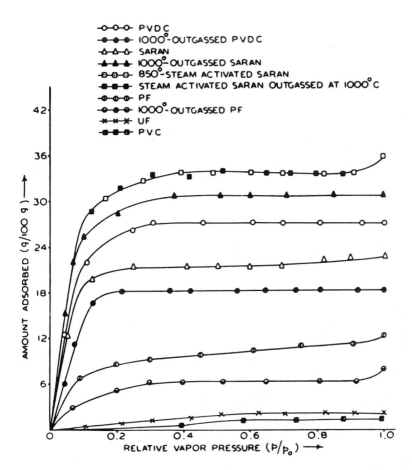

Figure 18 Adsorption isotherms of methanol vapors on different polymer carbons. PVDC = polyvinylidene chloride carbon; PF = poly-furfuryl alcohol carbon; PVC = polyvinyl chloride carbon; UF = urea formaldehyde resin carbon; Saran = a copolymer of 85% PVDC and 15% PVC. [From Bansal and Dhami (63).]

In addition, we note the difference of opinion regarding the adsorption of benzene vapors on active carbons.

4.2.3.1 Organic Vapors as Molecular Probes

Dacey and Thomas (60) studied the adsorption of several organic vapors on Saran charcoal prepared at 700°C. The charcoal adsorbed straight-chain hydrocarbons like n-pentane and flat molecules like benzene very rapidly, whereas neopentane, which is a larger molecule, was adsorbed very slowly. The difference in the rates of adsorption of benzene and neopentane was attributed to the existence of slit-shaped pores in the carbon. Walker and co-workers (61,62) measured the uptake of butane, isobutane, and neopentane on Saran charcoals as a function of the carbonization temperature. The surface areas covered by the different adsorbates were found to depend on the molecular dimensions of the adsorbate and the temperature of carbonization. The charcoals carbonized below 1200°C were associated with large-sized pores. At higher carbonization temperatures a shrinking of the micropores was noticed because the adsorption of neopentane was almost negligible. The adsorption of benzene and cyclohexane was almost 10 times more than that of neopentane.

Bansal and colleagues (63–67) investigated the molecular sieve behavior of polymer charcoals using several organic vapors of varying size and shape as molecular probes. They observed that the adsorption was determined by three factors: the size of the pores, the size and shape of the adsorbate molecules, and the chemical nature of the carbon surface. The shapes of the isotherms were generally type I with 90–95% of the total adsorption occurring at relative pressures below 0.2. This indicated that the charcoals were highly microporous with micropores only a few molecular diameters in width. The adsorption of polar molecules such as methanol and ethanol was influenced by the presence of acidic surface groups (Fig. 18). The adsorption of these alcohols decreased while those of benzene and other organic molecules increased on gradual elimination of the acidic oxygen complexes.

The surface areas calculated from the adsorption isotherms of organic vapors generally decreased with increase in the molecular diameter of the adsorbate. Thus for PVDC, Saran, and steam-activated Saran charcoals, the surface area accessible to larger isooctane (molecular diameter 0.68 nm) and α-pinene (molecular diameter 0.8 nm) were much smaller than the areas accessible to smaller molecules of benzene and n-hexane (Table 12). This indicated that these carbons contained an appreciable number of pores which were inaccessible to isooctane and α-pinene. Comparison of these areas with nitrogen and carbon dioxide surface areas showed (Table 12) that only about 13% and 7% of the nitrogen surface areas in the case of PVDC and between 10 and 25% in the case of steam-activated

Table 12 Surface Areas (m^2/g) of Various Polymer Carbons by Adsorption of Different Adsorbates

Sample	Nitrogen 77 K	Carbon dioxide 273 K	Benzene	n-Hexane	n-Heptane	Carbon tetrachloride	Isooctane	α-pinene
PVDC-6000	887	903	660	—	145	121	101	61
Saran-600	787	798	531	200	—	56	—	—
Saran-600°	1188	1138	749	282	—	195	115	324
PF-600	105	288	—	—	—	No adsorption	—	—
PF-900	106	163	—	—	—	No adsorption	—	—

Source: Bansal and Dhami (65).

Table 13 Pore Volume (ml/g) of Various Polymer Carbons Calculated by Adsorption at Relative Vapor Pressure 0.98

Sample	Nitrogen	Carbon dioxide	Water	Methanol	Ethanol	Benzene	n-Hexane	n-Heptane	Carbon tetra-chloride	Iso-Octane	α-Pinene
PVDC-600	0.40	0.37	0.34	0.35	0.35	0.33	0.11	0.09	0.10	0.07	0.06
Saran-600	0.39	0.28	0.29	0.29	0.29	0.33	0.10	0.09	0.08	0.06	0.02
Saran-600[a]	0.54	0.44	0.46	0.45	0.48	0.55	0.42	0.42	0.16	0.10	0.28
PF-600	0.04	—	0.14	0.15	0.14	0.10	0.01	0.03	—	—	—
UF-650	0.004	—	0.10	0.03	0.06	—	—	—	—	—	—
Polymer Carbons Outgassed at 1000°C											
PVDC-600			0.35	0.23	0.28	0.40	0.32	0.31	0.23	0.19	0.12
Saran-600			0.40	0.39	0.40	0.38	0.15	0.33	0.31	0.16	0.09
Saran-600[a]			0.84	0.47	0.41	0.27	0.19	0.27	0.27	0.11	0.08
PF-600			0.12	0.09	0.01	—	—	0.02	—	—	—
UF-650			0.10	—	—	—	—	—	—	—	—

[a]Steam activated at 850°C.
Source: Bansal and Dhami (65).

Saran charcoals respectively were accessible to isooctane and α-pinene. Thus surface area calculated from the adsorption of organic vapors has a limited significance in the case of highly microporous carbons.

The amounts of different organic vapors adsorbed at relative pressure of 0.98 (apparent pore volume) expressed as volume of liquid assuming the liquid to have normal liquid density (Table 13) showed that the apparent pore volume for a given charcoals was not the same for all the adsorbates, indicating that the Gurvitsch rule was not obeyed and that the pores in these charcoals were only a few molecular diameters in width. The results also indicated that the PVDC and Saran charcoals had an appreciable proportion of the pores with diameters less than 1 nm and the steam activation of the Saran charcoal enlarged some of the micropores. The pore volumes in the case of PVDC charcoal (Table 14) agree fairly well with the values obtained by other workers.

Table 14 Apparent Pore Volumes of Polyvinylidene Chloride Carbons

Adsorbate	Pore volume V (ml/g)		
	Dacey and Thomas (60)	Culver and Heath, quoted in Ref. 68	Bansal and Dhami (65)
Nitrogen	0.43	0.464	0.40
Water	0.37	—	0.34
Benzene		0.422	0.33
	0.403		
	0.390	—	—
Neopentane	0.34	—	—
n-Pentane	0.424	—	—
Toluene	0.432	—	—
Naphthalene	0.37	—	—
n-Hexane	0.425	—	0.11
Carbon tetraethyl	0.27	—	—
Ethyl chloride	—	0.445	—

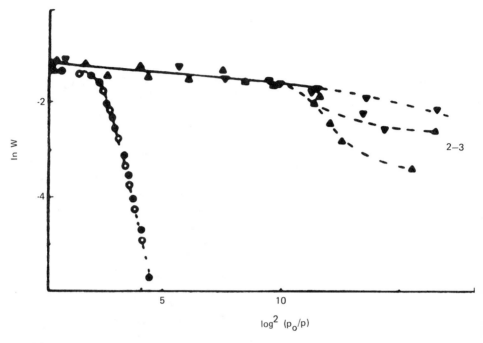

Figure 19 Pore size variation in carbon as a result of adsorption—desorption cycles of larger molecules (tertiobutyl benzene). [From Stoeckli et al. (33).]

Ainscough et al. (68), in the basis of studies of adsorption of several organic vapors on a PVDC charcoal, suggested that the total volume of an adsorbate adsorbed by a carbon could not be related to the relative size of the pores and the adsorbate molecules because of constrictions in the pores and the activated diffusion effects. The measured pore size will depend on the experimental condition. Stoeckli et al. (33) observed that successive adsorption—desorption cycles of large molecules such as tertiobutyl benzene and trimethyl phosphate on activated carbons can cause a substantial modification of the carbon micropore system. It was found that after three adsorption—desorption cycles of tertiobutyl benzene, the average pore size in an active carbon was shifted toward higher values (Fig. 19). At the same time an increase in heterogeneity was observed, which suggested that the pore size distribution shifted toward supermicropores under the influence of the larger molecule.

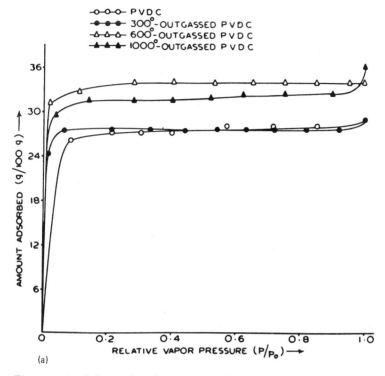

Figure 20   Adsorption isotherms of benzene on (a) PVDC carbon, (b) Spheron-6, and (c) Mogul before and after outgassing at different temperatures. [From Bansal and Dhami (63) and Puri et al. (69).]

4.2.3.2 Adsorption of Benzene Vapors

Adsorption of benzene to characterize microporous carbons has been studied by a number of investigators because of its smaller molecular dimensions and essentially nonpolar character. Dubinin and coworkers used benzene as a standard vapor for the derivation of their theories of physical adsorption. These workers are of the view that the adsorption of benzene involves purely dispersion interactions. However, Puri and co-workers (69), in their studies on the adsorption of benzene on sugar charcoals and carbon blacks, and Bansal and Dhami (66), on polymer carbons, observed that the presence of CO_2-complex, which is known to impart polar and hydrophilic character to the carbon surface, suppressed the adsorption of benzene, which is essentially nonpolar. With the elimination of most of this complex and the emergence of CO-complex as the only

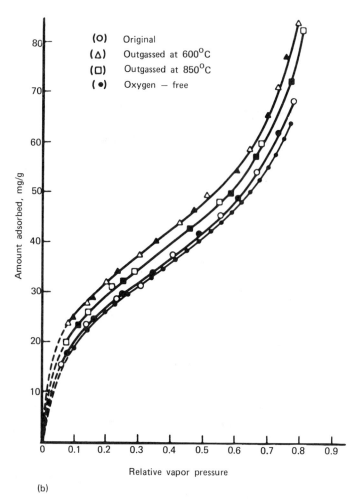

(b)

Figure 20 (continued)

predominant oxygen complex on the carbon surface, the adsorption
of benzene increased. The 600° outgassed carbon samples, which
were associated with relatively larger amounts of CO-complex,
showed higher adsorption of benzene at all relative pressures than
those outgassed at higher temperatures, even though their surface
areas were not significantly different (Fig. 20). This clearly indicated that the adsorption of benzene on CO_2-complex-free carbons
was a function not only of surface area but also of the CO-complex.

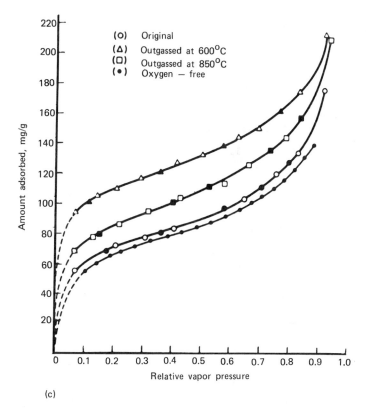

(c)

Figure 20 (continued)

It was further found that the additional adsorption of benzene at each relative pressure > 0.3 amounted roughly to 1 mol of benzene per mole of quinonic oxygen as determined by the sodium borohydride method (Table 15). This indicated the probability of interaction of π electron clouds of the benzene ring with the partial positive charge on the carbonyl carbon atom. The isosteric heats of adsorption for apparent monolayer coverages were also higher in the case of carbons containing quinonic oxygen (Table 16). The benzene adsorbed in the case of sugar charcoals could not be recovered completely even on prolonged evacuation. This was attributed to the complexing of benzene at the quinonic sites within the micropores and required high energy of activation to diffuse out. The adsorption of cyclohexane on the same charcoals and carbon blacks was not influenced by the associated oxygen, indicating that cyclohexane did

Table 15 Increased Benzene Adsorption in Relation to Quinonic Oxygen in the Case of 600°C-Outgassed Carbon Blacks

Relative vapor pressure	Increase in benzene adsorption on outgassing at 600°C (mmol/g)	Increased benzene adsorption / quinonic oxygen
	Mogul (quinonic oxygen = 0.658 mmol/g)	
0.30	0.564	0.86
0.40	0.577	0.88
0.50	0.596	0.89
0.60	0.615	0.94
0.80	0.621	0.94
	ELF-0 (quinonic oxygen = 0.239 mmol/g)	
0.30	0.269	1.16
0.40	0.263	1.10
0.50	0.282	1.18
0.60	0.307	1.28
	Spheron-4 (quinonic oxygen = 0.149 mmol/g)	
0.30	0.141	0.95
0.40	0.147	0.98
0.50	0.157	1.05
0.60	0.166	1.11
	Spheron-6 (quinonic oxygen = 0.0843 mmol/g)	
0.30	0.0772	0.92
0.40	0.0859	1.02
0.50	0.0897	1.06
0.60	0.0987	1.16

Source: Puri et al. (69).

Table 16 Isoteric and Net Heats of Adsorption of Benzene on Various Carbon Blacks

Sample (1)	Isoteric heat of adsorption (kcal/mol) (2)	Net heat of adsorption (kcal/mol) (3) = (2) − 7.3
Spheron-6		
Original	10.8	3.5
600°C-outgassed	14.8	7.5
850°C-outgassed	12.7	5.4
Oxygen-free	10.3	3.0
Mogul		
Original	10.6	3.3
600°C-outgassed	12.7	5.4
850°C-outgassed	11.3	4.0
Oxygen-free	10.2	2.9

Source: Puri et al. (69).

not interact with any of the surface oxygen complex. The interaction of benzene with quinone oxygen was therefore thought to be of a specific nature due to the presence of the π electrons of the benzene ring. These workers have thus suggested caution in measuring surface area from the adsorption isotherms of benzene.

4.3 CHARACTERIZATION BY ADSORPTION FROM SOLUTIONS

Adsorption by activated carbons from solutions was recognized with the early studies of Freundlich and Heller (70) on lower fatty acids and of Kipling (71) on weak electrolytes. However, a theoretical analysis of adsorption from solution and the derivation of a suitable adsorption equation have been comparatively difficult because both the components of a solution compete with each other for the available surface. Moreover, the thermal motion of the molecules in the liquid phase and their mutual interactions are much less well understood. It is therefore difficult to correctly assess the nature of the adsorbed phase whether monomolecular or multimolecular. This is usually determined by the porosity and the chemical nature of the adsorbent, the nature of the components of the solution, the concentration of the solution, and the mutual solubility of the components. The adsorption of a nonpolar solute will be higher on a

nonpolar adsorbent. But since there is competition between the solute and the solvent, the solvent should be polar in nature for the
solute to be adsorbed preferentially. Moreover, since the difference
in the polarities of the solute and the solvent determines their
mutual solubilities, the solvent should be as polar as possible for a
better adsorption of the nonpolar solute on the nonpolar adsorbent
such as activated carbon. The other factor that also determines the
adsorption from solutions is the steric arrangement, or chemical
structure, of the molecule. Since active carbons have a highly
microporous structure, some of the pores may be inaccessible to
larger molecules of the adsorbate. Thus the experimentally simple
technique of adsorption from solutions can be developed into a
method to characterize microporous carbons in terms of surface
area, microporosity, oxygen content, and hydrophobicity of the
carbon surfaces. This receives further attention because of the
growing importance of environmental control involving purification
of waterways by adsorption methods using active carbons.

Adsorption from solutions can be classified into adsorption of
solutes which have a limited solubility (i.e., from dilute solutions)
and adsorption of solutes which are completely miscible with the solvent in all proportions. In the former case the adsorption of the
solvent is of little consequence and is generally neglected as the
solutions are very dilute. In the latter case, however, the adsorption of both the compenents of the solution plays its part and has
to be considered. The importance of adsorption from solutions can
be judged by the fact that several monographs have been written on
the subject (72–75). However, our knowledge of the various factors
influencing adsorption from the solution phase is still inadequate.
We briefly consider in this section the important adsorption results
from the point of view of characterization of carbons.

4.3.1 Adsorption from Dilute Solutions

Early studies on the adsorption of homologous series of organic
compounds from aqueous solutions showed that the adsorption increased regularly with increase in the length of the carbon chain.
However, when the solvent was changed for a less polar one and
the adsorbent was changed for a more polar one, Traube's rule was
no longer obeyed. This indicated that the nature of the carbon as
well as the nature of the solvent influenced adsorption. Several
workers (76–78) attempted to calculate the surface areas of the
carbons from the adsorption of fatty acids from aqueous solutions
but failed to find values comparable to the BET values. In fact the
values of surface area decreased with increase in the molecular
dimensions of the acid. In the case of carbon blacks, which are

essentially nonporous, the surface areas calculated from the adsorption of acetic acid were comparable to their BET areas (79).

Puri and Arora (77,78) studied the isotherms of stearic acid from its solutions in benzene and carbon tetrachloride on a number of active carbons, carbon blacks, and graphon and observed that the ultimate amount of stearic acid adsorbed from the two solvents differed appreciably on active carbons and carbon blacks but was of the same order on graphon. Furthermore, the amount of stearic acid adsorbed was less in the case of carbon blacks compared to that on graphon, although the carbon blacks had much larger surface areas. This was attributed to the sensitivity of the adsorption of stearic acid to oxygen surface structures present on carbon blacks. It was postulated that the stearic acid molecules undergo dimerization in organic solvents and the dimer behaved as a nonpolar molecule. Consequently, its adsorption on polar oxygen group sites was inhibited. The adsorption of stearic acid increased when the surface oxygen groups were eliminated by evacuating at 1000°C (Table 17).

Adsorption of cationic and anionic dyes from aqueous solutions has been used to characterize carbons for their surface area, microporous structure, and polarity (77,78,80). For example, Graham (80) studied the adsorption of two dyes of opposite character but of approximately the same molecular dimensions (methylene blue and metanil yellow) from aqueous solutions on a number of active carbons, measuring separately the influence of pore size and the acidic groups. Graphon, which essentially has a uniform surface and is free of surface heterogeneities, was used as standard model substance. The surface available for adsorption was only 46% of the total in the case of methylene blue and only 25% in the case of metanil yellow, although the surface area occupied by pores with diameters greater than those of these dyes was about 75% (Table 18). This was attributed to the presence of acidic oxygen groups, which modified the adsorption characteristics toward these dyes. It was found that the available surface area increased when these acidic groups were removed by heat treatment at 900°C in nitrogen. In general, the adsorption of metanil yellow, which is an acidic dye, was less than that of methylene blue. The adsorption of metanil yellow also showed a linear relationship with the amount of acidic surface groups (Fig. 21). The results indicate that the acidic groups on the carbon surface tend to reduce the capacity of the carbon surface for anionic adsorbates, roughly in proportion to the concentration of these groups. Puri and Arora (77,78) determined the adsorption isotherms of methylene blue and Rhodamine B on graphon and on a number of commercial-grade active carbons. The adsorption isotherms were type I (Fig. 22), indicating a completion of the monolayer. However, when surface areas were calculated, the values agreed with the BET areas for graphon but differed appreciably for

Table 17 Maximum Surface Coverage of Various Carbons of Different Oxygen Contents by Stearic Acid Adsorbed from Carbon Tetrachloride Solution

Carbon	Oxygen content (g/100 g)	Nitrogen surface area (m^2/g)	Maximum surface coverage (%)
Activated carbons			
A	0.92	1096	20
B	0.91	988	17
C	0.72	924	19
D	1.41	528	14
E	1.32	898	16
Carbon blacks			
Graphon	0	86	100
Mogul	7.8	308	23
Spheron-C	3.2	254	25
Spheron-4	4.1	153	30
Spheron-6	3.1	120	28
Spheron-6 outgassed at 600°	1.3	115	40
Spheron-6, outgassed at 1000°	0.5	100	74
Spheron-6, outgassed at 1400°	0	91	100
ELF-0	4.1	171	27
ELF-0, outgassed at 600°	2.1	189	54
ELF-0, outgassed 15 1000°	0.6	178	77
ELF-0, outgassed at 1400°	0	142	100

Source: Arora (77).

Table 18 Adsorption of Methylene Blue and Metanil Yellow in Relation to Surface Area and Surface Acidity

| Carbon | Total surface area (m²/g) | Methylene blue | | Metanil yellow | | Ratio of accessible areas of metanil yellow and methylene blue | Acidity of carbon surface (mEq/100 m²) |
		Accessible area (m²/g)	Accessible area (%) of total surface	Accessible area (m²/g)	Accessible area (%) of total surface		
Graphon	83.9	83.9	100	83.9	100	1.00	0
1	1120	757	68	721	64	0.95	0.12
2	1130	836	74	629	56	0.76	0.45
3	1300	912	70	880	68	0.97	0.06
4	1300	602	46	329	25	0.55	0.94
4	1300	950	73	721	55	0.76	0.80
5	600	430	72	323	54	0.75	0.70
6	580	374	64	294	51	0.79	0.60

Source: Graham (80).

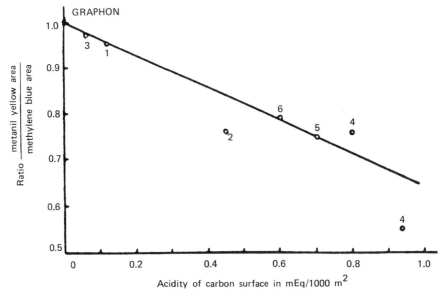

Figure 21 Influence of surface acidity of a carbon surface on its adsorption for metanil yellow. [From Graham (80).]

active carbons (Table 19). It was suggested that some of the micropores were not accessible to larger dye molecules.

The adsorption of phenols in the case of active carbons and carbon blacks was found to depend on the surface area of the carbon (77,78,81,82), as well as being influenced by the presence of oxygen complexes. The presence of oxygen resulted in the irreversible binding of phenol by the interaction of π electrons of the benzene ring system with the partial positive charge on carbonyl or quinonic carbon atoms (63,69). But when the irreversibly adsorbed amount was subtracted from the total adsorbed amount, the adsorption isotherms were type I on active carbons and type II on carbon blacks. The surface areas calculated agreed very closely with the nitrogen and carbon dioxide surface areas (Table 20). Coughlin et al. (83) examined the influence of chemisorbed oxygen in carbons on the adsorption of phenol, nitrobenzene, and sodium benzene sulfonate from aqueous solutions on a channel carbon black and two activated carbons. The presence of chemisorbed oxygen considerably reduced the capacity of the carbons for adsorption, in particular from dilute phenol solutions. These workers are of the view that chemisorbed oxygen depletes the electronic π band of graphitelike layers and reduces the dispersion forces of adsorption. In addition, the water

Table 19 Specific Surface Areas as Obtained by Dye Adsorption and Nitrogen Adsorption (BET) Methods and Relative Proportions of Ultramicropores in Various Activated Carbons

Activated carbon	N_2 adsorption (77° K)	Surface area (m^2/g) from dye adsorption		Average Dye Value	Area Constituted by ultramicropores
		Methylene blue	Rhodamine B		
A	109	916	896	906	190
B	988	797	763	780	208
C	924	638	613	625	299
D	528	564	548	346	182
E	898	232	228	623	275

Source: Arora (77).

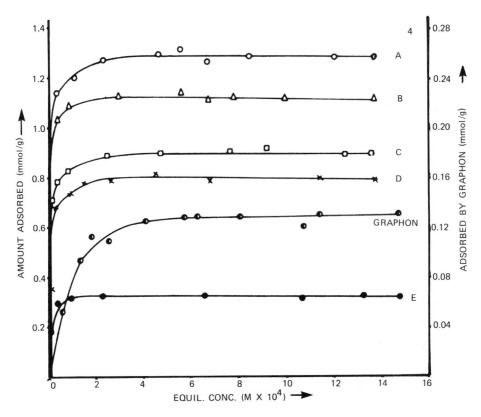

Figure 22 Adsorption isotherms of methylene blue from aqueous solutions on activated carbons and graphon. [From Arora (77).]

molecules cluster around the polar surface oxygen groups, again reducing the adsorption capacity. This latter suggestion was also supported by the work of Walker and Janov (84), who observed a proportionality between extent of water adsorption and extent of surface coverage by chemisorbed oxygen. Mattson et al. (85), however, suggested the formation of a donor—acceptor complex between the phenol and the surface carbonyl groups followed by complexation with the basal planes. Marsh and Campbell (81), during their adsorption studies of p-nitrophenol, benzoic acid, and oxalic acids from aqueous solutions on carbons prepared from polyfurfuryl alcohol and PVDC, observed that transitional pore filling occurred by these adsorbates. The adsorption of p-nitrophenol, particularly at low c/cs values of 10^{-3}, was extremely sensitive to microporous

Table 20 Specific Surface Areas of Carbons of Different Oxygen Contents as Obtained by Various Methods

Carbon	Oxygen content (g/100 g)	Surface area (m^2/g) obtained from:				
		Adsorption in gaseous phase		Adsorption in aqueous phase		Retention
		N (77 K) BET	CO_2 (273 K) D–P equation	Phenol	p-Nitrophenol	Retention of ethylene glycol
Activated carbons						
A	0.92	1096	1148	1154	1030	1194
B	0.91	988	1005	1020	1009	964
C	0.72	924	955	932	1018	1022
D	1.41	528	537	508	518	500
E	1.32	898	902	895	887	901
Sugar charcoal	17.8	412	388	—	382	424
Carbon blacks						
Spheron-C	3.2	254	288	253	247	268
Spheron-4	4.1	153	172	153	135	—
Spheron-6	3.1	120	138	120	108	121
Spheron-9	3.2	116	123	127	114	133
Philblack-A	0.5	46	46	50	46	48
Philblack-E	1.3	135	132	137	131	129
Philblack-I	1.3	117	107	110	109	109
Mogul	7.8	308	324	307	308	313
Mogul-A	7.1	228	252	233	254	262
ELF-0	4.1	171	209	173	202	182
Graphon	0	86	91	77	85	27

Source: Arora (77).

structure. Its adsorption resembled that of carbon dioxide and that at higher c/cs values resembled that of nitrogen at 77 K as well. The adsorption of p-nitrophenol was thus the most informative in terms of the characterization of the microporous structure of carbons. Oda et al. (86) observed that the presence of acidic oxygen structures retarded the adsorption both of phenol and benzoic acid from aqueous solutions; the effect was larger for phenol than for benzoic acid. The effect of surface acidity was also less pronounced in the case of competitive adsorption when both phenol and benzoic acid were present as an equimolecular mixture in aqueous solution. This was attributed to the change in solubility of both the components in the water. Benzoic acid, which was initially adsorbed, could not be substituted by phenol but phenol could be substituted by benzoic acid.

The adsorption of higher n-alkanols from their dilute solutions in n-heptane (87) exhibited a pronounced step, indicating a strong cooperative adsorption mechanism leading to a close-packed monolayer of alkanol molecules oriented with their chain axis parallel to the graphite basal planes. This unusual adsorption behavior in dilute nonaqueous solutions was attributed to an order—disorder transition of the pore alkanols at the liquid—carbon interface. Parkash (88) obtained type I adsorption isotherms for cationic pesticides Diquat and Paraquat from aqueous solutions on active carbons. The adsorption involved an intraparticle transport rate control mechanism. Competitive adsorption of the two pesticides from their equimolecular mixtures showed that the adsorption involved two different mechanisms.

The adsorption isotherms of ionic surfactants like dodecyl ammonium chloride, dodecyl pyridenium bromide, and sodium dodecyl sulfate from aqueous solutions on carbons were s-shaped with two maxima (89). The equilibrium concentration for the second maximum coincided with the CMC of the surfactants. The adsorption involved two steps, monomolecular adsorption of the Langmuir type and multilayer adsorption of the BET type. The polar sites on polar adsorbents attracted polar groups of the surfactant molecules by electrostatic attractive forces and formed oriented monomolecular layers and then multilayers between nonpolar chains of the surfactant molecules. The adsorption results clearly showed the influence of adsorbate and adsorbate—adsorbent interaction forces during adsorption from solutions (89—93). The adsorption of cationic polymers (94), such as nonyl phenol (NP) diethylamine, NP triethyltetramine, NP tetraethylpentamine, octyl phenol (OP) diethylenetriamine, OP tetraethylenepentamine from aqueous solutions on an activated carbon also showed a two-stage adsorption. At low adsorptions the molecules were postulated to be flat on the surface. A reorientation of the molecules occurred in the second stage as the adsorption

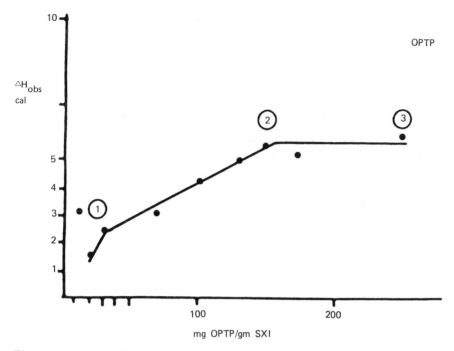

Figure 23 Heat of adsorption as a function of amount of octylphenol tetraethylene pentamine (OPTP) on Norit-activated carbon. [From Capelle (94).]

increased until at saturation the molecules were almost perpendicular to the surface. The variation of heats of adsorption with surface coverage showed two inflections (Fig. 23), corresponding to these two stages of adsorption.

Freundlich and Langmuir isotherm equations were employed to explain the earlier results of adsorption from solutions. However, the Freundlich isotherm lacks a theoretical basis and the Langmuir isotherm assumes a constant energy of adsorption over the entire surface, which is certainly not true in the case of carbons Kipling and Tester (95), Hansen and Fackler (96), and Manes and co-workers (97–103) applied Polanyi potential theory to the adsorption of organic compounds from aqueous and nonaqueous solutions. This model is better than the Langmuir model because it regards the surface as heterogeneous with respect to adsorption energies, and carbon surfaces are truly heterogeneous in nature. The Polanyi potential theory, however, has its limitations. It assumes that all the pores of the adsorbent are available to the adsorbate molecules, which is not completely true. Some of the micropores in carbons are not accessible to larger adsorbate molecules. Furthermore, the Polanyi potential approach does not recognize specific or chemical interactions between the adsorbate molecules and the surface oxygen structures, which are invariably found on carbon surfaces. Everett and co-workers (104–109) developed thermodynamic treatments, which have also been found to be inadequate to explain all the parameters involved in adsorption from solutions.

Urano et al. (110) tried to combine the generalizations of the Polanyi potential theory and the Freundlich isotherm equations to derive an equation which these workers claim could successfully predict the adsorption isotherm for several organic compounds from aqueous solutions on active carbons. Their adsorption equation is represented as

$$Q = (\frac{W_0}{bV}) \; (\frac{c}{c_0})^{1/\gamma n}$$

where Q is the equilibrium molar adsorption amount, V is the molar volume of the adsorbate, $1/n$ the exponential parameter of the modified Freundlich equation for a standard adsorbate, and b a constant which was approximated to 1.0 for liquid adsorbates and 1.5 for solid benzene-substitued compounds. These workers claim that the equation parameters W_0, n, and γ can be calculated from the properties of the activated carbon and the adsorbate as

$$W_0 = 1.9 \; V_3 - 0.03$$

$$n = \frac{1}{0.37 \; (V_3/V_{20}) - 0.05}$$

and

$$\gamma = \frac{V}{V_s}$$

where V and V_s are the molar volumes of the adsorbate and the standard adsorbate and V_3 and V_{20} represent volumes of the pores with diameters less than 3 and 20 nm. These workers compared the experimental adsorption data using a number of benzene derivatives, phenols and their derivatives, and aliphatic compounds on five different commercial activated carbons with the data calculated using this equation and found the agreement to be very close (Fig. 24),

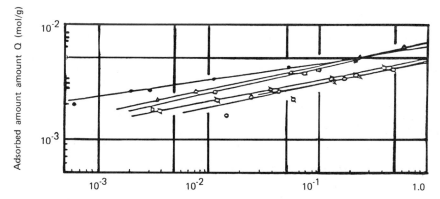

Figure 24 Comparison of theoretical and experimental adsorption isotherms of benzene derivatives. (Lines represent calculated isotherms, points experimental.) ⊙ , benzoic acid; △ , benzaldehyde; ⊡ , benzonitrile; ● , phenol; ⊐ , m. toluic acid; ⵕ, m. amino benzoic acid. [From Mano (110).]

thus making it possible to predict the adsorption capacities of different organic compounds from their aqueous solutions for any activated carbon whose pore size distribution is known.

Abe et al. (111) used the Freundlich isotherm equation to determine the adsorption data on a number of saccharides and polyhydric alcohols on an activated carbon and observed that the constants in the Freundlich equation were related to certain simple physical constants such as parachor, or molecular refractivity of the organic compounds. These workers also found that the adsorption isotherm for saccharides from aqueous solutions could be predicted from the number of carbon and oxygen atoms in the molecule.

4.3.2 Adsorption from Solutions at Higher Concentrations

When the two components of a solution are miscible with each other in all proportions, then neither of the two components can be regarded as a solvent over the whole range of concentrations. The change as a result of adsorption in such a system will be the resultant of the adsorption of both the components of the solution. The adsorption from such solutions can be represented in the form of composite isotherms, which are combinations of the isotherms for the individual components. Kipling (72) presented a composite isotherm equation, which is expressed as

$$\frac{n_0 \, \Delta x}{m} = n_1^s \, x_2 - n_2^s \, x_1$$

where Δx is the decrease in the mole fraction of component 1 in the bulk liquid which occurs when n_1^s moles of component 1 and n_2^s moles of component 2 are transferred from the solution onto m g of the solid in contact with n_0 moles of the solution. x_1 and x_2 are the mole fractions of components 1 and 2, and $n_0 \, \Delta x/m$ represents the surface excess or adsorption which, when plotted against mole fraction of a component, gives the composite isotherm. This composite isotherm equation can also be expressed in terms of weights and weight fractions as

$$\frac{W_0 \, \Delta c}{m} = W_1^s \, (1 - C) - W_2^s \, C$$

where W_0 is the initial weight of the solution, C is the equilibrium weight fraction of the solution, and W_1^s and W_2^s are the weights of the two components in the adsorbed phase. The two equations are related to each other as

$$\frac{n_0 \, \Delta x}{m} = \frac{W_0 \, \Delta C}{m} \, [\frac{1}{M_2 C + M_1 \, (1 - C)}]$$

where M_1 and M_2 are the molecular weights of the two components.

The relationship between the size of the composite isotherm and the adsorption of the individual components of the solution was illustrated by Ostwald and de Izaguirre (112) (Fig. 25). It can be seen clearly that negative adsorption is possible only when the solvent is also adsorbed and that maximum in the composite isotherm can occur even if there is no adsorption of the solvent. Nagy et al. (113,114), on the basis of a large number of composite isotherms, classified them into five types, two of which are U-shaped and three S-shaped (Fig. 26). The S shape of the isotherm indicates that the surface shows a preference for both the components of the solution but over different concentration ranges. The U-shaped isotherm, on the other hand, indicates a preference for only one component of the solution over the entire concentration range.

Gasser and Kipling (115) studied the adsorption of cyclohexane, methyl-, ethyl-, n-propyl, and n-butyl alcoholds from their solutions in benzene on Spheron-6 and graphon. The composite isotherms were S shaped in the case of alcohol—benzene solutions on

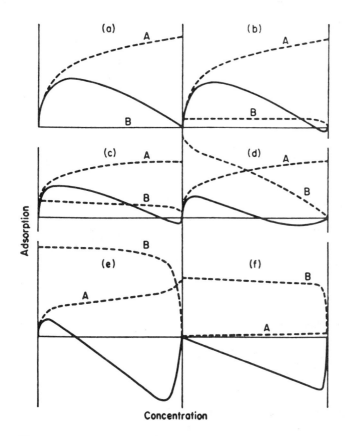

Figure 25 Composite and individual isotherms for binary systems
(solid line, composite; dashed line, individual for components).
(From Ostwald and de Izaguirre (112).]

Spheron-6 (Fig. 27a,d), indicating a preference for alcohols—the
most polar components of the solution—and U shaped for graphon
(Fig. 27c), indicating a preference for benzene—the less polar component of the solution. This was attributed to the presence of oxygen
structures on the surface of Spheron-6 and virtual absence of these
complexes on graphon surface. The marked preference for benzene by
Spheron-6 from benzene—cyclohexane solutions (Fig. 27b) was ascribed
to the interaction between polar oxygen complex on the surface with the
π electron clouds of the benzene molecule. A certain amount of methanol was chemisorbed from its solutions in benzene on oxygen-containing
sites. In the case of solution of *n*-butyl alcohol from benzene (Fig.
27c), the benzene was preferred over a wider range of concentrations,
although it is less polar than alcohol. This was attributed partly to

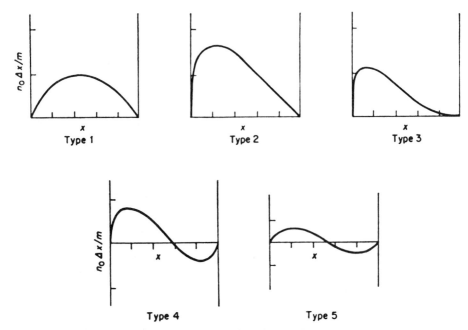

Figure 26 Classification of composite adsorption isotherms according to Nagy and Schay (113).

the fact that a benzene molecule can present six groups to the surface, whereas butyl alcohol can present only five groups, and partly to the possibility of π electron interactions. In addition, the hydrocarbon chain in n-butyl alcohol is of comparable importance with the hydroxyl group in determining the extent of adsorption (116).

Puri et al (117), while investigating the influence of carbon—oxygen surface complexes on selective adsorption of methanol—benzene, methanol—carbon tetrachloride, and ethanol—cyclohexane mixtures on a carbon black, observed that 1000°C outgassed sample, which was essentially free of any combined oxygen, showed a strong preference for benzene at all concentrations from benzene—methanol solutions, giving a typical U-shaped isotherm. However, when the same sample was degassed at a temperature of 700°C, where it loses all its oxygen present as CO_2-complex but retains most of its oxygen present as CO-complex, the sample showed even stronger preference for benzene. The original as well as the 400° outgassed samples showed more preference for methanol due to the presence of the polar CO_2-complex. The preference for methanol was enhanced when the CO_2-complex was enhanced by oxidation

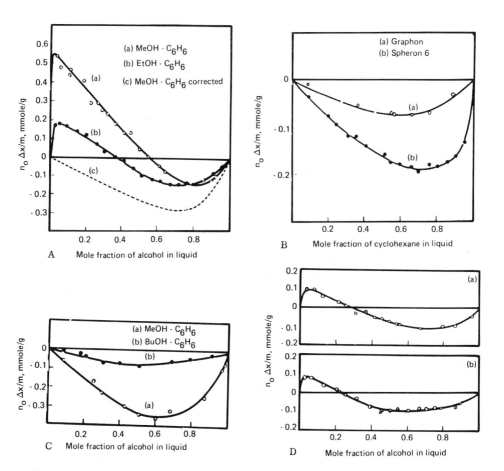

Figure 27 (A) Composite isotherms on Spheron-6 from binary solutions of alcohols and benzene. (B) Composite isotherms on Spheron-6 and graphon from binary solutions of cyclohexane and benzene. (C) Composite isotherms on graphon from binary solutions of alcohols in benzene. (D) Composite isotherms on Spheron-6 from binary solutions of alcohols in benzene for (1) *n*-propyl alcohol in benzene and (2) *n*-butyl alcohol in benzene. [From Gasser and Kipling (115)].

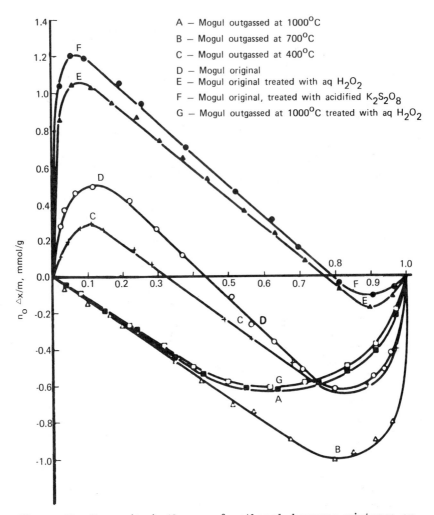

Figure 28 Composite isotherms of methanol—benzene mixtures on Mogul before and after various treatments. [From Puri et al. (117).]

with hydrogen peroxide or potassium persulfate. The change in the shape of the isotherms from S- to U-shaped as the temperature of outgassing was increased (Fig. 28) clearly indicated that while a part of the combined oxygen present as acidic CO_2-complex renders the surface polar and enhances its interaction with the more polar component of the solution, the presence of CO-complex enhances the preference of the surface for benzene, which is due to the interaction

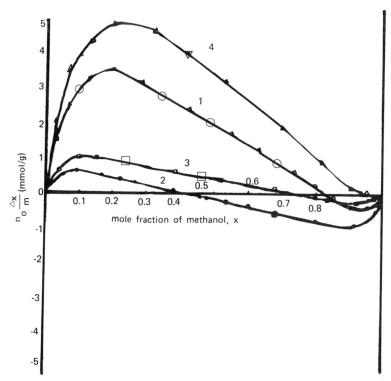

Figure 29 Composite isotherms from methanol—benzene solutions on various charcoals. ⊙ 1, original sugar charcoal; ● 2, 1000°-outgassed sugar charcoal; ⊡ 3, PVC charcoal; △ 4, H_2O_2-treated sugar charcoal. [From Bansal and Dhami (118).]

of the π electrons of the benzene ring with the partial positive charge on the carbonyl groups. In the case of methanol—carbon tetrachloride and ethanol—cyclohexane systems the interaction of carbon tetrachloride or of cyclohexane with the carbon surface was much weaker as compared to benzene and consequently the carbon surface preferred methanol or ethanol over wide ranges of concentration. The preference for methanol or for ethanol increased with increase in the oxygen content of the carbon surface on oxidation.

Bansal and Dhami (118) studied the adsorption from benzene—methanol solutions on a number of charcoals containing varying amounts of combined oxygen. The composite isotherms were S-shaped except for the charcoal oxidized with hydrogen peroxide, in which case the isotherm had a U shape. The extent of preference for methanol decreased when the charcoals were outgassed at 1000°C

Table 21 Thickness of the Adsorbed Layer on Various Charcoals

Sample	Thickness of the adsorbed layer		
	Methanol	Benzene	Total
Original sugar charcoal	1.30	0.68	1.98
1000°-outgassed charcoal	0.18	1.00	1.18
H_2O_2-treated charcoal	1.62	0.32	1.94
PVC charcoal	0.90	0.70	1.60

Source: Bansal and Dhami (118).

(Fig. 29), indicating the influence of the polar oxygen structures. The composite isotherms obtained by these workers showed linear regions which, according to Kipling and Tester (119) and Schay et al. (113,114,120), indicated a constant composition of the adsorbed phase along the range of concentrations covered by this linear region. These linear regions were extrapolated to get n_1^s and n_2^s from the intercepts at the ordinates through $x = 0$ and $x = 1$. The thickness t of the adsorbed layer was then calculated using the relationship

$$\frac{n_1^s}{(n_1^s)m} + \frac{n_2^s}{(n_2^s)m} = t$$

where $(n_1^s)m$ and $(n_2^s)m$ are the respective monolayer capacities. The calculated thickness of the adsorbed layer (Table 21) was more than unimolecular in the case of the charcoals associated with combined oxygen and approximately unimolecular for the oxygen-free sample. The higher thickness of the adsorbed layer in the case of the oxygen-containing carbons was attributed to the chemisorption of methanol on oxygen-containing sites where it could be held more strongly.

Bansal and Dhami (118) redetermined composite isotherms from benzene—methanol solutions on the charcoals after covering the chemisorption sites with methanol. It was found that in the case of oxygen-containing samples the preference for methanol decreased after the pretreatment but remained almost unchanged for the oxygen-free sample. The two isotherms before and after treatment with methanol in the case of the outgassed sample superimposed almost completely (Fig. 30). The break up of methanol into

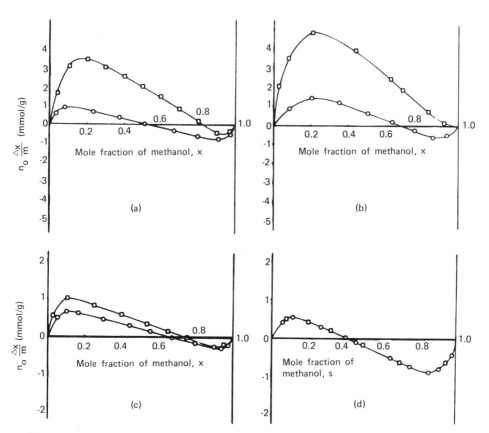

Figure 30 Composite isotherms from methanol—benzene solutions on carbons before and after methylation. (a) Original sugar charcoal; (b) H_2O_2-treated sugar charcoal; (c) PVC charcoal; (d) 1000°-outgassed sugar charcoal. [From Bansal and Dhami (118).]

chemisorption and physisorption as calculated from the composite isotherms before and after covering the chemisorption site (Table 22) showed that appreciable amounts of methanol were chemisorbed. The amounts n_1^s for methanol and n_2^s for benzene obtained from the composite isotherms of the charcoal samples after methylation were used to calculate the surface area. The values agreed fairly with the BET nitrogen and Dubinin—Polanyi carbon dioxide values. Thus the composite isotherms can also be used to calculate surface areas in cases where a linear region is obtained.

Jankowska et al. (121) studied adsorption from binary solutions of benzene and ethanol on a wood charcoal modified by heat treatment

Table 22 Breakup of Methanol into Physisorption and Chemisorption

| Sample | Methanol | | Benzene | | Methanol chemisorbed (vapor adsorption isotherm) (mmol/g) |
	Chemisorbed (mmol/g)	Physisorbed (mmol/g)	Chemisorbed (mmol/g)	Physisorbed (mmol/g)	
Original sugar charcoal	3.90	1.20	0	1.10	3.6
1000°-outgassed charcoal	0	0.80	0	1.20	0
H_2O_2-treated charcoal	5.0	2.20	0	1.0	—
PVC charcoal	0.64	0.52	0	0.35	0.58

Source: Bansal and Dhami (118).

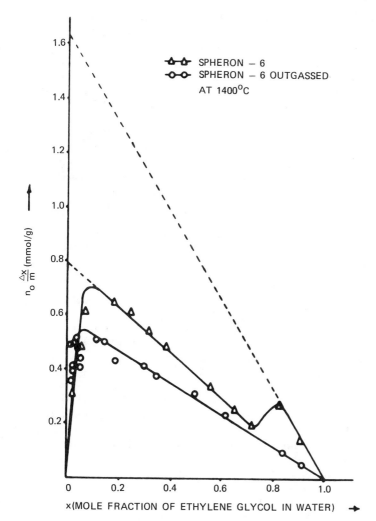

Figure 31 Composite adsorption isotherm from glycol—water binary solutions on Spheron-6. (From Puri et al., personal communication.)

in vacuum and by oxidation with nitric acid. The presence of acidic oxygen enhanced the preference for the more polar component of the solution and its removal enhanced the preference for the less polar component. These workers suggested that in addition to these polar and nonpolar interactions, the adsorption from solutions may also be influenced by energetic inhomogeneities of the carbon surface, which include defects, unsaturated sites, and free radicals.

Certain departures from the usual composite isotherm shapes have also been observed. Puri and co-workers studied the adsorption on Spheron-6 from ethylene glycol—water solutions in which both the components were polar in character. The isotherms were U shaped. The isotherm on the oxygen-free Spheron-6 showed one peak, whereas that on the oxygen-containing Spheron-6 showed two distinct peaks (Fig. 31). The composite isotherms with two peaks were termed "stepped isotherms" by Kipling et al. (72,122). When the two linear regions were extrapolated to zero concentration, the amounts adsorbed were found to be in the ratio of 1:2, indicating the possibility of the formation of a second layer.

4.4 ACTIVE SITES IN CARBONS

Active carbons have a stucture consisting of sheets of aromatic condensed ring systems stacked in nonpolar layers. These sheets have limited dimensions and therefore present edges. In addition, these sheets are associated with defects, dislocations, and discontinuities. The carbon atoms at these places have unpaired electrons and residual valencies and are richer in potential energy. Consequently, these carbon atoms are highly reactive and constitute active sites or active centers. A considerable amount of research has been aimed at understanding the number and nature of these active sites in view of the surface and catalytic reactions of carbons. Since the tendency of carbons to chemisorb oxygen is greater than their tendency to chemisorb any other species, much of our understanding of these active sites comes from the chemisorption of oxygen.

The first suggestion that the carbons are associated with different types of active sites came from the work of Rideal and Wright (123) on the oxidation of carbon surfaces with oxygen gas. These workers suggested three different types of sites which behaved differently at different oxygen pressures. The rates of oxygen chemisorption at 200°C were different at the three types of sites. Allardice (124), while studying the kinetics of chemisorption of oxygen on brown charcoal at temperatures between 25 and 300°C in the pressure range 100—700 torr, observed a two-step adsorption, which he attributed to the presence of two different types of sites. Dietz and McFarlane (125), while studying the adsorption of oxygen

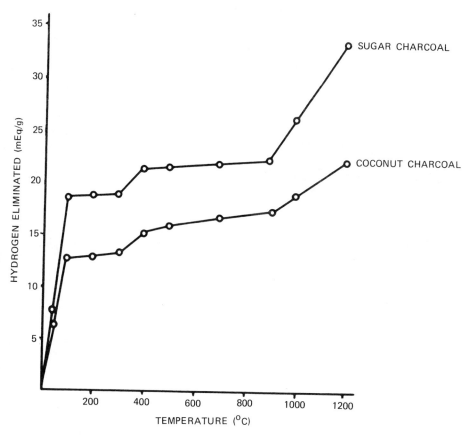

Figure 32 Elimination of chemisorbed hydrogen from charcoals by chloridation at different temperatures. [From Puri and Bansal (128a).]

on evaporated carbon films of high surface area at temperatures between 100 and 300°C and at oxygen pressures of the order of 100 millitorr, observed a rapid initial adsorption followed by a much slower adsorption. Carpenter and co-workers (126,127), during the intial stages of oxidation of different varieties of coals at temperatures of 65, 85, and 105°C, found that the chemisorption of oxygen obeyed the Elovich equation only in the first 5-min period. As the time period of oxidation increased, the quantity of oxygen chemisorbed exceeded the amounts predicted by the Elovich equation. This was attributed to the creation of fresh adsorption sites by the desorption of oxidation products such as CO_2, CO, and H_2O. Puri and Bansal (128,128a), while studying the chlorination of sugar and coconut

Table 23 Chemisorption of Oxygen on Activated Graphon at Different Temperatures

Experiment using the same sample	2.9% burnoff		19.9% burnoff	
	Temperature of chemisorption (C°)	Chemisorption capacity (μmol/g)	Temperature of chemisorption (C°)	Chemisorption capacity (μmol/g)
1	300	16.6	300	54.4
2	400	18.6	300	55.3
3	450	27.6	500	120.1
4	500	44.2	300	60.5
5	300	16.8	550	180.6
6	500	49.1	300	112.0

Source: Lussow et al. (129).

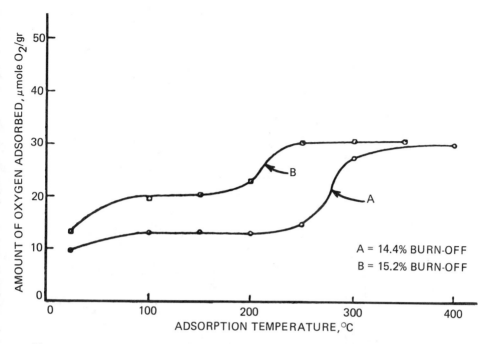

Figure 33 Saturation amounts of chemisorbed oxygen on activated graphon at different temperatures. [From Hart et al. (130).]

charcoals at different temperature between 30 and 1200°C, observed that the chemisorbed hydrogen was eliminated in a number of steps depending on the temperature of the treatment (Fig. 32). This was attributed to the fact that hydrogen in charcoals was bonded at different types of sites associated with varying energies of activation.

Lussow et al. (129) studied the kinetics of chemisorption of oxygen on graphon activated to eight different burnoffs between 0 and 35% in the temperature range of 450–675°C. The activated graphon samples were cleaned by evacuation at 950°C in a vacuum of 10^{-5} torr before chemisorbing oxygen. The saturation amounts of oxygen increased sharply at temperatures above 400°C, the amounts being almost two to three times greater. The adsorption of oxygen at 300°C after the first chemisorption at 500°C (Table 23) showed that the additional amount of chemisorption at temperatures above 400°C was not due to any additional activation of the graphon caused by the chemisorption above 400°C. It was also found that when the pressure of oxygen was increased from 0.5 to 700 torr the amount of oxygen chemisorbed was almost doubled. These results clearly

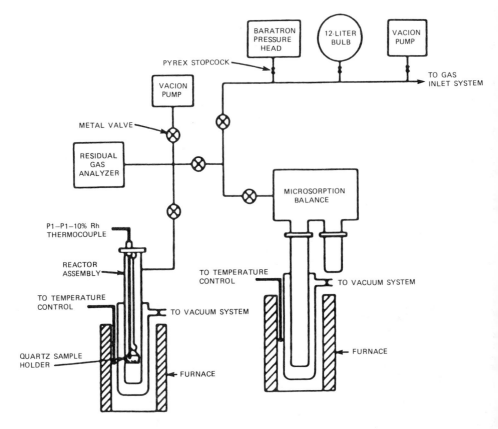

Figure 34 Schematic of ultraclean vacuum system for study of chemisorption of oxygen. [From Bansal et al. (132).]

showed that more than one type of site existed on the activated graphon surface. Hart et al. (130,131), while studying the rates of oxygen chemisorption on activated graphon (14.4% burnoff) in the temperature range 300–625°C at an oxygen pressure up to 0.5 torr using a mass spectrometer, observed two different rates of adsorption, one below and the other above 250°C. When the chemisorption at 300°C was studied for short time intervals, the saturation value was found to correspond to the value obtained below 250°C. However, when the adsorption was continued for longer periods another saturation value was obtained. This sharp increase in the saturation amount of oxygen at temperature above 250°C or at longer time intervals was attributed to the presence of two types of sites, which differed in their activation energies of adsorption (Fig. 33). The activation energy of adsorption at relatively more active sites was found to be 7.4 kcal/mol. The maximum amount of oxygen chemisorbed on these sites was estimated to occupy 2.8 m^2/g or about 2.6% of the BET surface area.

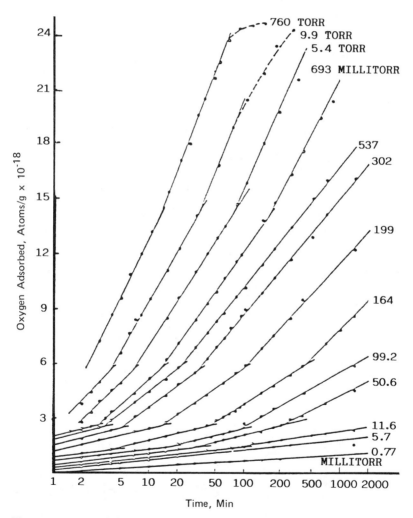

Figure 35 Elovitch plots of oxygen chemisorption on graphon at different pressures and temperatures. [From Walker et al. (133).]

Table 24 Instantaneous Adsorption Rates Calculated Midway in Each Linear Region; Rates Normalized with Respect to Pressure

Pressure (mtorr)	Rate (atoms/g · min · mtorr)				
	I ($Q = 0.6 \times 10^{18}$)	II ($q = 1.8 \times 10^{18}$)	III ($q = 4.5 \times 10^{18}$)	IV ($q = 9.6 \times 10^{18}$)	V ($q = 18.0 \times 10^{18}$)
0.77	2.1×10^{15}	—	—	—	—
5.76	2.5	—	—	—	—
11.6	2.7	—	—	—	—
22.9	2.0	2.6×10^{14}	—	—	—
50.6	2.0	2.9	—	—	—
99.2	—	2.9	—	—	—
164	—	3.0	—	—	—
199	—	4.0	2.2×10^{14}	—	—
302	—	3.9	2.0	5.2×10^{13}	—
537	—	—	2.2	6.5	—
693	—	—	2.3	6.9	8.7×10^{12}
5438	—	—	2.0	4.5	7.6
9930	—	—	2.0	4.9	8.9
760×10^3	—	—	—	1.0	3.1

Source: Walker et al. (133).

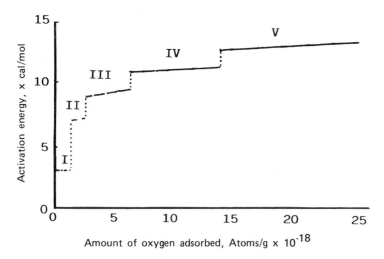

Figure 36 Variation of activation energy as a function of oxygen chemisorption on activated graphon. [From Bansal et al. (132).]

Bansal and colleagues (132–134) characterized ultraclean surfaces of graphon by studying the kinetics of chemisorption at low temperatures in the range 70–160°C and at oxygen pressures between 7.7×10^{-4} and 760 torr. The activated graphon sample (16.6% burnoff) was cleaned by heat treatment at 1000°C in a vacuum of the order of 10^{-9} torr. The apparatus used for cleaning the graphon sample and for studying chemisorption kinetics was a stainless steel system that conformed to all the requirements of an ultraclean system. The apparatus essentially consisted of a gas inlet system; a vacuum assembly, which contained a vacion pump; and an adsorption unit, which included a baratron differential manometer, a residual gas analyzer, and a Cahn microsorption balance (Fig. 34).

The results of these kinetic measurements when plotted according to the Elovich equation (q vs. log t) showed linear regions; the number of linear regions and the time of appearance of a linear region depended on the initial oxygen pressure or the temperature of the adsorption (Fig. 35). For example, the plots showed only one linear region at oxygen pressures lower than 10 mtorr and two linear regions between 50 and 683 mtorr pressure and at adsorption temperatures between 0 and 125°C. The number of these linear regions decreased at higher pressures or at higher temperatures. The instantaneous rates of adsorption calculated midway in each linear region when normalized with respect to oxygen pressure were found essentially proportional to the first power of oxygen pressure,

Table 25 Comparison of Experimental and Theoretical Values of Activation Energy for Chemisorption of Hydrogen on Graphite

Experimental		Theoretical	
Site	E (kcal/mol)	C−C distance (nm)	E (kcal/mol)
I	5.7	0.362	9
II	8.3	0.335	11
III	18.4	0.284	19
IV	30.4	0.246	28
V	—	0.142	50

Source: Bansal et al. (134).

which was varied widely, Furthermore, the rate of oxygen chemisorption decreased sharply in advancing from stage I to stage V (Table 24), where it was 250-fold less than that for stage I. These linear regions were thus postulated to represent different kinetic stages of the same chemisorption process involving adsorption at different types of sites. This received further support from the fact that each of the kinetic stages appeared after the adsorption of a definite amount of oxygen, although the time and the temperature of appearance and the temporal range of existence of any particular kinetic stage was determined by the initial pressure of the gas or the temperature of adsorption. In general, any stage appeared earlier and lasted for a shorter time as the pressure or the temperature was increased. A lower kinetic stage disappeared almost completely at higher pressures or at higher temperatures because the rates of adsorption were so rapid that the lower kinetic stage passed in the time period between the exposure of the carbon to the gas and the first measurement, so that the amounts adsorbed under these conditions in the very first measurement were larger than the amounts chemisorbed in the lower stages of adsorption. The activation energies of adsorption calculated from the Arrhenius plots of instantaneous rates at different coverages were found to be independent of surface coverage on any one group of sites, although the activation energies were different for adsorption on different groups of sites (Fig. 36). In all, five different groups of sites were observed in these studies. The activation energies varied between 3.1 and 12.4 kcal/mol as the chemisorption proceeded from the most active to the least active sites. Bansal et al. (134) also studied the kinetics of chemisorption of hydrogen on the same sample of activated graphon as a function of hydrogen pressure and adsorption temperature. The rates of hydrogen chemisorption were very low compared to those of oxygen. Only four types of active sites could be observed with an adsorption temperature of 600°C. Chemisorption experiments were not carried out at higher temperatures because the graphon surface showed burning, producing gaseous species which vitiated the kinetic measurements.

The existence of these discrete types of sites has been attributed to the difference in the geometrical arrangement of the surface carbon atoms (132—134). As the carbon surface presents several carbon—carbon distances to the incoming gas molecules, the activated complex formed between the gas atoms and the two surface carbon atoms would be expected to have different potential energy configurations, depending on these spacings between the carbon atoms, resulting in a variation of the activation energies of chemisorption. Sherman and Eyring (135) made theoretical calculations of the energy of activation for dissociative chemisorption of hydrogen on a carbon surface and found the values to vary with the carbon—carbon spacings. A comparison of the theoretical values of Sherman and Eyring (135) and the experimental values obtained by Bansal et al. (134) is given in Table 25. The carbon—carbon distances selected in this table are those which the hydrogen molecule would most likely enounter when approaching the carbon surface. The spacing 0.246 nm represents the configuration terminating in (10$\bar{1}$ 1) face, the spacings 0.142 and 0.284 nm for (11$\bar{2}$ 1) termination, and the spacings of 0.335 and 0.362 nm for distances between edge carbon atoms in adjacent basal planes. Agreement between the experimental and the theoretical values supports the concept that these carbon spacings act like discrete types of sites on which the chemisorption of gases generally can occur.

Puri et al. (136—138) obtained fairly convincing evidence for the presence of certain highly active sites of an entirely different kind, which they called unsaturated sites. These unsaturated sites can be determined by interacting carbons with the aqueous solution of bromine in potassium bromide. These sites are produced when carbons associated with oxygen are outgassed at high temperatures, the optimum temperature being between 600 and 700°C. The concentration of these sites was shown to vary with the combined oxygen which comes off as carbon dioxide (Table 26). One mole of unsaturated sites was created by the chemisorption of two moles of oxygen as carbon dioxide.

The concentration of the active sites on a carbon surface has been measured in terms of the active surface area (ASA) by Laine and co-workers (139—141). The active surface, according to these workers, is an index of the reactivity of a carbon surface and can be determined from the amount of oxygen chemisorbed at 300°C in

Table 26 Active Sites as Determined by Bromine Adsorption from Aqueous Solutions by Carbons

Carbon	Sugar charcoal		Cotton stalk charcoal	
	Unsaturated active sites (mmol/g)	Decrease in CO_2-complex (mmol/g)	Unsaturated active sites (mmol/g)	Decrease in CO_2-complex (mmol/g)
Original	—	—	—	—
Outgassed at				
300°	0.80	1.69	0.27	0.56
400°	1.34	2.59	0.49	1.06
500°	1.62	3.31	0.67	1.41
700°	1.85	3.63	0.89	1.69
900°	1.83	3.63	0.88	1.69
1000°	1.87	3.70	0.86	1.65
1200°	1.85	3.64	0.89	1.69

Source: Puri et al. (136).

Table 27 Comparison of Active Surface Area Calculated by Chemisorption of Oxygen at 300°C and the Most Active Site Area Obtained from Chemisorption at Higher Temperatures

Burnoff (%)	ASA (Laine method) (μmol O/g)	Most active site area (μmol O/g)
0	5.0	5.0
0	4.4	53.9
3.3	17.8	40.4
6.4	27.6	27.6
7.7	30.9	28.8
14.4	44.4	45.2
14.4	44.4	45.2
14.4	44.4	44.8
14.4	44.4	50.3
18.5	52.0	49.5
18.5	52.0	47.8
18.5	52.0	48.2
18.5	52.0	50.4
18.5	52.0	46.0
18.5	52.0	50.2
18.5	52.0	47.0
18.0	52.0	50.7
18.5	52.0	52.7
18.5	52.0	53.4
19.9	54.4	50.0
25.8	65.0	64.6
25.8	65.0	65.0
34.9	82.0	76.0

Source: Lussow et al. (129).

24 hr at an initial oxygen pressure of 0.5 torr. Assuming that one oxygen atom is adsorbed at each carbon atom and that each carbon atom occupies an area of 0.083 nm^2, the oxygen chemisorbed can be converted into active surface area. These workers activated graphon to seven different burnoffs in order to create varying amounts of active surface and observed that ASA increased with the degree of burnoff and could be related to the reactivity of the graphon toward oxygen. However, in later work (129,130) on the chemisorption of oxygen at higher temperatures between 300 and 675°C, they observed that there existed more than one type of active site which differed in reactivities toward oxygen. The rate of chemisorption of oxygen on the less active sites was not appreciable at low oxygen pressures until the chemisorption temperature was 400°C or above. Consequently they suggested that Laine's method, which involved chemisorption at 300°C, determined only the surface area covered by the most active groups of active sites. These workers (129) calculated the area covered by the most active group of sites from the low coverage part of the reactivity data, when the reactivity of the less active sites was supposedly negligible, and found the values to be in good agreement with Laine's values (Table 27). Thus these workers pointed out the necessity of obtaining reactivity data under proper conditions to ensure that the observed reactivity was not being unduly influenced by the reaction on less active sites.

Hoffman et al. (142), while studying the chemisorption of several hydrocarbons (e.g., propylene, ethylene, propane) and methane on activated graphon, observed that the adsorption of each hydrocarbon increased with burnoff due to increase in the active surface area ASA of the graphon. However, the ASA covered by these hydrocarbons was much less compared to the ASA covered by oxygen at all degrees of burnoff (Table 28). Furthermore, propylene covered a larger ASA compared to ethylene or methane or n-butane. This may be attributed partly to the larger size of these hydrocarbon molecules, which, when adsorbed on an active site, are likely to shield some of the neighboring sites to make them unavailable for adsorption. Thus ASA has a meaning only with respect to the chemisorption of a particular species. Whereas Laine's ASA can measure reactivity toward oxygen, it fails to measure the reactivity of the graphon toward these hydrocarbons.

Dentzer et al. (143) examined the adorption and decomposition of silver diamine complexes from ammoniacal solutions on a graphitized carbon black, Vulcan 3, activated to different degrees of burnoff. These workers observed that the amount of silver adsorbed increased with increase in the degree of burnoff. A linear relationship (Fig. 37) was observed between the amount of silver adsorbed and the ASA as determined by Laine's method. This was attributed to

Table 28 Active Surface Area Occupied by the Chemisorption of Different Gases on Graphon at 300°C

Sample	Burnoff (%)			Oxygen ASA (m^2/g)	Propylene (ASA (m^2/g)	Ethylene ASA (m^2/g)	n-Butane ASA (m^2/g)
	723 K	1223 K	Total				
1	0	0	0	0.264	0.056	0.033	0.012
2	0	0.61	0.61	0.94	0.44	0.32	0.18
3	4.9	0.85	5.75	2.25	1.06	0.72	0.46
4	24.1	0.40	24.5	5.00	1.89	0.19	0.87

Source: Hoffman et al. (142).

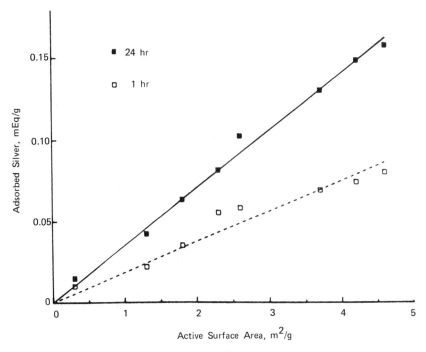

Figure 37 Amount of silver adsorbed in relation to active surface area of carbons. [From Dentzer et al. (143).]

the specific reductive interaction of the silver diamine with the carbon-active sites producing metallic silver, which was chemisorbed on the active sites.

REFERENCES

1. Leslie, J., *Tilloch's Phil. Mag.*, *14:* 20 (1802).
2. Chessick, J. J. and Zettlemoyer, A. C., *Adv. Catal. 11:* 263 (1959).
3. Zettlemoyer, A. C., Iyengar, R. D., and Scheidt, P., *J. Colloid Interace Sci.*, *22:* 172 (1966).
4. Wade, W. H. and Hackerman, N., *J. Phys. Chem.*, *64:* 1196 (1960); *65:* 1681 (1961).
5. Wade, W. H., Teranishi, S., and Durham, J. L., *J. Colloid Interface Sci.*, *20:* 838 (1966).
6. Whalen, J. W., *J. Phys. Chem.*, *65:* 1676 (1961).

7. Rozouk, R. L., Nashed, S. H., and Mourad, W. E., *Carbon*, *2:* 539 (1964).
8. Wade, W. H., *J. Colloid Interface Sci.*, *31:* 111 (1969).
9. Brusset, H., Martin, J. J. P., and Mendelsbaum, H. G., *Bull. Soc. Chim. France*, *7:* 2346 (1967).
10. Stoeckli, H. F. and Kraehenbuehl, F., *Carbon*. *19:* 353 (1982).
11. Barton, S. S., Boulton, G. L., and Harrison, B. H., *Carbon*, *10:* 391 (1972).
12. Barton, S. S. and Harrison, B. H., *Carbon*, *10:* 745 (1972).
13. Barton, S. S. and Harrison, B. H., *Carbon*, *13:* 47 (1975).
14. Zettlemoyer, A. C. and Chessick, J. J., in *Advances in Chemistry*, N°. 43, American Chemical Society, Washington D.C., 1974, p. 58.
15. Puri, B. R., Singh, D. D., and Sharma, L. R., *J. Phys. Chem.*, *62:* 756 (1958).
16. Puri, B. R., in *Chemistry and Physics of Carbon* P. L. Walker, Jr., Ed., Marcel Dekker, New York, 1970.
17. Youssef, A. M., *Carbon*, *13:* 449 (1975).
18. Bangham, D. H. and Razouk, R. I., *Proc. Roy. Soc. A, 166:* 572 (1938).
19. Maggs, F. A. P. and Robins, G. A., in *Characterisation of Porous Solids*, S. J. Gregg, K. S. W. Sing, and H. F. Stoeckli, Eds., Soc. of Chem. Ind., 1979, p. 59.
20. Atkins, D., McLeod, A. I., Sing, K. S. W., and Capon, A., *Carbon, 20:* 339 (1982).
21. Zettlemoyer, A. C., Pendleton, P., and Micale, F. J., in *Adsorption from Solution*, R. H. Ottewill, C. H. Rochester, and A. L. Smith, Eds., Academic Press, New York, 1983, p. 113.
22. Hagiwara, S., Tsutsumi, K., and Takahashi, H., *Carbon, 19:* 107 (1981).
23. Hagiwara, S., Tsutsumi, K., and Takahashi, H., *Carbon, 9:* 693 (1971).
24. Hagiwara, S., Tsutsumi, K., and Takahashi, H., *Nippon Kagaku Kaishi, 8:* 1369 (1973).
25. Zettlemoyer, A. C., *Ind. Eng. Chem.*, *57:* 27 (1965).
26. Stoeckli, H. F., Kraehenbuhl, F., and Morel, D., *Carbon, 21:* 589 (1983).
27. Dubinin, M. M., in *Progress in Surface and Membrane Service*, Vol. 9, D. A. Cadenhead, Ed., Academic Press, New York, 1975, pp. 1–70.
28. Radeke, K. H., *Carbon, 22:* 473 (1984).
29. Stoeckli, H. F. and Kraehenbuel, F., *Carbon, 22:* 297 (1984).
30. Kraehenbuehl, F., Ph.D. disseration, University of Neuchâtel, 1983.

31. Stoeckli, H. F., Kraehenbuehl, F., Lavanchy, A., and Huber, U., *J. Chem. Phys.*, *81:* 785 (1984).
32. Barton, S. S., Evans, M. J. B., Holland, J., and Koresh, J. E., *Carbon*, *22:* 265 (1984).
33. Stoeckli, H. F., Perret, A., and Mena, P., *Carbon*, *18:* 443 (1980).
34. Pierce, C. and Smith, R. N., *J. Phys. Colloid Chem.*, *54:* 784 (1950).
35. Pierce, C., Smith, R. N., Wiley, J. K., and Corder, H. *J. Am. Chem. Soc.*, *73:* 4551 (1951).
36. Dubinin, M. M., Zaverina, E. D., and Serpinski, V. V., *J. Chem. Soc.*, 1760 (1955).
37. Dubinin, M. M., *Carbon*, *18:* 355 (1980).
38. Dubinin, M. M. and Serpinski, V. V., *Carbon*, *19:* 402 (1981).
39. Puri, B. R., Murari, K., and Singh, D. D., *J. Phys. Chem.*, *65:* 37 (1961).
40. Puri, B. R., *Carbon*, *4:* 391 (1966).
41. Youssef, A. M., *Carbon*, *12:* 433 (1974).
42. Youssef, A. M., Ghazy, T. M., and Nabarawy, T. El, *Carbon*, *20:* 113 (1982).
43. Wiig, E. O. and Juhola, A. J., *J. Am. Chem. Soc.*, *71:* 561 (1949).
44. Juhola, A. J. and Wiig, E. O., *J. Am. Chem. Soc.*, *71:* 2069 (1949).
45. Juhola, A. J. and Smith, S. B., *J. Am. Chem. Soc.*, *74:* 61 (1952).
46. McDermot, H. L. and Arnell, A. C., *J. Phys. Chem.*, *58:* 492 (1954).
47. Kiselev, A. V. and Kovaleva, N. V., *Zh. Fiz. Khim.* *30:* 2775 (1956).
48. Avgul, N. N., Dzhigit, D. M., Kiselev, A. V., and Scherbakova, K. D., *Dokl. Akad. Nauk. SSSR*, *92:* 105 (1953).
49. Avgul, N. N., Dzhigit, O. M., Kiselev, A. V., and Scherbakova, K. D., *Dokl. Akad. Nauk. SSSR*, *101:* 285 (1955).
50. Bansal, R. C., Dhami, T. L., and Parkash, S., *Carbon*, *16:* 389 (1978).
51. Dubinin, M. M., in *Chemistry and Physics of Carbon*, Vol. 2, P. L. Walker, Jr., Ed., Marcel Dekker, New York, 1966, pp. 51—120.
52. Marsh, H. and Rand, B., *J. Colloid Interface Sci.*, *33:* 101 (1970).
53. Dubinin, M. M., *Zh. Fiz. Khim.*, *39:* 1305 (1965).
54. Zukal, A. and Kadlec, O., Collect. C. Zech *Chem. Commun.*, *38:* 321 (1973).

55. Kadlec, O., *Chem. Zvesti* (Chem. Papers), *29:* 660 (1975).
56. Rozwadowski, M. and Wojsz, R., *Carbon, 19:* 383 (1981).
57. Rozwadowski, M., Siedlowski, J., and Wojsz, R., *Carbon, 17:* 411 (1979).
58. Wolff, H., in *The Hydrogen Bond,* Vol. 3, North-Holland, Amsterdam, 1976, p. 1225.
59. Wolff, H. and Landeck, H., *Ber. Bunsenges. Phys. Chem., 81:* 1054 (1977).
60. Dacey, J. R. and Thomas, D. G., *Trans. Faraday Soc., 50:* 740 (1954).
61. Walker, P. L., Jr., Austin, L. G., and Nandi, S. P., in *Chemistry and Physics of Carbon,* Vol. 2, P. L. Walker, Jr., Ed., Marcel Dekker, New York, 1966, pp. 257–371.
62. Walker, P. L., Jr., Lamond, T. G., and Metcalfe, J. E., in *2nd Conf. Ind. Carbon and Graphite,* London, 1965, p. 7.
63. Bansal, R. C. and Dhami, T. L., *Carbon, 18:* 137 (1980).
64. Bansal, R. C. and Dhami, T. L., *Carbon, 18:* 297 (1980).
65. Bansal, R. C. and Dhami, T. L., *Indian J. Tech., 23:* 92 (1985).
66. Bansal, R. C. and Dhami, T. L., in *First Indian Conf. Carbon,* December 1982, Proc. Indian Carbon Soc. Publ.
67. Bansal, R. C., Parkash, S., and Dhami, T. L., *Indian J. Chem., 19A:* 1116 (1980).
68. Ainscough, A. N., Dollimore, D., and Heal, G. R., *Carbon, 11:* 189 (1973).
69. Puri, B. R., Kaistha, B. C., Vardhan, Y., and Mahajan, O. P., *Carbon, 11:* 329 (1973).
70. Freundlich, H. and Heller, W., *J. Am. Chem. Soc., 61:* 228 (1939).
71. Kipling, J. J., *J. Chem. Soc., 1839:* 1483.
72. Kipling, J. J., *Adsorption from Solutions of Nonelectrolytes,* Academic Press, New York, 1965.
73. Mattson, J. S. and Mark, H. B., Jr., *Activated Carbon: Surface Chemistry and Adsorption from Solution,* Marcel Dekker, New York, 1971.
74. Suffet, I. H. and McGuire, M. J., *Activated Carbon Adsorption,* Vols. I, II, Ann Arbor Science Publishers, Ann Arbor, Mich., 1981.
75. Ottewill, R. H., Rochester, C. H., and Smith, A. L., Eds., *Adsorption from Solutions,* Academic Press, New York, 1983.
76. Lemeiur, R. U. and Morrison, J. L., *Can. J. Res., 25 B:* 440 (1947).
77. Arora, V. M., Ph.D. dissertation, Panjab University, Chandigarh, 1977.

78. Puri, B. R., in *Activated Carbon Adsorption*, I. H. Suffet and M. J. McGuire, Eds., Ann Arbor Science Publishers, Ann Arbor, Mich., 1981, p. 353.
79. Smith, H. A. and Hurley, R. B., *J. Phys. Colloid Chem. 53:* 1409 (1949).
80. Graham, D., *J. Phys. Chem. 59:* 896 (1955).
81. Marsh, H. and Campbell, H. G., *Carbon, 9:* 489 (1971).
82. Puri, B. R., Gupta, M., and Singh, D. D., *J. Indian Chem. Soc., 52:* 26 (1975).
83. Coughlin, R. W., Ezra, F. S., and Tan, R. N., *J. Colloid Interface Sci., 28:* 386 (1968).
84. Walker, P. L., Jr., and Janov, J., in *Hydrophobic Surfaces*, F. M. Fowkes, Ed., Academic Press, New York, 1969, p. 107.
85. Mattson, J. S., Mark, H. B., Jr., Malbin, M. D., Weber, W. J., and Crittenden, J. C., *J. Colloid Interface Sci., 31:* 116 (1969).
86. Oda, H., Kishida, M., and Yokokawa, C., *Carbon, 19:* 243 (1981).
87. Furdenegg, G. H., Koch, C., and Liphard, M., in *Adsorption from Solutions*, R. H. Ottewill, C. H. Rochester, and A. L. Smith, Eds., Academic Press, New York, 1983, pp. 87.
88. Parkash, S., *Carbon, 12:* 483 (1974).
89. Tamamush, B., in *Adsorption from Solutions*, R. H. Ottewill, C. H. Rochester, and A. L. Smith, Eds., Academic Press, New York, 1983, p. 79.
90. Tamamush, B. and Tamaki, K., in *Proc. 2nd Int. Cong. Surface Activity 1957*, Vol. 3, Butterworth, London, 1959, p. 449.
91. Tamamushi, B. and Tamaki, K., *Trans. Faraday Soc., 55:* 1007 (1959).
92. Eda, K., *J. Chem. Soc. (Japan), 80:* 343, 347, 349, 461, 465, 708 (1960).
93. Yamada, H., Fukumura, K., and Tamamushi, B., *Bull. Chem. Soc. Japan, 53:* 3054 (1980).
94. Capelle, A., in *Activated Carbon: A Fascinating Material*, A. Capelle and F. deVooys, Eds., Norit N. V., Netherlands, 1983, p. 191.
95. Kipling, J. J. and Tester, D. A., *J. Am. Chem. Soc., 1922:* 4123.
96. Hansen, R. S. and Fackler, W. V., Jr., *J. Phys. Chem., 57:* 6346 (1953).
97. Manes, M. and Hoffer, L. J. E., *J. Phys. Chem., 73:* 584 (1969).
98. Wohleber, D. A. and Manes, M., *J. Phys. Chem., 75:* 61 (1971).

99. Wohler, D. A. and Manes, M., *J. Phys. Chem.*, *75:* 3720 (1971).
100. Chion, C. C. T. and Manes, M., *J. Phys. Chem.*, *77:* 811 (1973).
102. Schenz, T. W. and Manes, M., *J. Phys. Chem.*, *79:* 604 (1975).
103. Rosen, M. R. and Manes, M., *J. Phys. Chem.*, *80:* 953 (1976).
104. Everett, D. H., *Trans. Faraday Soc.*, *60:* 1803 (1964).
105. Everett, D. H., *Trans. Faraday Soc.*, *61:* 2478 (1965).
106. Ash, S. G., Everett, D. H., and Findenegg, G. H., *Trans. Faraday Soc.*, *64:* 2639 (1968).
107. Ash, S. G., Brown, R., and Everett, D. H., *J. Chem. Thermodyn.* *5:* 239 (1973).
108. Ash, S. G., Brown, R., and Everett, D. H., *Trans. Faraday Soc.*, *70:* 123 (1974).
109. Everett, D. H., in *Adsorption from Solutions*, R. H. Ottewill, C. H. Rochester, and A. L. Smith, Eds., Academic Press, New York, 1983, p. 1.
110. Mano, K., Koichi, Y., and Yamamoto, E., *J. Colloid Interface Sci.*, *86:* 43 (1982).
111. Abe, I., Hayashi, K., and Kitagawa, M., *Carbon, 21:* 189 (1983).
112. Ostwald, W. and de Izaguirre, R., *Kolloid Z.*, *30:* 279 (1922).
113. Nagy, L. G. and Schay, G., *Magyar Kem. Foly.*, *66:* 31 (1960).
114. Schay, G., Nagy, L. G., and Szekrengery, T., *Period. Polytech.*, *4:* 95 (1960).
115. Gasser, C. G. and Kipling, J. J., in *Proc. 4th Conf. Carbon, University of Buffalo*, Pergamon Press, New York, 1960, p. 55.
116. Gasser, C. G. and Kipling, J. J., 13th ACS national meeting, Chicago, 1958.
117. Puri, B. R., Singh, D. D., and Kaistha, B. C., *Carbon, 10:* 481 (1972).
118. Bansal, R. C. and Dhami, T. L., *Carbon, 15:* 153 (1977).
119. Kipling, J. J. and Tester, D. A., *J. Chem. Soc. (London)*, *1952:* 4123.
120. Schay, G. and Nagy, L. G., *J. Chem. Phys.*, *1961:* 149.
121. Jankowska, H., Swiatowsi, A., Oscik, J., and Kusak, R., *Carbon, 21:* 117 (1983).
122. Cornford, P. V., Kipling, J. J., and Wright, E. H. M., *Trans. Faraday Soc.*, *58:* 74 (1962).
123. Rideal, E. K. and Wright, M. W., *J. Am. Chem. Soc.*, *1925:* 1347.
124. Allardice, D. J., *Carbon, 4:* 255 (1966).

125. Deitz, V. R. and McFarlane, E. F., in *Proc. Fifth Conf. Carbon*, Vol. 2, Pergamon Press, New York, p. 219.
126. Carpenter, D. L. and Giddings, D. W., *Fuel, 43:* 375 (1964).
127. Carpenter, D. L. and Sergeant, G. D., *Fuel, 45:* 311 (1966).
128. Puri, B. R., Malhotra, S. L., and Bansal, R. C., *J. Indian Chem. Soc., 40:* 179 (19   ).
128a. Puri, B. R. and Bansal, R. C., *Chem. Ind. London, 1963:* 574.
129. Lussow, R. O., Vastola, F. J., and Walker, P. L., Jr., *Carbon, 5:* 591 (1967).
130. Hart, P. J., Vastola, F. J., and Walker, P. L., Jr., *Carbon, 5:* 363 (1967).
131. Walker, P. L., Jr., Vastola, F. J., and Hart, P. J., in *Proc. Symp. Fundamentals of Gas Surface Interactions, San Diego, Calif.*, Academic Press, New York, 1967, p. 307.
132. Bansal, R. C., Vastola, F. J., and Walker, P. L., Jr., *J. Colloid Interface Sci., 32:* 187 (1970).
133. Walker, P. L., Jr., Bansal, R. C., and Vastola, F. J., in *The Structure and Chemistry of Solid Surfaces*, Wiley, 1969, pp. 81-1–81-16.
134. Bansal, R. C., Vastola, F. J., and Walker, P. L., Jr., *Carbon, 9:* 185 (1971).
135. Sherman, A. and Eyring, H., *J. Am. Chem. Soc., 54:* 2661 (1932).
136. Puri, B. R., Sandle, N. K., and Mahajan, O. P., *J. Chem. Soc. (London), 1963:* 4880.
137. Puri, B. R. and Bansal, R. C., *Carbon, 3:* 523 (1966).
138. Bansal, R. C., Dhami, T. L., and Parkash, S., *Carbon, 18:* 395 (1980).
139. Laine, N. R., Ph.D. dissertation, Pennsylvania State University, 1982.
140. Laine, N. R., Vastola, F. J., and Walker, P. L., Jr., *J. Phys. Chem., 67:* 2030 (1963).
141. Laine, N. R., Vastola, F. J., and Walker, P. L., Jr., in *Proc. Fifth Conf. Carbon*, Vol. 2, Pergamon Press, New York, 1963, p. 211.
142. Hoffman, W. P., Vastola, F. J., and Walker, P. L., Jr., *Carbon, 22:* 585 (1984).
143. Dentzer, J., Ehrburger, P., and Lahaye, J. *Colloid Interface Science, 112:* 170–177 (1986).

5

Surface Modification of Carbons

Carbons are almost invariably associated with appreciable amounts of oxygen. The oxygen is chemisorbed even on mere exposure to air or oxygen preferably at 400–500°C. The oxygen is fixed firmly and comes off only as oxides of carbon on high-temperature heat treatment in vacuum or in an inert atmosphere. Similarly, it is well known that all microcrystalline carbons contain chemically bonded hydrogen, the amount depending on the history of its formation. The hydrogen is held so firmly that it is not given off completely even on outgassing at 1200°C. These carbons can also fix nitrogen on treatment with ammonia; sulfur on treatment with hydrogen sulfide, carbon disulfide, or sulfur; chlorine on treatment with the gas; and bromine on treatment in the gaseous or solution phase. These treatments give rise to stable carbon–nitrogen, carbon–sulfur, carbon–chlorine, or carbon–bromine surface structures (surface compounds) respectively. There is also evidence that the carbons can adsorb certain molecular species such as phenols, amines, nitrobenzene, surfactants, and several other cationic species.

X-ray diffraction studies have shown that these heteroatoms or molecular species are bonded or retained at the edges and corners of the aromatic sheets or to carbon atoms in defect positions of the aromatic sheets, or they can be incorporated within the carbon layer, forming heterocyclic ring systems. These carbon atoms have unsaturated valencies and have a tendency to reduce their potential energy by forming bonds with these heteroatoms. Since these edges constitute the main adsorbing surface, the presence of these heteroatoms or molecular species is expected to modify the surface characteristics and surface behavior of these carbons. Similar surface compounds in carbon blacks determine their application in the rubber, plastic, and paint industries and they determine the lubricating properties in graphites as well as their use as moderators in atomic reactors. In the case of carbon fibers these very surface compounds determine their adhesion to plastic matrices and consequently determine their use in composites.

5.1 MODIFICATION OF CARBONS BY OXIDATION

Carbons are always associated with varying amounts of chemisorbed oxygen unless special care is taken to eliminate it. In fact this combined oxygen has often been found to be the source of the property by which a carbon becomes useful or effective in certain respects. Carbons also have a great tendency to extend this layer of bonded oxygen by chemisorbing oxygen, and many of their reactions arise because of this tendency. For example, carbons are capable of decomposing oxidizing gases such as ozone (1—4), oxides of nitrogen (5,6) chemisorbing oxygen in each case. They also decompose aqueous solutions of silver salts (7), halogens (8—10), ferric chloride (11), potassium and ammonium persulfates (12—13), sodium hypochlorite (14,15),·potassium permanganate (16,17), potassium dichromate (17), sodium thiosulfate (18), hydrogen peroxide (19,20), and nitric acid (17,21,22). In each case there is chemisorption of oxygen and a buildup of the oxide layer on the carbon surface.

Carbons can also be oxidized by heat treatment in air, carbon dioxide, or oxygen. The nature and the amount of surface oxides formed on treatment with oxygen depends on the nature of the carbon and the history of its formation, its surface area, and the temperature of treatment. The reaction of carbons with oxygen can proceed in several ways depending on the temperature at which the reaction is carried out:

$$C + O \longrightarrow C(O) \quad \text{formation of oxygen surface compound}$$
surface compound

$$C + O \longrightarrow CO + CO_2 \quad \text{gasification}$$

$$C(O) \longrightarrow CO + CO_2 \quad \text{decomposition of surface compound}$$

At temperatures below 400°C, the chemisorption of oxygen and formation of the carbon—oxygen surface compounds dominate, whereas at temperatures above 400°C, the decomposition of the surface compounds and the gasification of the carbon are the dominating reactions. In case of oxidative treatments in solutions, the major reaction is the formation of the surface compounds, although some gasification may also occur depending on the strength of the oxidative treatment and the severity of the experimental conditions. The formation of the surface oxygen compounds using various types of carbon and using different oxidative treatments in gaseous and liquid phases has been studied by a large number of investigators and has been well reviewed (23–25). Thus we do not discuss this aspect in detail, but merely say that carbons have a tendency to pick up oxygen at least to some extent under all conditions when in contact with an oxidative reagent.

As mentioned earlier, the oxygen in carbons is present at sites which are generally responsible for their adsorption characteristics, surface reactions, and surface behavior. Thus its presence will greatly modify the surface properties of carbons. In fact it is assumed that most of the properties of carbons are the properties of these oxygen structures. Consequently, their influence on carbon surface behavior is felt and discussed in almost every chapter of the book in one form or another. We therefore do not discuss the modification of carbons by these surface compounds separately in this chapter.

5.2 MODIFICATION OF ACTIVE CARBONS BY SURFACE IMPREGNATION

The impregnation of carbons at once brings to mind the impregnation by metals, a subject that has been widely studied and extensively reviewed because of its importance in heterogeneous catalysis. Metals and their oxides, dispersed as small particles on high surface area carbons and other supports, have been used widely and are being used as catalysts for various industrial applications. The primary role of the carbon support is to favor the formation of a highly dispersed metal phase and to physically separate the metal crystallites, thus increasing their stability on sintering. The study of supported small metal crystallites started in the late 1970s and has grown significantly since. The impregnation of metals in carbonaceous materials also modifies the gasification characteristics and varies the porous structure of the final carbon product. Thus the procedure has been applied to obtain carbons with a given microporous structure. The subject of this section, however, is not these developments but a study of the modification of the surface behavior and adsorption characteristics of carbons by chemical reagents which are present as a part of the adsorbed phase.

Activated carbons impregnated with potassium iodide and similar compounds (26) and with amines (27,28) including several pyridines have been widely used in the nuclear industry for the retention of radioactive iodine compounds from coolant release and ventilation systems. These carbons are required to remove very low levels of iodine and its compounds with high efficiency from gas streams, which in some cases could be of very high humidity. The performance of a carbon has been found to vary with manufacture and by aging. Billinge et al. (26) compared the efficiency of potassium iodide—impregnated coconut and coal-based charcoals for trapping radioactive methyl iodide from gas-cooled nuclear reactors. The coal-based impregnated charcoal was better than the impregnated coconut charcoal, which in turn was better than the impregnated activated charcoal. The trapping of methyl iodide was related to the associated oxygen in carbons. The oxygen evolved as carbon dioxide on out-gassing reduced the retention efficiency more than the oxygen evolved as carbon monoxide. The impregnated potassium iodide reacted with the oxygen groups on carbons and modified their desorption behavior, thereby improving the efficiency to retain radioactive methyl iodide.

Impregnation with pyridine and picoline has been used as a treatment to keep ASC Whetlerites from degrading in the presence of moisture (29). Activated carbons impregnated with Cu^{2+}, Ag^+, Cr^{6+}, NH^{4+}, and CO_3^{2-} were found to be efficient adsorbates for arsene, hydrogen cyanide, cynogen chloride, chloroform, and phosgene. Baker and Poziomek (27) modified adsorptive properties of coal-based active carbons by impregnating them with pyridine, 4-vinylpyridine, 4-aminopyridine, 4-cynopyridine, and 4-n-propyl-pyridine. The carbons were impregnated by volatile impregnants in a rotary evaporator and the amounts impregnated determined by the difference between the initial and the final weight of the carbon. The nonvolatile impregnants were adsorbed on the carbon surface from alcohol solutions. The adsorption capacity of the impregnated charcoals were compared for carbon tetrachloride and cynogen chloride under dynamic conditions and for methyl iodide under equilibrium conditions.

Impregnation of a small percentage of pyridines decreased the adsorption of carbon tetrachloride by a small percentage (Table 1). Increasing amounts of impregnants decreased the adsorption of carbon tetrachloride. In the case of cynogen chloride, the reactivity of the carbons increased with increase in the amount of a given impregnant. The reactivity, however, varied from one pyridine to another but not necessarily in the order of their expected neucleophilicities or basicities (Table 2). The reactivity of the charcoals impregnated with the five pyridines toward methyl iodide did not parallel their reactivity toward cynogen chloride but was rather in the following order:

Pyridine > 4-vinylpyridine > 4-aminopyridine >

4-cynopyridine > 4-n-propylpyridine

(Table 2). When the same charcoal sample was impregnated with 4-aminopyridine and 4-vinylpyridine to different degrees, the retention of methyl iodide increased (from 53 to 75%) with increase in the

Table 1 Influence of Impregnation of Carbon on the Relative Adsorption of Carbon Tetrachloride

Impregnant	Loading (mmol/g)	Relative CCl_4 adsorption (%)
None	0	100
Pyridine	0.253	96
4-Cyanopyridine	0.192	95
4-Vinylpyridine	0.190	93

Source: Baker and Poziomek (27). Reproduced with permission from Pergamon Press.

concentration of 4-aminopyridine on the surface of the carbon but remained more or less unchanged (around 71%) with increase in the concentration of 4-vinylpyridine on the carbon surface (Table 3).

Baker and Poziomek (27) observed that the charcoal impregnated with 4-vinylpyridine showed different behavior in many respects. First, it showed higher adsorption of cynogen chloride (0.605 mmol/g) compared to pyridine-impregnated charcoal (0.482 mmol/g), although 4-vinylpyridine is less basic than pyridine. Second, the 4-vinylpyridine-impregnated charcoal failed to result in any weight loss on heating at 150°C under 3 torr pressure for many hours compared to a 5% weight loss in the case of 4-n-propylpyridine (which has a boiling point similar to that of 4-vinylpyridine). Third, 4-vinylpyridine could be removed from the charcoal surface only to the extent of 66% against 86% for n-propylpyridine when subjected to soxhlet extraction with ethanol for 6 hr. On the basis of this evidence, these workers (27) suggested that 4-vinylpyridine undergoes adsorption polymerization when adsorbed on activated charcoal. To further substantiate their views these workers studied the adsorption of cynogen chloride on an active carbon impregnated with poly-4-vinylpyridine. This sample adsorbed 0.073 mmol CNCl/g, which was comparable with the control carbon sample (0.085 mmol/CNCl/g), indicating that the nitrogens of the polymer were not available for reaction with cynogen chloride. This contrasted sharply with the results of cynogen chloride adsorption on the carbon sample impregnated with 4-vinylpyridine (0.200 mmol CNCl/g), which indicated that the nitrogens of the adsorption polymer were available for reaction. The 4-vinylpyridine—impregnated carbon could adsorb only two-thirds of the amount of water adsorbed by activated carbon or any of the other impregnated carbons.

Table 2 Influence of Impregnation of Carbon on the Adsorption of Cynogen Chloride

Impregnant	Amount impregnated (mmol/100 g)	Saturation adsorption capacity for CNCl₃ (mmol/100 g)	Ratio of additional CNCl adsorbed to impregnant (mmol)
None	0	85	0
4-Aminopyridine	212	272	0.882
Pyridine	253	207	0.482
4-Vinylpyridine	190	200	0.605

Source: Baker and Poziomek (27). Reproduced with permission from Pergamon Press.

Table 3 Influence of Impregnation of Charcoal on the Retention of Methyl Iodide

Impregnant	Concentration of impregnant (mmol/g)	Methyl iodide retained (%)
4-Vinylpyridine	0.048	76.5
	0.096	73.3
	0.190	71.3
	0.286	77.0
	0.475	77.7
4-Aminopyridine	0.053	53.5
	0.106	57.5
	0.212	61.7
	0.318	69.2
	0.530	75.5
Pyridine	0.253[a]	80.2
4-Cynopyridine	0.192[a]	54.8
4-n-Propylpyridine	0.165[a]	48.7
4-Vinylpyridine	0.190[a]	71.3
4-Aminopyridine	0.212[a]	61.7

[a]Corresponds to a 2% by weight loading on the charcoal.
Source: Baker and Poziomek (27). Reproduced with permission from Pergamon Press.

Barnir and Aharoni (30) compared the adsorption of cynogen chloride on active carbon before and after impregnation with Cu^{2+}, Cr^{6+}, Ag^+, and NH_3 in a given ratio. The adsorption of cynogen chloride, which was reversible in the case of active carbon, became irreversible after impregnation, although the adsorption capacity did not show any increase. Thus whereas cynogen chloride was the main gas evolved on heating the active carbon treated with cynogen chloride, carbon dioxide was mainly desorbed from the impregnated charcoal. The presence of moisture in the carbon also hindered the desorption of cynogen chloride and enhanced the desorption of carbon dioxide. It was postulated that the cynogen chloride was adsorbed both physically and chemically on the surface of the carbon as well as on the additive (impregnant) surface. The cynogen chloride chemisorbed on the impregnant surface reacted faster with water sorbed on the carbon or linked with the impregnating material producing carbon dioxide and ammonium chloride:

$$CNCl + 2H_2O \rightarrow CO_2 + NH_4Cl$$

This process regenerated the surface of the impregnant where more cynogen chloride could be chemisorbed.

Reucroft and Chion (31,32) compared the adsorption behavior of BPL-activated carbon with ASC Whetlerite (obtained by impregnating BPL-activated carbon with Cu^{2+}, CrO_4^{2-}, and Ag^+) and ASB carbons (prepared by impregnating BPL with Cu^{2+} and BO_3 in different mole ratios) for chloroform, phosgene, cynogen chloride, and hydrogen cyanide using a gravimetric adsorption system. The adsorption isotherms were analyzed in terms of the Dubinin–Polanyi equation. A comparison of the affinity coefficients (β_{ex}) calculated from the slopes of the experimental isotherms with theoretical values (β_{th}) for adsorption of chloroform and phosgene (Table 4) showed about 17% higher adsorption than predicted on a theoretical basis for all carbons except ASC Whetlerite—impregnated carbons. In the case of the latter carbon the isotherm was not linear in the low-pressure region and consequently (β_{ex}) and (β_{th}) for adsorption of cynogen chloride for the BPL and impregnated BPL—activated carbon (Table 5) were about 20% higher than predicted for each of the three ASB-impregnated carbons. The ASC Whetlerite carbon isotherm did not obey the Dubinin–Polanyi equation and therefore the (β_{ex}) could not be obtained. The impregnation of the carbon generally lowered the micropore volume (W_0) in all cases, which was attributed to some pores being blocked or occupied by impregnating species.

The impregnated carbons showed both chemisorption and physisorption, the chemisorption being more pronounced in case of phosgene, cynogen chloride, and hydrogen cyanide on ASC and ASB carbons compred to BPL-activated carbon. Both ASC and ASB

Table 4 Adsorption Parameters of Dubinin–Polanyi Equation for Nonimpregnated and Impregnated Activated Carbons

$\beta_{th} = 0.78$

Adsorbent	CHCl₃ isotherm		COCl₂ isotherm		$\dfrac{\beta \exp K_{CHCl_3}}{(K_{COCl_2})^{\frac{1}{2}}}$
	W_0 (cm^3/g)	$k \times 10^8$ $(cal^{-2}mol^2)$	W_0 (cm^3/g)	$k \times 10^8$ $(cal^{-2}mol^2)$	
BPL-activated carbon	0.43	2.50	0.43	3.02	0.913
ASC Whetlerite	0.35	2.90	0.36	—	—
ASB (0.1 H₃BO₃:1)	0.35	2.25	0.36	2.67	0.918
ASB (0.5 H₃BO₃:1)	0.34	2.26	0.35	2.68	0.918
ASB (1 H₃BO₃:1)	0.34	2.25	0.35	2.68	0.916
ASB (2 H₃BO₃:1)	0.32	2.27	0.32	2.67	0.922

Source: Chion and Reucroft (31). Reproduced with permission from Pergamon Press.

Table 5 Dubinin–Polanyi Adsorption Equation Parameters for Nonimpregnated and Impregnated Active Carbon

Adsorbent	CNCl $\beta_{th} = 0.53$			HCN $\beta_{th} = 0.305$		
	W_0 (cm^3/g)	$k \times 10^8$ (cal^{-2}mol^2)	β exp, $\dfrac{K_{CHCl_3}}{(K_{CNCl})^{\frac{1}{2}}}$	W_0 (cm^3/g)	$k \times 10^8$ (cal^{-2}mol^2)	β exp, $\dfrac{K_{CHCl_3}}{(K_{HCN})^{\frac{1}{2}}}$
BPL-activated carbon	0.42	6.11	0.640	0.42	14.9	0.410
ASC Whetlerite	0.36	—	—	0.35	—	—
ASB (0.1 H$_3$BO$_3$:1)	0.36	6.80	0.685	—	—	—
ASB (1 H$_3$BO$_3$:1)	0.35	4.71	0.691	0.14	11.4	0.444
ASB (2 H$_3$BO$_3$:1)	0.32	4.81	0.687	0.33	11.3	0.448

Source: Reucroft and Chion (32). Reproduced with permission from Pergamon Press.

carbons retained appreciable amounts of the three adsorbates after evacuation at 150°C, the amount retained being more in ASC Whetlerite. The chemisorbed amount of cynogen chloride depended on the highest initial exposure pressure of cynogen chloride, whereas the adsorption of hydrogen cyanide was almost independent of this pressure. The adsorptive capacity of these carbons decreased in the following order:

ASC Whetlerite > ASB carbons > BPL-activated carbon

This indicated that CrO_4^{2-} (or a complex form of Cr^{2-} with other ions) or Ag^+ in ASC Whetlerite produced a greater degree of chemisorptive interactions and was primarily responsible for the higher affinity toward these adsorbates. It was also found that the chemisorption increased in the order of the decreasing size of the adsorbate molecule:

Phosgene chemisorption > cynogen chloride chemisorption >

hydrogen cyanide chemisorption

for all the carbons.

The influence of temperature on the adsorption—desorption isotherms of hydrogen cyanide on BPL-activated and ASC Whetlerite-impregnated carbons was studied by Freeman et al. (33). The Dubinin—Polanyi equation gave 0.42 and 0.35 ml/g as the micropore volume for BPL and ASC Whetlerite carbons at all temperatures between 0 and 122°C. The adsorption decreased with increasing temperature for both carbons but was greater for ASC Whetlerite at lower pressures at all temperatures. This was attributed to adsorption at the active sites provided by the ionic impregnants. At higher pressure the larger surface area of the BPL-activated carbon was the dominant factor in adsorption. The Whetlerite showed chemisorption, the average value of chemisorption (0.00102 mol HCN/g C) being almost equivalent to the number of moles (0.00131 mol/g) of the impregnant, which clearly indicated that the additional adsorption at the lower pressure and the chemisorption in case of Whetlerite were due to the ionic impregnants. The heats of adsorption calculated using the Clausius—Clapeyron (Fig. 1) equation coverages at high both for BPL and Whetlerite carbons were in good agreement with the heat of liquefaction of hydrogen cyanide ($\tilde{=}$ 6.5 kcal/mol). The heats of adsorption for the Whetlerite, however, were higher at low surface coverages, indicating chemisorption of hydrogen cyanide on the ionic impregnant sites.

The adsorption of water vapor and its binary mixtures with hydrogen cyanide on ASC Whetlerite—impregnated and BPL-activated carbons (34) showed that ASC Whetlerite adsorbed 10 times more

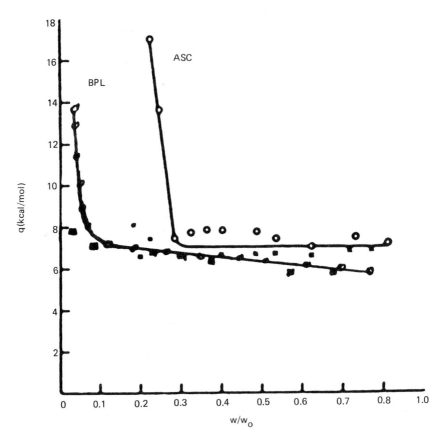

Figure 1 Isoteric heats of sorption for HCN on BPL-activated carbon and ASC Whetlerite carbon. [From Freeman et al. (33). Reproduced with permission from Pergamon Press.]

water than BPL-activated carbon at low pressures (Fig. 2), but the isotherms crossed each other at a relative pressure of about 0.7 and the saturation adsorption was more for the BPL-activated carbon. This was attributed to the adsorption being determined by the greater affinity of impregnant ions at lower relative pressures and to pore volume considerations at higher relative pressures. In ASC Whetlerite a part of the pore volume is occupied by the impregnant ions and consequently the saturation adsorption capacity is less than that for BPL-activated carbon. The adsorption of hydrogen cyanide was greater for both carbons when they had preadsorbed water and this amount exceeded the amounts of water and hydrogen cyanide

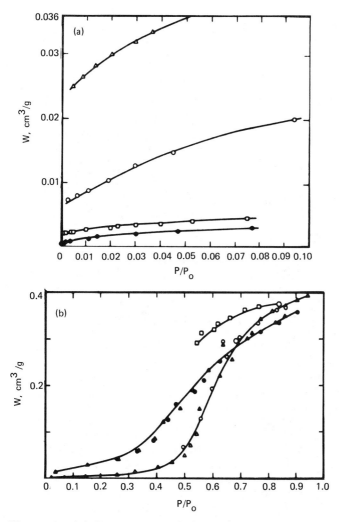

Figure 2 (a) Low pressure adsorption and desorption isotherms of water on BPL-activated and ASC Whetlerite—impregnated carbons. Open triangle, desorption of H_2O on ASC Whetlerite; open circle, adsorption of H_2O on ASC Whetlerite; open square, desorption of H_2O on BPL; closed circle, adsorption of H_2O on BPL. (b) High-pressure adsorption and desorption isotherms of water on BPL-activated and ASC Whetlerite—impregnated carbons. Open triangle, adsorption run No. 1 of H_2O on BPL; open circle, adsorption run No. 2 of H_2O on BPL; open circle, adsorption run No. 3 of H_2O on BPL; open square, desorption of H_2O on BPL; closed triangle, adsorption run No. 1 of H_2O on ASC Whetlerite; closed circle, adsorption run No. 2 of H_2O on ASC Whetlerite. [From Freeman and Reucroft (34). Reproduced with permission from Pergamon Press.]

Table 6 Adsorption Capacity of Impregnated Carbons for H_2S, SO_2, and HCN

	Weight uptake ($\mu g/cm^2$)		
Impregnant	H_2S	SO_2	HCN
None	50	220	120
$AgNO_3$	210	—	—
$Na_2CR_2O_7$	310	480	120
Cu (5%)	360	470	320
Cu (5%) + $Na_2CR_2O_7$	760	960	890

Source: Capon et al. (35).

adsorbed separately. The chemisorption of hydrogen cyanide on ASC Whetlerite was also higher in the presence of water.

An activated charcoal cloth prepared by carbonization of viscose rayon in the presence of $CNCl_2$ when impregnated with oxidizing agents such as $KMnO_4$, Na_2CR_2O, and ClO_2, with an organic tertiary amine, triethylenediamine (TEDA), and $AgNO_3$ from aqueous solutions, showed enhanced adsorption capacity (reactivity) toward low-boiling pollutant gases such as SO_2, NO_2, H_2S, HCN, and CNCl (35). The exact nature of the reaction was found to depend on the nature of the impregnant and the oxidizing agent (Table 6). The presence of $CuCl_2$ as the impregnant enhanced the adsorption capacity for all gases while the presence of organic amine (TEDA) enhanced considerably the adsorption of CNCl (Table 7). The adsorption capacity of the cloth for physically adsorbed molecules, however, was not significantly impaired after impregnation.

Jonas (36) suggested that the removal of gas or vapor from a flowing stream by a bed of chemically impregnated activated carbon granules, in which both physical adsorption and chemical reaction occur, could be represented as a series of seven consecutive steps. This sequence of steps, which is initiated when the flowing gas air stream approaches the first layer of carbon in the bed, includes mass transfer, surface diffusion, intragranular diffusion, physical adsorption, chemical reaction, and surface renewal. Jonas also deduced mathematical equations and applied logic analysis to find the rate-determining step in the sequence.

An upsurge in the use of activated carbons for treatment of wastewater and effluent from industries and the regeneration of the spent-up carbon by steam activation resulted in a considerable

Table 7 Adsorption Capacity of Impregnated Carbons for NO_2 and $CNCl$

Impregnant	Weight uptake of NO_2 ($\mu g/cm^2$)	Impregnant	Weight uptake of $CNCl$ ($\mu g/cm^2$)
None	10	None	60
$KMnO_4$	30	Cu (5%)	160
Cu (14%)	110	Cu (5%) + $Na_2Cr_2O_7$ + pyridine	320
Cu (14%) + $KMnO_4$	160	Cu (5%) + $Na_2Cr_2O_7$ + TEDA (1%)	860
Cu (14%) + ClO_2	520	Cu (5%) + $Na_2Cr_2O_7$ + TEDA (3%)	1530

Source: Capon et al. (35).

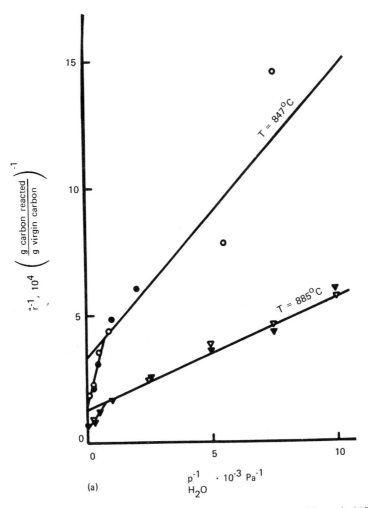

Figure 3 Steam—carbon reaction data for (a) 847 and 885°C (phenol loaded sample, ● ▼; untreated sample, ○ ▽) and (b) 915 and 950°C (phenol-loaded sample, ▲ ■; untreated sample, △ □). [From Krebbs and Smith (41). Reproduced with permission from Pergamon Press.]

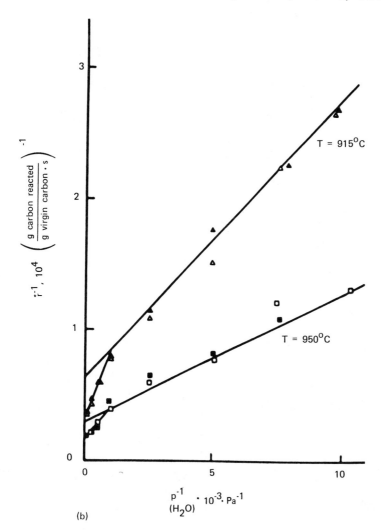

(b)

Figure 3 (continued)

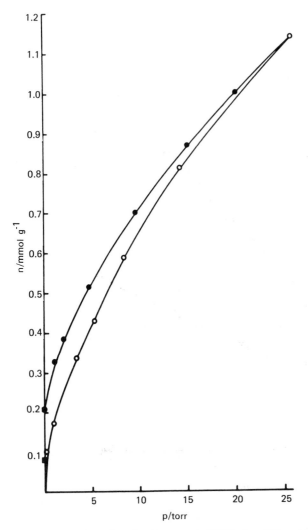

Figure 4 Sorption isotherm at 25°C for PH_3—carbon cloth (sample outgassed at 110°C). Open circle, sorption; closed circle, desorption; close square, desorption at 110°C. [From Hall et al. (42). Reproduced with permission from Pergamon Press.]

amount of research relating to modification of activated carbons by the adsorbate residues left on the surface of these carbons. Umehra et al. (38.39) used sodium dodecyl benzene sulfonate (DBS) as the adsorbate and observed that after thermal regeneration the DBS residue increased the rate of steam—carbon reaction by one order of magnitude, which was attributed to the catalytic effect of metal atoms. Chihara et al. (37) studied activated charcoals having sucrose adsorbed on them, but the residue left after thermal regeneration was so small that it was not possible to conclude if the residue modified the subsequent steam regeneration. Reichenberger (40) adsorbed phenol on an activated carbon and examined the rate of steam regeneration in the temperature range 700—800°C and at steam pressures 10^4—10^5 Pa and observed a 25% increase in carbon gasification. Krebbs and Smith (41) used TGA to study the kinetics of steam activation over a wide range of steam pressures (10_2—10_4 Pa) and temperatures between 850 and 950°C using an activated carbon before and after adsorption of phenol. The rates of steam—carbon reaction were essentially the same for virgin and phenol-loaded carbons (Fig. 3). These workers were of the view that when the residue from thermal decomposition was purely organic, it did not affect the steam gasification rates.

Hall et al. (42) recently carried out detailed and comprehensive investigations into the adsorption of phosphine (PH_3) and water vapor on an activated carbon cloth before and after impregnation with $AgNO_3$ and $Cu(NO_3)_2 \cdot 3H_2O$ from aqueous solutions. The techniques used included adsorption by gravimetric measurements, surface characterization using XPS, SEM (scanning electron microscopy), ED-X-ray analysis (energy dispersive X-ray analysis), and IR spectroscopy to study the corresponding gas phase. The adsorption isotherms in the pressure range 0—25 torr (Fig. 4) indicated that raw cloth had strong affinity for PH_3. Impregnation enhanced the sorption capacity (Fig. 5) of the cloth above a certain pressure limit, which depended on the impregnant content of the carbon cloth, but apparently reduced the adsorption capacity at lower pressures. The isotherms exhibited adsorption—desorption hysteresis with the desorption curve failing to meet the adsorption curve even when the pressure approached zero. However, a second run on the untreated cloth was reversible and did not show hysteresis. The failure of the hysteresis loop to close even after prolonged evacuation indicated that the raw cloth as well as the impregnated samples retained an appreciable amount of PH_3, which was bonded strongly. The amount of PH_3 retained compared to the raw cloth was as much as five times greater when the impregnant was $AgNO_3$ and three times greater when it was $Cu(NO_3)_2 \cdot 3H_2O$. The enhancement in the adsorption of PH_3 was more when the impregnant was $Cu(NO_3)_2$ than when it was $AgNO_3$ with the same molar concentrations of the metal

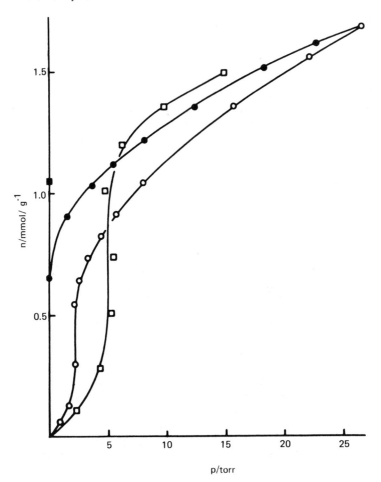

Figure 5 Sorption isotherms at 25°C for $PH_3-CU(NO_3)_2 \cdot 3H_2O$ freshly impregnated carbon cloth (sample outgassed at 25°C) and PH_3-AgNO_3 freshly impregnated carbon cloth (>5% Ag, sample out-gassed at 25°C). For Cu: open circle sorption; closed circle, desorption. For Ag: open square, sorption; closed square, desorption. [From Hall et al. (42). Reproduced with permission from Pergamon Press.]

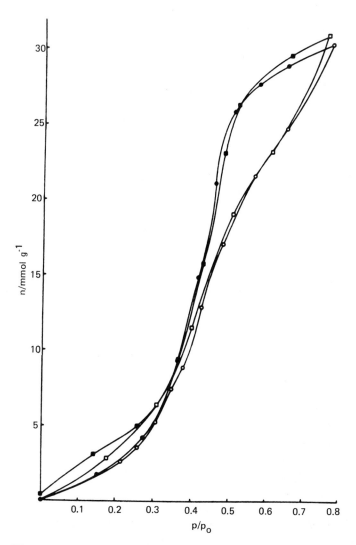

Figure 6 Adsorption isotherms at 25°C for H_2O on raw and on $AgNO_3$-impregnated carbon cloth (= 5% Ag samples outgassed at 25°C). For raw cloth: open circle, adsorption; closed circle, desorption. For Ag impregnated cloth: open square, adsorption; closed square desorption. [From Hall et al. (42). Reproduced with permission from Pergamon Press.]

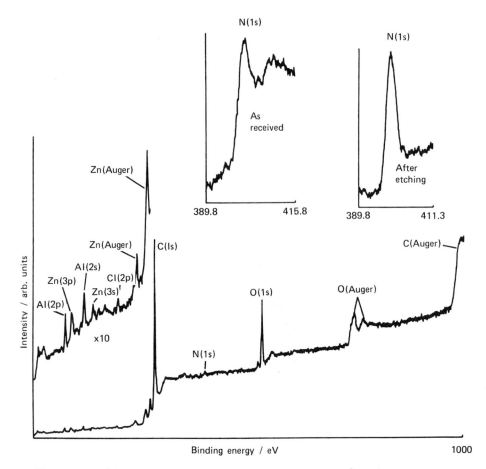

Figure 7 Widescan X-ray photoelectron spectrum of carbon cloth, plus insets of a high-sensitivity scan of the 0–280 eV region with a high resolution scan of the N (1s) signal before and after 25 min etching. [From Hall et al. (42). Reproduced with permission from Pergamon Press.]

ions. Dubinin–Astakhov plots (when $a = 1$) were linear for raw carbon cloth over all pressure ranges but showed deviations from linearity in the low-pressure region for the impregnated carbon cloth. This indicated different mechanisms in adsorption at low- and high-pressure regions.

Water vapor adsorption isotherms at 25°C were type V of the BET classification on raw carbon cloth with a marked hysteresis

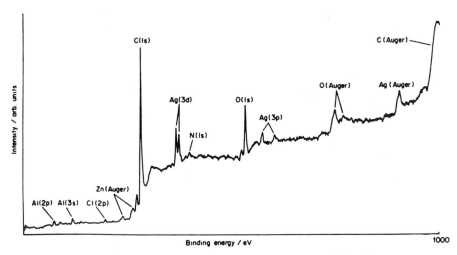

Figure 8 Widescan X-ray photoelectron spectrum of $AgNO_3$-impregnated carbon cloth (= 5% Ag) after 1 min etching. [From Hall et al. (42). Reproduced with permission from Pergamon Press.]

loop. Impregnation with $AgNO_3$ had little effect on the relative pressures studied except below about 0.3 where the uptake was enhanced by as much as 25% during adsorption and even more during desorption. The isotherm also showed a second hysteresis loop (Fig. 6), which failed to close as the pressure approached zero. The nonclosure of the hysteresis loop indicates the amount of water adsorbed irreversibly. The sorption capacity of the $AgNO_3$-impregnated carbon cloth for PH_3 was reduced markedly by the presence of water when exposed to a mixture of PH_3 and water vapors.

Infrared studies of the gaseous species showed the presence of N_2O when the $AgNO_3$-impregnated carbon cloth was exposed to PH_3, although no such IR-active species were observed when raw carbon cloth was exposed to PH_3. The presence of water reduced the amount of N_2O produced in the case of the impregnated cloth. Such species were also observed when bulk $AgNO_3$ or $Cu(NO_3)_2 \cdot 3H_2O$ was reacted with PH_3. This indicates that PH_3 reacts chemically with the impregnant on the carbon cloth surface. SEM and ED-X-ray analysis showed nodular growth of virtually pure silver together with $AgNO_3$ on the surface of the $AgNO_3$-impregnated carbon cloth. After exposure to PH_3, phosphorus was largely confined to areas away from these nodules. XPS spectra (Figs. 7 and 8) showed a layer of $AgNO_3$ together with some metallic silver on the $AgNO_3$-impregnated

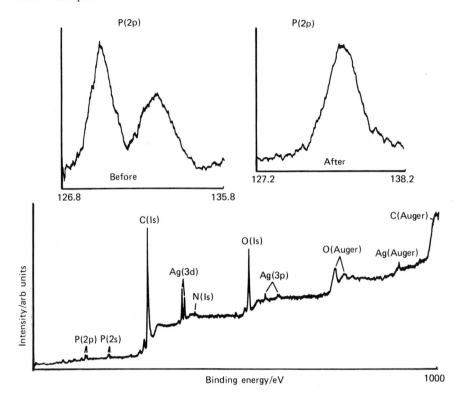

Figure 9 Widescan X-ray photoelectron spectrum of AgNO$_3$-impregnated carbon cloth ($\simeq$ 5% Ag) after exposure to PH$_3$, plus insets of the P(2p) signal before and after 5 days of air exposure. [From Hall et al. (42). Reproduced with permission from Pergamon Press.]

cloth. After exposure to PH$_3$ the formation of at least two phosphorous compounds, one of which was air sensitive, was indicated (Fig. 9).

Kaistha and Bansal (42a) impregnated an activated carbon with sodium ions by treating the oxidized carbon with appropriate quantities of 0.1 M solutions of NaHCO$_3$. The treatments resulted in the impregnation of appreciable amounts of sodium ions, the amount being larger in the case of NaOH treatment. The adsorption isotherms of aqueous oxalic acid on the oxidized and the ion-impregnated carbon (42a) showed a considerable increase in the case of the impregnated carbons, the extent of increase depending on the increase in the degree of impregnation. For example, the amount of oxalic acid adsorbed increased from 0.75 mmol/g for the oxidized carbon to 1.52 and 1.80 mmol/g after impregnation. Furthermore, the amount of oxalic acid adsorbed on the impregnated carbons was very close to the amount of sodium ions impregnated on the carbon surface, indicating that the adsorption of oxalic acid involved an exchange mechanism. This received further support from the observation that when the sodium ions were removed from the carbon surface by treatment with acid and then washing repeatedly with hot distilled water, the adsorption of oxalic acid superimposed almost completely the adsorption on the nonimpregnated carbon at all concentrations. The impregnated carbons had smaller surface area than the original oxidized carbon. When the amounts of oxalic acid adsorbed were calculated per unit surface area, it was found that the impregnated carbons showed a much more enhanced adsorption of oxalic acid, indicating that impregnated carbons have a better efficiency for the removal of oxalic acid from aqueous solutions.

5.3 MODIFICATION OF CARBONS BY HALOGENATION

The treatment of activated carbons, charcoals, and carbon blacks with halogens has been studied by several investigators with a view to modify their surface character. The adsorption of the halogens is both physical (43,44) and chemical and proceeds through several mechanisms, including addition at the unsaturated sites (45—55), exchange with chemisorbed hydrogen (48—55), and surface oxidation of the carbon (45,46,56), depending on the nature of the carbon surface, the oxygen and hydrogen contents of the carbon, the experimental conditions, and the nature of the halogenating species. The halogen fixed on the surface of carbons in the form of carbon—halogen surface compounds is thermally highly stable and can be eliminated on heat treatment in vacuum up to 1000°C if the carbon has no residual hydrogen. However, a part of the halogen could be exchanged with OH groups on treatment with alkali hydroxides and with NH_2 groups on heat treatment with ammonia gas.

Reyerson and co-workers (44,57,58), Alekseeviskii and Likharev (59), and Emmett (60) treated activated carbons and charcoals with chlorine and bromine in the gaseous states and observed that appreciable amounts of the halogens were fixed irreversibly on the carbon surface. The carbon—chlorine or the carbon—bromine complex was very stable and could not be decomposed on high-temperature treatment in vacuum or by refluxing with alkalis. Alekseeviskii and

Table 8 Amount of Chlorine Fixed by Different Carbons

Carbon	Oxygen content (%)	Hydrogen content (%)	Chlorine fixed (meq/g)
Sugar charcoal			
Original	32.52	3.38	6.70
400°-outgassed	16.86	3.37	5.35
500°-outgassed	12.76	3.07	4.21
700°-outgassed	7.48	2.58	2.87
1000°-outgassed	1.42	1.15	2.70
1200°-outgassed	0	1.02	2.57
Coconut shell charcoal			
Original	15.45	1.24	11.35
700°-outgassed	7.25	1.05	5.63
1000°-outgassed	0.52	0.62	3.50
1200°-outgassed	0	0.35	2.31

Source: Puri et al. (51).

Likharev, however, found that the carbons activated in chlorine were better than many of the commercially available materials and suggested that treatment with chlorine gas could be used as an activating process.

The interaction of halogens with carbon blacks was studied by Ruff (61), Boehm et al. (62), and Rivin and Aron (63). Boehm et al. (62) observed that the chlorine fixed on treatment at 400–500°C was roughly equivalent to the initial hydrogen content of the carbon black and was available for reactions with ammonia at elevated temperatures as well as with potassium hydroxide and sodium cyanide on fusion. Rivin and Aron (63), however, found that the reactivity as well as the stability of the carbon–halogen surface compounds formed on treatment with gaseous halogens or from solutions was in the order Cl > Br > I. Thus while all of the iodine could be dissociated, only 80% of the bromine and 60% of the chlorine could be dissociated in the presence of suitable neucleophiles. Puri and Bansal (52), during their interaction of carbon blacks with chlorine, found that only a part of the chemisorbed chlorine could be recovered on boiling under reflux with sodium hydroxide solution (2.5 N) or on heat treatment in vacuum at 1000°C.

Puri and co-workers (51, 64) carried out systematic investigations into the formation and properties of carbon–chlorine surface compounds formed on sugar and coconut charcoals associated with varying amounts of oxygen and hydrogen, in the temperature range 35–600°C and at varying pressures of the chlorine. The interaction resulted in the fixation of appreciable amounts of chlorine, the magnitude depending on the temperature of treatment up to 450°C and the oxygen and hydrogen contents of the charcoals. The maximum amounts of chlorine fixed at 450°C were 24% in sugar and 40% in coconut charcoal, and this amount decreased gradually and appreciably on outgassing the charcoals at gradually increasing temperatures (Table 8). The kinetic measurements showed that the reaction between carbon and chlorine was first order.

Puri and Bansal (46) showed that the fixation of chlorine occurs partly by addition at the unsaturated sites vacated by the combined oxygen (47,48) and partly in exchange (substitution) for hydrogen (Table 9). Since original charcoals which had not been outgassed contained few unsaturated sites, the bulk of the chlorine was fixed by exchange with hydrogen (Table 9), as observed by Boehm et al. (62). However, when these charcoals were outgassed, the hydrogen content decreased and unsaturated sites were created by the evolution of oxygen; the chlorine fixed now was more by addition than by substitution. In the case of the chlorination of 1000° outgassed charcoals, which—although they retained about 30% of the total hydrogen content of the original charcoals—were not able to fix chlorine by substitution (Table 9), almost the entire amount of chlorine was

Table 9 Fixation of Bromine by Various Samples of Charcoal and Carbon Black before and after Fixation of Chlorine (values in meq/g)

Sample	(a) H$_2$ present in sample	(b) Cl$_2$ fixed at 450	Unsaturation as determined by Br$_2$ uptake from aqueous solution		(e) Cl$_2$ fixed by addition [(c−d)]	(f) Cl$_2$ fixed by substitution [(b) − (e)]
			(c) Before chlorination	(d) After chlorination		
Sugar charcoal						
Degassed at 400°	33.78	6.35	3.12	1.68	1.44	4.91
Degassed at 700°	25.80	3.35	3.76	0.63	3.13	0.23
Degassed at 1000°	11.50	3.09	3.88	0.85	3.03	0.06

Degassed at 1200°	10.20	3.07	3.80	0.86	2.94	0.13
Coconut charcoal						
Degassed at 400°	20.10	10.28	2.40	0.65	1.75	8.53
Degassed at 700°	10.10	5.63	3.16	0.93	2.23	3.4
Degassed at 1000°	6.40	3.39	3.14	1.14	2.00	1.39
Degassed at 1200°	6.10	2.31	3.16	1.24	1.92	0.39
Carbon black						
Philblack-A	3.50*	3.57	0.30	0	0.30	3.27
Spheron-9	5.30	3.31	0.43	0	0.43	2.88
Spheron-6	4.90	3.97	0.62	0	0.62	3.35
Spheron-4	3.80	3.14	0.45	0	0.45	2.69
Spheron-C	2.50	2.65	0.70	0	0.70	1.95
ELF-0	3.80	4.40	0.12	0	0.12	4.34
Mogul	3.00	3.15	0	0	0	3.15

Source: Puri and Bansal (46).

fixed by addition at the unsaturated sites. This is due to the fact
that this hydrogen, which could not be desorbed on evacuation at
1000°C, was present in charcoals at less active sites (65) dispersed
in the body of the carbon surface or held up in the extremely fine
microcapillary pores and was not readily available for exchange with
chlorine. This view was further supported by the fact that coconut
charcoals, which have relatively coarser porous structure than sugar
charcoals, could fix more chlorine by substitution. The carbon
blacks, which were essentially nonporous, fixed almost all of their
chlorine by exchange with hydrogen (Table 9). These views regarding the fixation of chlorine were substantiated by the chlorination of
anthracites by Walker et al. (66) and by Puri and co-workers by
their reactions of charcoals with chlorine water (45), a mixture of
HCl and H_2O_2 (67), and a mixture of oxygen and chlorine at 400°C
(68).

Tobias and Soffer (69) carried out an interesting study of the
stepwise chlorination of a carbon cloth (TCM 128), a carbon black
(Continex N-110), and a graphitized carbon black (Carbopak B)
after outgassing them at 1000°C to eliminate interference of chemisorbed oxygen to chlorine adsorption. The chlorination was found to
involve fixation of chlorine and the liberation of hydrochloric acid
(Table 10). These workers are of the view that the interaction between chlorine and carbon involved three processes: addition at the
olefinic double bonds, exchange with chemisorbed hydrogen, and dehydrogenation of the carbon. The ratio of chlorine fixed to hydrochloric acid formed (R = Cl_2 uptake/HCl formed) was taken as a
measure of the particular process which dominates in a given step.
Thus in the initial stages when the ratio R was higher the most preferred process was the addition of chlorine at the double bond sites
followed by exchange with hydrogen and the least favored dehydrogenation. Assuming that when $R > 1$, which was the situation at the
first chlorination, the dehydrogenation is negligible and assuming
that addition is negligible when $R < 1$, these workers calculated the
amounts of chlorine fixed by addition, by exchange, and by dehydrogenation processes (Table 12). The results clearly showed that
although clorine addition was the first to occur in both carbon black
and active carbon, the exchange reactions were more predominant in
the case of carbon blacks (46). The dehydrogenation reactions which
occurred to a negligible extent in carbon black were high in active
carbon.

To verify that chlorine exchanged with hydrogen these workers
(69) carried out a number of chlorination—hydrogenation cycles,
measuring the amounts of chlorine fixed and hydrochloric acid formed
after each cycle. Significant amounts of hydrochloric acid were produced during both the chlorination and hydrogenation steps together
with the fixation of the corresponding gaseous species (Table 12).
This indicated that there was an exchange of C—H bond by C-Cl

Table 10 Stepwise Chlorination at 773 K of Carbons after Degassing the Activated Charcoal TCM 128 and the Graphitized Black-Carbolak-8 after Degassing at 1273 K

Step	Cl_2 adsorption (mEq/g)	HCl release (mEq/g)	Ratio of Cl_2 uptake to HCl reformed
Continex N-110			
1	0.225	0.048	4.69
2	0.225	0.203	1.11
3	0.152	0.149	1.03
4	0.140	0.144	0.97
5	0.124	0.122	1.02
6	0.100	0.103	0.97
7[a]	0.135	0.126	1.07
8[a]	0.133	0.143	0.98
9[a]	0.141	0.134	1.05
Total			
TCM 128			
1	0.43	0.32	1.50
2	0.30	0.17	1.76
3	0.18	0.21	0.86
4	0.06	0.05	1.20
5	—	0.02	—
Total	1.02	0.77	1.32
Carbopack-B			
1	0.10	0.01	10
2	0.02	0.02	1
Total	0.12	0.03	4

[a]The cell temperature was cycled once between 773 and 473 K at 1.5 K/min. This enhanced the reaction rate considerably.
Source: Tobias and Soffer (69). Reproduced with permission from Pergamon Press.

bond during chlorination and exchange of C-Cl bond by C-H bond during hydrogenation. Furthermore, after some initial irreversible changes, the system behaved almost reproducibly upon repeated chlorination and hydrogenation cycles. Such reproducibility in fact was attained starting from the second cycle (Table 12), indicating that it was purely an exchange process.

The interaction of carbons with bromine involved the same mechanisms as the interaction with chlorine. Puri and co-workers (47,48) reacted charcoals and carbon blacks with bromine in aqueous solutions. The bromine uptake of the carbons increased with increase in

Table 11 Contributions of Various Reactions Responsible for Cl_2 Interaction with Carbons (data derived from Table 10)

Step	Addition (mEq/g)	Exchange (mEq/g)	Dehydrogenation (mEq/g)
Continex N-110 as received			
1	0.177	0.048	—
2	0.022	0.203	—
3	0.005	0.147	—
4	—	0.136	0.004
5	0.002	0.122	—
6	—	0.100	0.003
7	0.009	0.126	—
8	—	0.173	0.010
9	0.007	0.134	—
Total	0.222	1.153	0.017
TCM 128 as received			
1	0.16	0.32	—
2	0.13	0.17	—
3	—	0.18	0.03
4	0.01	0.05	—
5	—	0.02	0.02
Total	0.30	0.74	0.05
Carbopack-B			
	0.09	0.01	

Source: Tobias and Soffer (69). Reproduced with permission from Pergamon Press.

the temperature of degassing up to 700°C and remained unchanged thereafter. The uptake of bromine was attributed to addition at the ethylenic double bond sites, which were created by the elimination of that part of the associated oxygen that was evolved as carbon dioxide on evacuation (CO_2-complex). Since the entire amount of this complex was desorbed at temperatures below 800°C, the uptake of bromine remained more or less unchanged after this heat treatment temperature. Similar results were obtained by Stearns and Johnson (70) in their reaction of channel blacks with aqueous solutions of bromine.

Bansal et al. (56) reacted polymer carbons with aqueous solutions of bromine and observed that the reaction involved the fixation of bromine as well as the formation of hydrogen bromide (Table 13). The amount of bromine converted into hydrogen bromide was related to surface acidity (Fig. 10) while the amount fixed depended on the

Table 12 Chlorination—hydrogenation Cycles of Activated Carbon Cloth TCM-128, after Degassing at 1273 K (chlorination performed at 778 K; hydrogenation at 1073 K)

Treatment	Chlorine fixed (mEq/g)	Hydrogen fixed (mEq/g)	HCl released (mmol/g)
First cycle			
Chlorination	1.62		1.14
Hydrogenation		2.138	1.26
Second cycle			
Chlorination	2.26		1.92
Hydrogenation		2.14	1.47
Third cycle			
Chlorination	1.94		1.93
Hydrogenation		2.23	1.47
Fourth cycle			
Chlorination	2.13		1.93
Hydrogenation		2.14	1.47

Source: Tobias and Soffer (69). Reproduced with permission from Pergamon Press.

nature of the char and its history of formation. PVDC and Saran chars were prepared from polymers containing no oxygen as a part of their chemical structure, and consequently contained very little associated oxygen with them. These chars therefore had very little unsaturation of the type suggested by Puri et al. (47,48) and Stearns and Johnson (70). On the other hand, these chars were highly microporous; thus a large portion of the bromine was adsorbed in the microcapillary pores. PF (polyfurfuryl alcohol) and UF (urea formaldehyde) chars, which were prepared by the carbonization of oxygen-containing polymers, had unsaturated sites so that bromine could be fixed by addition. These views were supported by the fact that about 30% of adsorbed bromine could be released from brominated PVDC and Saran chars on heat treatment at 50°C (the adsorption was carried out at 30°C), whereas only less than 1% could be recovered from PF and UF chars (Table 14).

Brooks and Spotswood (71), during their reaction of chars obtained from bituminous coal with bromine dissolved in carbon tetrachloride, observed three different types of adsorbed bromine. A part of the adsorbed bromine which could be removed with boiling water or alcohol was attributed to addition compounds of phenanthrene or

Table 13 Hydrogen Bromide Formed and Bromine fixed on Treatment of Polymer Charcoals with Aqueous Solutions of Bromine

Sample	HBr formed (mEq/g)	bromine fixed (mEq/g)	NaOH neutralized (mEq/g)
PVDC-600	18.06	13.02	2.12
Saran-600	13.12	16.10	1.43
Saran-600 (steam activated at 850°C)	3.10	18.24	0.50
PF-140	15.24	21.70	1.83
PF-400	8.42	18.72	0.97
PF-600	2.06	10.04	0.42
PF-900	0.23	5.23	0.16
UF-400	6.07	17.01	0.76
UF-650	4.12	9.22	0.78
UF-850	1.10	5.54	0.07

Key: PVDC = polyvinylidene chloride carbon prepared at 600°C; Saran = saran carbon prepared at 600°C; PF = polyfurfuryl alcohol carbon prepared at temperature indicated; UF = urea formaldehyde resin carbon prepared at temperature indicated.
Source: Bansal et al. (56). Reproduced with permission from Pergamon Press.

anthracene type. The small amount of bromine which was not removed by water but could be recovered by hydrolysis with sodium hydroxide was ascribed to residual aliphatic or alicyclic structures. The rest of the adsorbed bromine, which was difficult to remove except under strongly alkaline conditons, was considered to be substituted for hydrogen in polycyclic ring systems. Watson and Parkinson (72) working with carbon blacks, and Puri et al. (49), working with charcoals and carbon blacks, observed that the interaction with bromine from carbon tetrachloride solutions was partly reversible and partly irreversible (Tables 15,16). The amount of bromide adsorbed reversibly was related to the specific surface area of the carbons (Fig. 11). The irreversibly adsorbed bromine involved both addition at the unsaturated sites and a partial exchange with the chemibound hydrogen. The interaction of charcoals and carbon blacks with bromine vapors (53) was also found to involve addition of bromine at

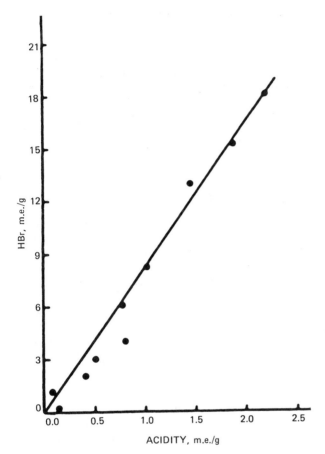

Figure 10 Surface acidity of polymer carbon in relation to hydrobromic acid formed. [From Bansal et al. (56). Reproduced with permission from Pergamon Press.]

the unsaturated sites, exchange with chemibound hydrogen, and dehydrogenation of the carbons in the case of charcoals, but only an
exchange reaction in the case of carbon blacks. The amount of bromine fixed was as high as 38% in the original sugar charcoal and
about 31% in Spheron-9.

The treatment of active carbons, charcoals, and carbon blacks
with iodine solutions in aqueous (73,74) and nonaqueous media does
not involve chemisorption of iodine or formation of carbon—iodine
surface structures. Consequently, iodine may not be in a position to
modify carbon surface properties. It may, however, be mentioned
that the adsorption of iodine, being purely physical, has been used

Table 14   Desorption of Bromine on Heat Treatment at Different
Temperatures

Sample identification	Bromine fixed (mEq/g)	Bromine desorbed on heat treatment at (mEq/g):			
		50°C	100°C	150°C	200°C
PVDC-600	18.02	5.41	6.10	7.01	7.52
Saran-600 (steam activated at 850°C)	18.24	5.04	6.91	7.41	8.82
PF-600	10.04	0.22	0.58	0.84	1.10
UF-650	9.22	0.14	0.57	0.72	0.95

Key: PVDC = polyvinylidene chloride carbon; PF = polyfurfuryl
alcohol carbon; UF = urea formaldehyde resin carbon.
Source: Bansal et al. (56). Reproduced with permission from Pergamon
Press.

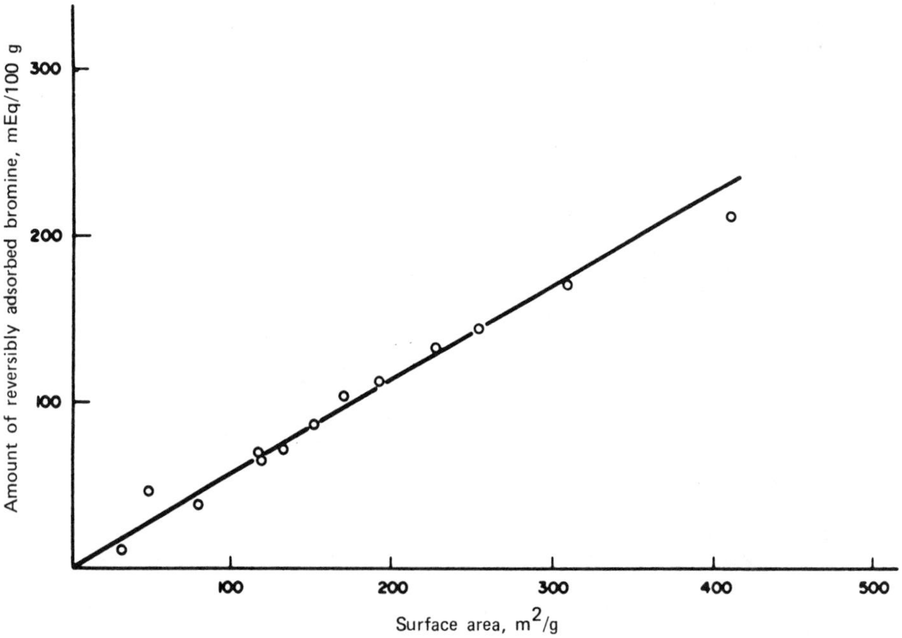

Figure 11   Reversibly adsorbed bromine against surface area.
[From Puri et al. (49). Reproduced with permission from Pergamon
Press.]

Table 15 Reversible and Irreversible Adsorption of Bromine by Carbon Blacks and Sugar Charcoal of Different Specific Surface Areas, Surface Unsaturations, and Hydrogen Contents

(1) Carbon	(2) Surface area (m²/g)	(3) Hydrogen content	(4) Bromine adsorbed reversibly	(5) Bromine adsorbed irreversibly	(6) HBr formed (bromine fixed by substitution)	(7) Surface unsaturation (bromine fixed by addition at unsaturated sites)	(8) Bromine unaccounted for [5 - (6 + 7)]
Mogul	308	480	171	132	23	16	93
Mogul-A	228	510	132	131	18	14	99
Elf-0	171	471	105	124	25	13	86
Spheron-C	253	332	144	148	19	70	59
Spheron-4	153	472	86	117	21	41	55
Spheron-9	116	620	65	118	23	42	53
Phiblack-A	46	350	26	32	11	11	10
Phiblack-I	116	242	68	57	16	21	20
Phiblack-E	135	310	72	61	13	22	26
Vulcan-SC	194	140	111	122	11	64	48
Phiblack-O	80	310	38	45	12	18	15
Kosmos-40	31	350	11	24	8	9	7
Carbolac	839	470	226	183	32	96	55
Sugar	412	1984	210	393	140	46	207

Source: Puri et al. (49). Reproduced with permission from Pergamon Press.

Table 16 Effect of Outgassing Carbon Blacks and Sugar Charcoal on Reversible and Irreversible Adsorption of Bromine[a]

(1) Carbon	(2) Surface area (m/g)	(3) Bromine adsorbed reversibly	(4) Bromine adsorbed irreversibly	(5) HBr formed (bromine fixed in substitution for hydrogen)	(6) Surface unsaturation (bromine fixed by addition at unsaturated sites)	(7) Bromine unaccounted for [4 - 5 (5 + 6)]
Mogul						
Original	308	171	132	23	16	93
Outgassed at 600°C	335	183	267	13	76	178
Outgassed at 1000°C	328	173	221	0	74	147
Carbolac						
Original	839	226	183	32	96	55
Outgassed at 600°C	815	232	291	20	168	103
Outgassed at 1000°C	760	201	272	11	170	91

	(2)	(3)	(4)	(5)	(6)	(7)
Spheron-C						
Original	253	144	148	19	70	59
Outgassed at 600°C	290	173	183	15	88	80
Outgassed at 1000°C	281	160	174	10	82	82
Vulcan-SC						
Original	194	111	122	11	64	47
Outgassed at 600°C	202	118	154	0	80	74
Outgassed at 1000°C	198	104	152	0	76	76
Sugar charcoal						
Original	412	210	393	140	46	207
Outgassed at 600°C	637	282	557	119	380	58
Outgassed at 1000°C	388	201	433	0	375	58

[a]Columns (2) – (7) expressed in milliequivalents per 100 grams.
Source: Puri et al. (49). Reproduced with permission from Pergamon Press.

as a measure of surface area by several workers (73—76), although opinions differ as to the way the iodine is adsorbed on carbons. Whereas Hill and Marsh (76) consider adsorption of iodine from aqueous solutions a process of micropore filling and multilayer adsorption, Puri and Bansal (73) view it as a monolayer adsorption from both aqueous and nonaqueous solutions of iodine. These workers obtained surface areas of charcoals and carbon blacks which were comparable to BET (N_2) surface areas. The only important factor in determining surface area by adsorption of iodine is to work under standardized conditions of iodine concentration and the time of contact. According to these workers contacting 1 g of a carbon with 100 ml of 0.15 M iodine solution in 2.1 M potassium iodide for a period of 72 hr or with 0.3 N solution of iodine in benzene or chloroform for 20 days gives reasonable values of surface area. Juhola (74) also observed that adsorption of iodine from aqueous solutions on active carbons was unimolecular while absorption from the vapor phase involved pore filling of micropores (less than 30 Å diameter).

5.3.1 Modification of Surface Properties

The foregoing perusal of the literature on the reactions between carbon and chlorine or bromine shows that carbon—chlorine or carbon—bromine surface structures are formed by addition at the unsaturated sites and/or by exchange of chemibound hydrogen. In addition, these elements may penetrate deep into the microporous structure and require sufficiently high temperature to desorb them. Consequently these chemisorbed or physically adsorbed species may completely modify the surface properties and surface reactions of carbons. For example, the chemisorption of chlorine and/or bromine may produce a polar but nonhydrogen bonding adsorbent similar to the abundant polar carbons that are associated with oxygen surface structures. The carbon—chlorine or carbon—bromine bond may be exchanged with other functional groups to obtain new kinds of surface-modified carbon adsorbents and electrodes. The microporosity and adsorption stereoselectivity of microporous carbons may be modified, giving rise to newer carbons with absolutely new adsorption characteristics and behavior.

Puri and Bansal (77) studied the surface characteristics and surface behavior of sugar and coconut charcoals modified by treatment with chlorine. The true and bulk densities of the charcoals increased linearly with increase in the amount of chlorine chemisorbed by the carbon. The pH of the charcoal, the acid adsorption capacity, and the base neutralization capacities, however, remain unaltered, indicating that the presence of carbon—chlorine surface compounds did not alter the status of surface acidity of the charcoals in any way.

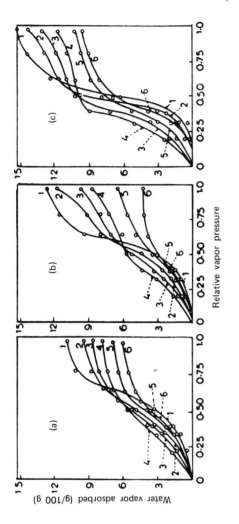

Figure 12 Water vapor adsorption isotherm on sugar charcoal. [From Puri and Bansal (77).]

Curve	Sample	Chlorine fixed (meq/g)		
		a	b	c
1	Untreated	0	0	0
2	Chlorinated at 100°	1.09	0.97	0.58
3	Chlorinated at 200°	1.67	1.32	1.70
4	Chlorinated at 300°	2.12	1.69	2.02
5	Chlorinated at 400°	3.50	3.35	3.07
6	Chlorinated at 500°	3.56	3.28	2.99

The adsorption isotherms of water vapor on sugar charcoals out-gassed at 500 (a), 700 (b), and 1000°C (c) containing varying amounts of associated chlorine are presented in Figure 12. These charcoals were essentially free of CO_2-complex so that no secondary effect due to interaction of this complex with water as reported in the literature was possible. The isotherms on the chlorinated samples intersected those on the untreated samples in the 0.5—0.6 relative vapor pressure range in every case, the chlorinated samples taking up more moisture at lower relative vapor pressures and less moisture at higher relative vapor pressures. The effect increased with the fixation of chlorine as long as the temperature of chlorination was 300°C or less. At higher temperatures of chlorination, the sorption values as above decreased (Fig. 12) at all relative vapor pressures in every sample. The increased sorption of water vapors in the lower part of the isotherm cannot be attributed to any chemical or quasi-chemical interaction (such as hydrogen bonding) of water with the chlorinated surface, as is well known for oxygenated carbon surfaces, because in that case the increase would have continued with increase in the chlorine content of the charcoal. These workers (77) are of the view that the fixation of chlorine results more in conditioning the pore structure and size distribution of capillary pores or in changing the location and frequency of active sites involved in the sorption of water vapor than in altering the chemical nature of the surface. This received further support from the shapes of the water isotherms on one of the carbon blacks (Spheron-6) before and after treatment with chlorine (Fig. 13). The isotherm changed shape from type II to type IV and then to type I, indicating a considerable narrowing down of the capillary pores as more and more chlorine was being fixed on the carbon black. The immersional heats of wetting in water of the chlorinated samples also increased only slightly (Table 17), indicating little or no interaction between adsorbed chlorine and water.

The treatment of sugar and coconut charcoals with bromine pro-duced a similar increase in adsorption of water vapor up to a rela-tive vapor pressure of 0.5 and a decrease in adsorption at higher relative vapor pressures. However, in the case of carbon blacks treated with bromine (48), the water vapor adsorption isotherms are shifted bodily upward (Fig. 14) after fixation of bromine, there be-ing almost a uniform increase in all ranges of vapor pressure. The average in adsorption at 0.5 relative vapor pressure in all the four samples was 2.33 mmol/g, while the average fixation of bromine cor-responded to 2.58 milliatoms/g, suggesting one molecule of additional water was being adsorbed at each site where a bromine atom had been fixed. The heat of immersion of the brominated samples in-creased significantly after fixation of bromine, almost in proportion to the amount of bromine fixed (Tables 18, 19).

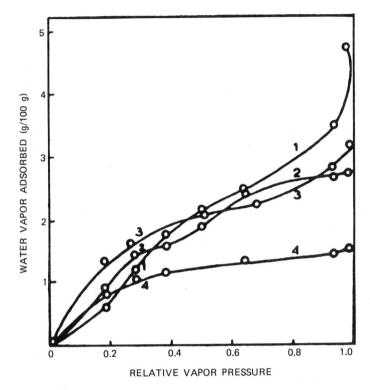

Figure 13 Water vapor adsorption isotherms of Spheron-6 before and after fixation of chlorine. Chlorine fixed: (1) untreated sample, 0; (2) 1–18 mEq/g; (3) 2–45 mEq/g; (4) 3–97 mEq/g. [From Puri and Bansal (77).]

Puri and co-workers (64,77) and Boehm et al. (20) also examined the availability of combined chlorine in carbons for substitution by other functional groups. Exhaustive boiling under reflux with 2.5 N sodium hydroxide could hydrolyze only a part of the combined chlorine but the amount hydrolyzed was independent of the amount of chlorine fixed by the carbon. However, the fraction of the chlorine hydrolyzed generally decreased with increase in the temperature of the treatment (Table 20). Almost the same amount of chlorine was eliminated as hydrochloric acid and recovered as ammonium chloride on treatment of the chlorinated samples with dry ammonia gas at 300°C. The amount of nitrogen fixed by the chlorinated sample as a result of ammonia treatment was determined by ultimate analysis and was found to be almost equivalent to the amount of chlorine displaced

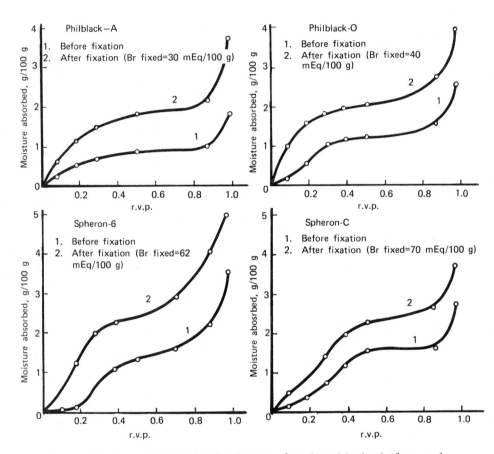

Figure 14 Water adsorption isotherms of carbon blacks before and after fixation of bromine. [From Puri and Bansal (48). Reproduced with permission from Pergamon Press.]

Table 17 Heats of Immersion in Water of Various Samples of Sugar and Coconut Charcoals before and after Fixation of Chlorine

Treatment temperature (°C)	1000°C-degassed		1200°C-degassed	
	Chlorine fixed (mEq/g)	Heat of immersion in water (cal/g)	Chlorine fixed (mEq/g)	Heat of immersion in water (cal/g)
Sugar charcoal				
—	—	5.26	—	4.65
100	1.37	5.51	0.58	4.89
200	1.99	5.97	1.70	5.06
300	2.39	6.51	2.02	5.36
400	3.09	6.45	3.07	5.31
Coconut charcoal				
—	—	3.96	—	2.57
100	1.76	4.26	1.15	2.79
200	2.56	4.65	2.06	3.10
300	3.10	5.15	2.10	3.33
400	3.39	5.21	2.31	3.36

Source: Puri and Bansal (77).

Table 18 Heats of Immersion of Carbon Blacks in Water in Relation to Bromine Fixed

Sample	Bromine fixed (mmol/100g)	Heat of immersion in water (cal/100g)
Philblack-A	0	83
	15	98
Philblack-A	0	195
	20	226
Spheron-6	0	196
	31	226
Spheron-C	0	306
	35	342

Source: Puri and Bansal (48). Reproduced with permission from Pergamon Press.

Table 19 Heats of Immersion of Carbon Blacks in
Water Before and After Chlorination

| Sample | Heat of immersion (cal/g) | |
	Before chlorination	After chlorination
Philblack-A	0.83	1.10
Philblack-E	2.57	2.84
Philblack-I	2.61	2.87
Spheron-9	2.27	2.61
Spheron-6	1.96	2.32
Spheron-4	2.95	3.53
Spheron-C	3.06	3.65
ELF-0	2.90	3.78

Source: Puri and Bansal (48). Reproduced with permission from
Pergamon Press.

(Table 21), indicating the replacement of chlorine by amino groups.
The presence of an amino group imparted a basic character to the
carbon surface. The pH value of the aqueous suspension was raised
from 8.4 to above 9.1 and there was a noticeable increase in the acid
adsorption capacity (Table 21). Equivalence between chlorine eliminated,
nitrogen fixed and increase in basicity (i.e., increase in adsorption
of HCl) on ammonia treatment of the chlorinated carbons clearly indicated that a part of the combined chlorine was available for exchange
reactions with other groups. Boehm et al. (62), however, observed
that in the case of chlorinated carbon black CK-3, although all of the
chlorine could be removed on fusion with potassium hydroxide and
about 70% could be removed on treatment with dry ammonia, the increase
in the oxygen content in the first treatment and in the nitrogen content
in the second treatment was far lower then would be expected for
exchange by -OH or $-NH_2$ groups respectively. Reaction of the chlorinated black with sodium cyanide or copper cyanide increased nitrogen
content corresponding to decrease in chlorine content, but the hydrolysis of the cyanide groups did not yield carboxyl groups.

Table 20 Chlorine Fixed by Charcoal and Recovered as Chloride Ion on Refluxing with Sodium Hydroxide[a]

Chlorination temperature (°C)	Original		Degassed at 700°		Degassed at 1000°		Degassed at 1200°	
	A	B	A	B	A	B	A	B
Sugar charcoal								
100	1.81	1.15	0.97	0.65	1.37	0.40	0.58	0.35
200	3.80	1.20	1.32	0.65	1.99	0.40	1.70	0.40
300	6.08	1.25	1.69	0.70	2.34	0.45	2.02	0.45
400	6.70	1.20	2.87	0.75	2.87	0.40	2.57	0.45
500	4.00	1.00	3.50	0.70	2.70	0.40	2.56	0.45
Coconut shell charcoal								
100	3.00	1.10	2.69	0.35	1.76	0.56	1.15	0.30
200	4.61	1.25	3.49	0.35	2.56	0.60	2.06	0.35
300	11.38	1.75	4.87	0.40	3.10	0.64	2.10	0.40
400	11.35	1.80	5.63	0.40	3.39	0.64	2.14	0.40

[a]Values refer to the amount of chlorine in milliequivalents per gram. A, fixed; B, hydrolyzed.
Source: Puri et al. (64).

Table 21 Interaction of Sugar and Coconut Charcoals with Ammonia at 300°C

Sample	Chlorine fixed (mEq/g)	Chlorine eliminated as NH_4Cl on treatment with NH_3 (mEq/g)	Nitrogen fixed on treatment with NH_3 (mEq/g)	Increase in the amount of acid adsorbed after treating chlorinated product with NH_3 (mEq/g)	Chlorine hydrolyzed on refluxing with 2.5 N NaOH (mEq/g)
1000°C-degassed sugar charcoal treated with chlorine at 400°C	3.09	0.65	0.68	0.71	0.40
1200°C-degassed sugar charcoal treated with chlorine at 400°C	3.07	0.49	0.52	0.53	0.45
1000°C-degassed coconut charcoal treated with chlorine at 400°C	3.39	0.64	0.62	0.59	0.64
1200°C-degassed coconut charcoal treated with chlorine at 400°C	2.31	0.38	0.36	0.40	0.40

Source: Puri and Bansal (77).

Brooks and Spotswood (71), while reacting their brominated chars with copper cyanide in pyridine at 200°C, observed that the combined bromine was almost completely replaced by nitrogen but the cyano groups failed to undergo hydrolysis and were unaffected by treatment with 80% sulfuric acid. Reactions of brominated chars with *n*-butyl lithium in tetrahydrofuran followed by carbonization eliminated bromine completely and increased the oxygen and hydrogen contents of the chars, but the number of carboxyl groups introduced was very small.

Tobias and Soffer (78) recently observed that chlorination of carbons followed by degassing at the same temperature as the chlorination or at higher temperatures creates double bonds which are different from the graphitic surface π bonds in their chemical reactivity. These double bonds, which were in addition to those already existing on untreated carbons, were available for adsorption of hydrogen and hydrochloric acid. These adsorbed species could be partially degassed by thermal cycling, indicating that at least part of these olefinic double bonds were thermally labile and might be available for exchange with other species. Thus carbons modified by chlorine treatment may be of importance in catalysis.

5.4 MODIFICATION OF CARBONS BY SULFUR SURFACE COMPOUNDS

The carbon—sulfur surface compounds have been reported on a wide variety of chars, activated carbons, carbon blacks, and coals. They are formed either during or subsequent to the formation of the carbon. However, their formation, their high thermal stability, the manner and the form in which they can be decomposed, the nature of the carbon surface structure, and their influence on the surface behavior of carbons are some of the factors which have intrigued the surface carbon chemist. These sulfur compounds are nonstoichiometric, falling within a wide range of composition depending on the nature of the carbon, the experimental conditions of its formation, and the magnitude of the surface of the carbon. They frequently contain appreciable amounts of sulfur, which may be as high as 40—50%, even when the contribution of the sulfur contents of inorganic impurities is excluded. These surface compounds can be neither extracted with solvents nor decomposed completely on heat treatment in vacuum at 1000°C but they can be removed completely as hydrogen sulfide on heat treatment in hydrogen between 500 and 700°C.

They are generally formed by heating a carbon in the presence of sulfur vapors (79—88, 90, 91, 102, 104, 105), or sulfurous gases such as hydrogen sulfide (90, 92—96), sulfur dioxide (97—100, 104, 105), and carbon disulfide (89, 103—105), or during carbonization of

organic compounds containing sulfur (83,109), or during carbonization of organic materials in the presence of elementary sulfur (80),
or materials yielding sulfurous pyrolytic products (110). The solid
sulfur surface compounds so formed by different processes show
many similarities with regard to their nonstoichiometric character and
chemical behavior, although their formation by the different methods
offers different possibilities of sulfur addition to, or substitution in,
the carbon lattice.

5.4.1 Formation of Sulfur Surface Structures

Wibaut (79—83) studied the interaction of several charcoals and carbon
blacks with sulfur between 100 and 1000°C and observed the chemisorption of appreciable amounts, varying between 18 and 25%, of
sulfur with the evolution of small amounts of hydrogen sulfide and
carbon sulfide. The sulfur surface compound was very stable and
could not be decomposed completely even on heat treatment in vacuum at 1100°C. Juja and Blanke (84), on the other hand, working in
the same temperature range, speculated that the fixation of sulfur
was partly due to the capillary condensation and physical adsorption
and partly due to chemisorption, depending on the nature of the
carbon and the experimental conditions. Heat treatment of charcoals
with sulfur in a rotating tube furnace at 400°C incorporated about
41% of sulfur in the charcoal, only 12% of which could be washed
with solvents (85). The fixation of sulfur lowered the hydrogen content of the charcoal, decreased porosity, and considerably lowered
the adsorption of water vapor below 0.5 relative vapor pressure.

Boehm et al. (86,87) could fix about 20% sulfur by heating
activated sugar charcoal with sulfur at 600°C in an evacuated tube.
The amount of sulfur fixed was equivalent to the amount of oxygen
fixed when the same charcoal was oxidized in oxygen at 400°C, indicating that oxygen and sulfur were bonded similarly on the carbon
surface. The sulfurized charcoal catalyzed the reaction of sodium
azide with iodine which is characteristic of sulfure in sulfide form.
Sykes and White reacted coconut charcoal with sulfur (88) and also
with carbon disulfide (89), at low pressure, in the temperature
range 627—927 K. The treatment of the carbon with sulfur evolved
carbon disulfide while the same sample when treated with carbon disulfide developed a pressure of sulfur vapors. This showed that sulfur and carbon disulfide were being adsorbed at the same set of
sites, giving rise to different sulfur-containing interconvertible surface structures. The interaction of the carbon with hydrogen sulfide,
(89) by a flow method in the same temperature range, fixed appreciable amounts of sulfur, evolving significant amounts of carbon disulfide. However, it could not be ascertained if the carbon disulfide

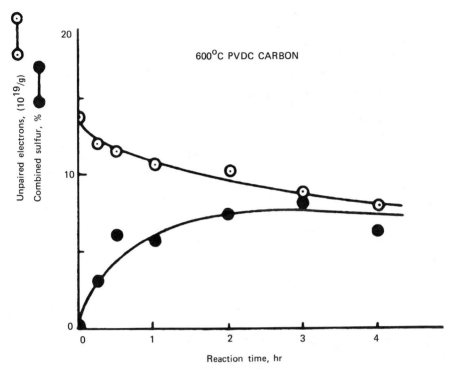

Figure 15 Comparison of sulfur retention and changes in unpaired electron concentration. (Sulfure vapor pressure, 15 mm; reaction temperature, 300°C.) [From Blayden and Patrick (91). Reproduced with permission from Pergamon Press.]

was produced by direct interaction between hydrogen sulfide and carbon or through the formation of some intermediate surface structures.

The interaction of sulfur and hydrogen sulfide with carbon blacks (90) in the temperature range 150–600°C involved two reactions: a rapid reaction, which was directly proportional to the amount of oxygen present as 1,4-quinone groups indicating addition of one sulfur atom for each quinone-type oxygen atom, and a slow reaction which was retarded by the presence of oxygen atoms. The amount of sulfur fixed in the slow reaction was related to the hydrogen content of the carbon, indicating an exchange for hydrogen.

Blayden and Patrick (91) studied the interaction of PVC, PVDC, and cellulose chars, prepared by carbonization at different temperatures between 300 and 500°C, with sulfur vapors at pressures

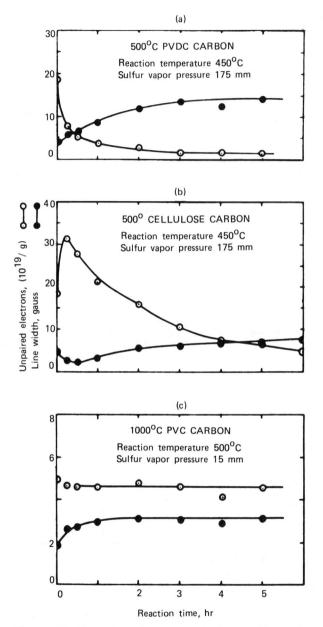

Figure 16 Reaction of representative carbons with sulfur vapor (typical changes of ESR with reaction time). [From Blayden and Patrick (91). Reproduced with permission from Pergamon Press.]

between 15 to 175 mm and with hydrogen sulfide gas. The amount of sulfur fixed depended on the nature of the carbon, the temperature of carbonization of the polymer, the reaction conditions, and the sulfurizing agent and the hydrogen content of the carbon. The reaction with sulfur vapors caused dehydrogenation of the carbon, forming hydrogen sulfide and the fixation of as much as 45% sulfur. The amount of sulfur retained by the carbon decreased with increase in the temperature of treatment. The interaction with hydrogen sulfide appeared to be more complex and less systematic and the sulfur content of the residual product was not more than 4—5%. The sulfur surface structures were stable to nitrous acid and hydrogen peroxide and the combined sulfur could not be extracted by solvents or by sodium hydroxide solution. ESR studies indicated a decrease in the spin concentration progressively with increasing sulfur content in low-temperature carbons sulfurized with sulfur vapors (Fig. 15). There was little or no change in spin concentration in the case of PVC char carbonized at 1000°C (Fig. 16). It was suggested that the sulfur was fixed by chemisorption including spin pairing of sulfur with spin centers and by exchange with combined hydrogen. No definite conclusions were drawn regarding the structure of the resulting surface compound, but it was suggested that surface thioethers and disulfides were the possible surface groups which formed peripheral heterocyclic structures on the layer lattice in the carbon surface or sulfur bridges between the adjacent layers. The presence of sulfide groups was based on the reaction of sodium azide with iodine and on certain reduction reactions.

The interaction of charcoals and carbon blacks with aqueous solutions of hydrogen sulfide (92—96) resulted in the fixation of about 6—7% of sulfur. The fixation of sulfur on oxygen-free carbons occurred by addition at the unsaturated sites (Table 22), whereas the fixation on the oxygenated carbons took place partly by addition at the unsaturated sites and partly by substitution in place of that part of associated oxygen which was evolved as carbon monoxide on evacuation (Table 22). The interaction of carbons with sulfur dioxide has also been studied by several workers (97—100). The reaction with an active carbon (100), between 50 and 650°C in the absence of oxygen in a flow system containing 0.5 vol % SO_2, showed that the chemisorption of sulfur was very small in the temperature range 50—300°C. However, sulfur dioxide reacted with the carbon at 600°C rapidly producing carbon dioxide, carbon monoxide, and elemental sulfur, which was partly deposited in the cooler portion of the reactor and partly fixed by the carbon. Young and Steinberg (101), however, observed that appreciable amounts of sulfur dioxide were chemisorbed by a coconut charcoal on heat treatment with the gas between 400 and 500°C. The IR spectroscopy of the treated charcoal

Table 22   Amount of Sulfur Fixed in Relation to the Degree of
Unsaturation in Carbon Blacks Evacuated at 1200°C

Carbon black	Amount of sulfur fixed (mmol/100 g)	Degree of unsaturation as determined by bromine adsorption (mmol/100 g)
Pelletex	6.1	7.0
Kosmos-40	10.2	9.1
Statex-B	6.6	4.9
Philblack-A	18.8	18.0
Philblack-O	22.4	24.0
Philblack-I	10.2	9.1
Spheron-9	30.6	29.1
Spheron-C	40.8	42.3
ELF-O	15.9	17.0
Mogul	33.2	31.3
Mogul-A	30.8	36.9

Source: Bansal and Gupta (96).

showed the formation of sulfate ion by bonding of sulfur dioxide to
the surface oxygen.

Puri and co-workers (102–105) carried out a systematic study of
the formation, stability, and mechanism of fixation of sulfur by
sugar and coconut charcoals associated with varying amounts of oxygen and hydrogen using sulfur vapors, H_2S, CS_2, and SO_2 as sulfurizing reagents at 600°C. Each carbon fixed an appreciable amount
of sulfur depending on the amount of associated oxygen and hydrogen of the carbon and the nature of the sulfurizing reagent (Table
25). The sulfur was fixed partly by addition at the unsaturated
sites and partly by exchange with certain oxygen structures, which
come off as carbon monoxide on high temperature evacuation (Fig.
17). As carbon monoxide was evolved by the decomposition of phenolic and quinonic structures, it was suggested that the exchange
with these groups leads to the formation of sulfide and hydrosulfide
groups.

The initial hydrogen content of the charcoals also played a significant role in determining the amount of sulfur fixed by a charcoal. It was suggested that nascent hydrogen, which was evolved

Table 23 Amount of Sulfur fixed in Relation to the Degree of Unsaturation in Carbon Blacks

Carbon black	Amount of sulfur fixed (mmol/100g)	Degree of unsaturation as determined by bromine adsorption (mmol/100g)
Furnace blacks		
Pelletex	15.0	2.5
Kosmos-40	20.5	4.6
Statex-B	21.5	11.0
Philblack-A	25.5	15.0
Philblack-O	36.2	21.2
Philblack-I	47.5	25.4
Vulcan-SC	56.0	29.2
Channel blacks		
Spheron-9	71.3	17.2
Spheron-C	83.5	27.2
ELF-O	71.0	2.0
Mogul	130.5	0
Mogul-A	133.5	0
Oxidized blacks		
Spheron-9	65.2	13.2
Philblack-I	45.1	19.8

Source: Bansal and Gupta (96).

from the carbon during the treatment at 600°C, reacted with the sulfurizing agent to produce monoatomic sulfur as

$$S_2 + 2H \longrightarrow H_2S + S$$

$$CS_2 + 2H \longrightarrow H_2S + C + S$$

$$SO_2 + 4H \longrightarrow 2H_2O + S$$

Since H_2S could not undergo such a reaction with hydrogen, the amount of sulfur fixed on treatment with this reagent was not influenced by the hydrogen content of the charcoals (Fig. 18).

Table 24 Fixation of Sulfur on Various Samples of Sugar and Coconut Charcoal on Treatment with Different Sulfurizing Reagents at 600°C

(1)	(2)	(3)	(4)	(5)	(6)
	Sulfur fixed (mEq/g)	Unsaturation as determined by Br$_2$ uptake		Sulfur fixed by addition (mEq/g) [(3) - (5)]	Sulfur fixed by a substitution process (mEq/g) [(2) - (5)]
Sample		Before fixation of sulfur	After fixation of sulfur		
Using sulfur as sulfurizing reagent					
Sugar charcoal					
Original	12.3	3.1	0.5	2.6	9.7
400°-outgassed	11.1	3.5	0.2	3.3	7.8
700°-outgassed	7.3	3.8	0	3.8	3.5
1000°-outgassed	5.2	3.8	0	3.8	1.4
Coconut charcoal					
Original	13.5	3.0	0	3.0	10.5
400°-outgassed	10.5	3.4	0	3.4	7.1
700°-outgassed	8.4	3.6	0	3.6	4.8
1000°-outgassed	4.8	3.6	0	3.6	1.2
Using hydrogen sulfide as sulfurizing reagent					
Sugar charcoal					
Original					
400°-outgassed	7.0	3.1	0.1	2.9	4.6
700°-outgassed	6.5	3.5	0	3.5	3.5
1000°-outgassed	5.5	3.8	0	3.8	2.7

Coconut charcoal

Original	7.2	3.0	0	2.9	4.3
400°-outgassed	7.1	3.4	0	3.4	3.7
700°-outgassed	5.9	3.6	0	3.6	2.3
1000°-outgassed	4.3	3.6	0	3.6	0.7

Using carbon disulfide as sulfurizing reagent

Sugar charcoal

Original	12.2	3.1	0	3.1	9.1
400°-outgassed	12.3	3.5	0	3.5	8.8
700°-outgassed	8.1	3.8	0	3.8	4.3
1000°-outgassed	5.0	3.8	0	3.8	1.2

Coconut charcoal

Original	14.2	3.0	0	3.0	11.2
400°-outgassed	14.2	3.4	0	3.4	10.8
700°-outgassed	11.3	3.6	0	3.6	7.7
1000°-outgassed	5.5	3.6	0	3.6	1.9

Using sulfur dioxide as sulfurizing reagent

Sugar charcoal

Original	17.0	3.1	0.2	2.9	14.1
400°-outgassed	16.6	3.5	0	3.5	13.1
700°-outgassed	10.8	3.8	0	3.8	7.0
1000°-outgassed	5.4	3.8	0.3	3.5	1.9

Coconut charcoal

Original	14.7	3.0	0	3.0	11.7
400°-outgassed	13.2	3.5	0.3	3.2	10.0
700°-outgassed	9.7	3.6	0	3.6	6.1
1000°-outgassed	4.2	3.6	0.3	3.3	0.9

Source: Puri and Hazra (105). Reproduced with permission from Pergamon Press.

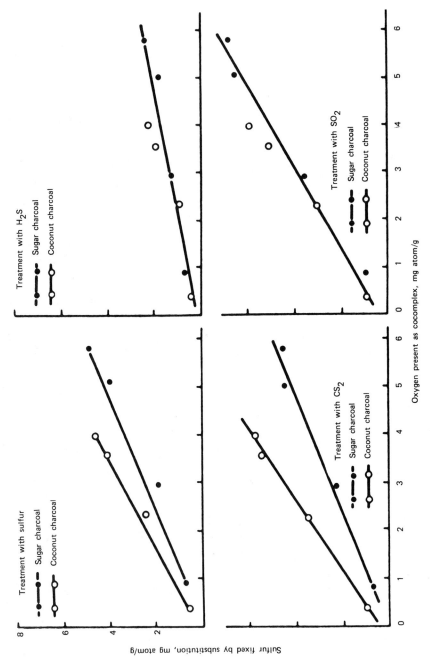

Figure 17 Amount of sulfur fixed by "substitution" in relation to oxygen content present as CO-complex. [From Puri and Hazra (105). Reproduced with permission from Pergamon Press.]

Table 25 Decomposition of Carbon-Sulfur Surface Complex formed on 400°-Outgassed Sugar Charcoal on Heating in a Current of Nitrogen at Different Temperatures

Temperature of heating in nitrogen (°C)	Sulfur evolved as (mEq/g):				Sulfur left in the residue (mEq/g)
	Free sulfur	H_2S	CS_2	SO_2	
	Complex formed on treatment with H_2S (initial sulfur content = 7.50 mEq/g)				
400°	0	0	0.39	0	6.85
500°	0	0.21	0.62	0	6.37
700°	0.10	1.50	0.69	0	4.52
900°	0.12	2.51	1.01	0	3.31
1000°	0.12	3.12	3.04	0	0.23
	Complex formed on treatment with SO_2 (initial sulfur content = 16.6 mEq/g)				
600°	0.12	1.23	0.40	0.95	13.15
800°	0.25	1.41	2.20	1.13	10.20
1000°	0.32	2.10	6.92	1.20	4.62
1200°	0.41	2.70	10.21	1.20	0.83

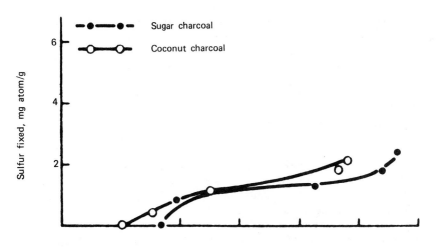

Figure 18 Amount of sulfur fixed by "substitution" in relation to hydrogen content of the various charcoals. [From Puri and Hazra (105). Reproduced with permission from Pergamon Press.]

The carbon—sulfur surface compound was highly stable and could not be recovered on boiling with 2.5 N sodium hydroxide solution under reflux for 12 hr (106) and could be recovered only partly on heat treatment in vacuum or in nitrogen at 1000°C (Table 25). However, it could be decomposed completely on heat treatment in hydrogen at 900°C. It was not possible to identify exactly the sulfur-containing surface structures, but Puri et al. (104) explored the probable nature of these groups by estimating the amount of sulfur recovered on treatment of the sulfurated carbons with oxidizing solutions such as aqueous chlorine, acidified potassium chlorate, and concentrated nitric acid. They observed that except in the case of surface compounds formed with sulfur dioxide, only a part of the bonded sulfur could be recovered even with the most drastic treatment with nitric acid at 110°C (Table 26). The amount of sulfur recovered varied up to 10% for the carbon disulfide—treated product and was almost 100% with the sulfur dioxide—treated product. This showed that different types of sulfur structures were formed by different treatments. These workers speculated that the treatment with sulfur dioxide produced readily oxidizable sulfoxides and sulfone structures as well as sulfide and hydrosulfide groups. The formation of sulfoxide and sulfone groups was evidenced by the chemisorption of appreciable amounts of oxygen during the reaction of carbons with sulfur dioxide.

When the surface compounds were formed by treatment with hydrogen sulfide or sulfur vapors, the amount of sulfur recovered by oxidation with nitric acid corresponded, fairly closely, to the amount of sulfur fixed by exchange (Table 24) and the amount not recovered to the amount fixed by addition at the unsaturated sites. The former was attributed to sulfides and hydrosulfides and the latter to highly stable sulfur-containing aromatic ring structures of the carbon layers. The presence of sulfide groups was evidenced by the catalysis of the sodium azide—iodine reaction by the sulfurized carbons. In the case of carbon disulfide—treated carbons, the sulfur recovered was very small ($= 6-8\%$ of the total fixed). Since some carbon also is deposited in this reaction, it was suggested that some new peripheral heterocyclic structures were condensed on the layer lattices in the carbon surface.

Chang (107) prepared surface sulfur compounds by heating activated carbons, carbon blacks, graphon, petroleum pitches, and polymer chars with carbon disulfide, H_2S, $SOCl_2$ (thionyl chloride), and SO_2, in the temperature range 800—900°C. Sulfur compounds containing greater than 30% sulfur were obtained. However, no detectable amount of sulfur could be fixed on graphon. Thermogravimetric analysis of the surface compounds showed the formation of three different surface structures for activated carbons (Fig. 19). A major portion of the combined sulfur was weakly held and could be recovered with carbon disulfide or on heat treatment at 280°C. The

Table 26 Recovery of Sulfur Fixed on Treatment of Charcoals with Various Sulfurizing Reagents on Treatment with Hot (110°) Nitric Acid

Sample	Total sulfur fixed (mEq/g)	Amount of sulfur recovered as H_2SO_4 (mEq/g)
400°-outgassed sugar charcoal		
Treated with SO_2	16.60	15.06
Treated with H_2S	7.00	3.18
Treated with S	11.10	4.12
Treated with CS_2	12.30	0.76
400°-outgassed coconut charcoal		
Treated with SO_2	13.20	11.90
Treated with H_2S	7.10	3.25
Treated with S	10.50	5.06
Treated with CS_2	14.20	0.82
1000°-outgassed sugar charcoal		
Treated with SO_2	5.40	4.90
Treated with H_2S	5.50	1.55
Treated with S	5.20	1.68
Treated with CS_2	5.00	0.56
1000°-outgassed coconut charcoal		
Treated with SO_2	4.20	3.90
Treated with H_2S	4.30	0.65
Treated with S	4.80	1.40
Treated with CS_2	5.50	0.32

Source: Puri et al. (104).

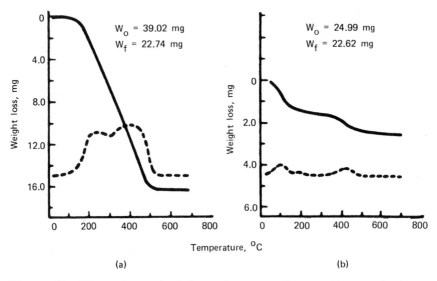

Figure 19   Thermal gravimetric analyses on the reaction product between Pittsburgh CPG and sulfur (a) before and (b) after CS_2 washing. [From Chang (107). Reproduced with permission from Pergamon Press.]

second type of surface structure was stable to carbon disulfide washing and could be recovered on heat treatment at about 450°C. The third sulfur structure, which these workers called as C_xS, was chemically stable to KOH (9 N, 72 hr) and after an initial loss of a small percentage of sulfur was stable even to strong oxidizing agents such as HNO_3 (5 N, 72 hr), HCl (5 N, 24 hr), and to heat treatment in Ar at 700°C. X-ray diffractometry and electron microscopy of C_xS surface compound showed that it had a disordered structure. ESCA spectra (Fig. 20) showed a carbon 1s binding energy of 284.3 eV and a sulfur $2p_{3/2}$ binding energy of 163.7 eV. Chang postulated that these binding energies and the FTIR absorption bands at 1180 and 1150 cm^{-1} (Fig. 21) suggest that the surface structures on C_xS were thiocarbonyls (> C = S) and thiolactones ($\overset{S}{\underset{C}{\diagdown}}C = s$) analogous to carbonyls and lactones existing on the surface of activated carbons. carbons.

5.4.2 Modification of Carbon Surface by Sulfur Surface Compounds

It is apparent that almost all types of carbons can fix appreciable amounts of sulfur-forming sulfur surface compounds. It is also recognized that sulfur is bonded to the peripheral carbon atoms,

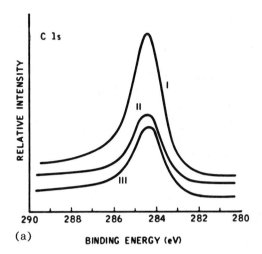

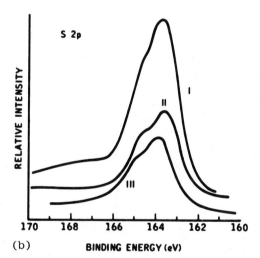

Figure 20 (a) XPS spectra of carbon 1s: (I) C_2S (PVDF); (II) mixture of PVDF charcoal and 24% sulfur; (III) C_2S (sucrose). (b) XPS spectra of sulfur 2p: (I) C_2S (sucrose); (II) mixture of PVDF charcoal and 24% sulfur; (III) C_2S (PVDF). [From Chang (107). Reproduced with permission from Pergamon Press.]

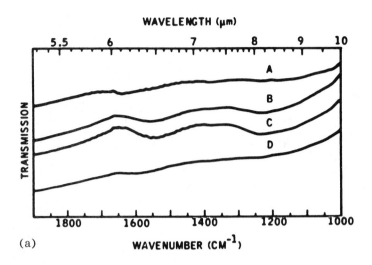

(a)

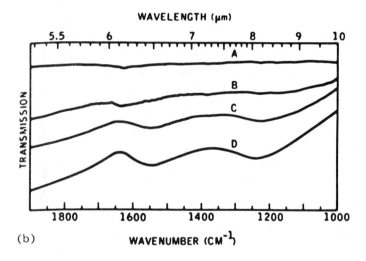

(b)

Figure 21   (a) FTIR spectra in the region 1000–1900 cm^{-1}: (A) PVDF charcoal vs. air; (B) C$_x$S (PVDF) vs. air; (C) C$_x$S (sucrose) vs. air; (D) (PVDF charcoal + 24.8% sulfur) vs. air; (b) FTIR spectra in the region 1000–1900 cm^{-1}: (A) KBr vs. air; (B) PVDF charcoal vs. air; (C) C$_x$S (PVDF) vs. air; (D) C$_x$S (PVDF) vs. PVDF charcoal. [From Chang (107). Reproduced with permission from Pergamon Press.]

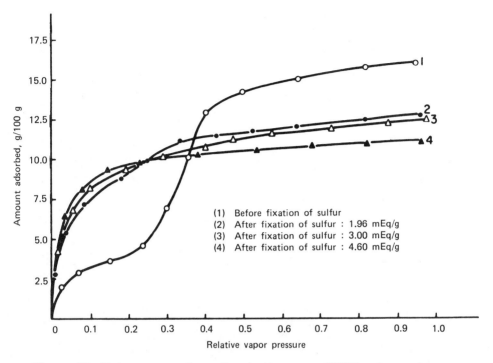

Figure 22 Water vapor adsorption isotherm on 700°C outgassed sugar charcoal before and after fixation of sulfur. [From Puri and Hazra (105). Reproduced with permission from Pergamon Press.]

penetrates into the porous structure, is added at the double bond sites, and is exchanged for hydrogen as well as oxygen associated with the carbon surface. Since the peripheral carbon atoms due to unsatisfied valencies, the unsaturated active sites, and the associated oxygen determine the surface reactions and adsorption characteristics of carbons, it is reasonable to believe that the formation of surface sulfur complexes should modify the behavior of carbons. The formation of carbon—sulfur surface compounds is well known but their influence in modifying carbon surface properties has not been studied systematically.

The influence of sulfur surface compounds on the adsorption behavior of carbons toward polar and nonpolar vapors of varying molecular dimensions was examined by Puri and Hazra (105). The adsorption of water vapors increased appreciably at relative vapor pressures lower than 0.4 and decreased at higher relative pressures (Fig. 22). The effect increased with increase in the amount of sulfur fixed and was attributed to the variation of the pore size distribu-

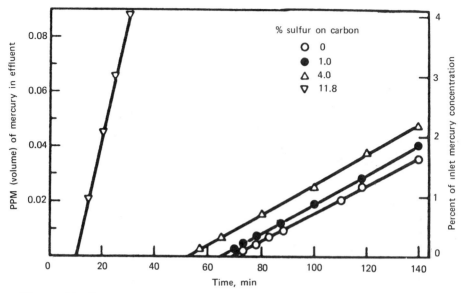

Figure 23 Breakthrough curves of mercury in a dry air stream at 25°C from Saran carbon beds containing varying amounts of sulfur. [From Sinha and Walker (108). Reproduced with permission from Pergamon Press.]

tion caused by the fixation of sulfur along the pore walls. A gradual change in the shape of the isotherm with increasing amounts of sulfur in the carbon indicated that the carbon had become highly microporous. Calculation of the pore size distribution by the Kelvin equation showed a decrease in the pore size from a very wide range to about 10–12 Å. The charcoal, however, could still adsorb appreciable amounts of water vapor. The adsorption isotherms of methanol and benzene vapors, which have higher cross-sectional areas (16.7 Å^2 and 41 Å^2 respectively) compared to water (cross-sectional area 10.5 Å^2), indicated that these larger molecules found smaller and smaller excess as more and more sulfur was being incorporated into the charcoals. There was very little adsorption of benzene and no adsorption of still bigger molecules such as ∝-pinene. The charcoal behaved as a molecular sieve after the fixation of sulfur. Thus fixation of sulfur could be developed as a method for preparing carbon molecular sieves.

Sulfurized Saran carbon carbonized at 900°C and loaded with varying amounts of sulfur between 1 and 12% was used by Sinha and Walker (108) for the removal of mercury vapors from the air stream using an atomic absorption spectrometer. The contaminated air stream

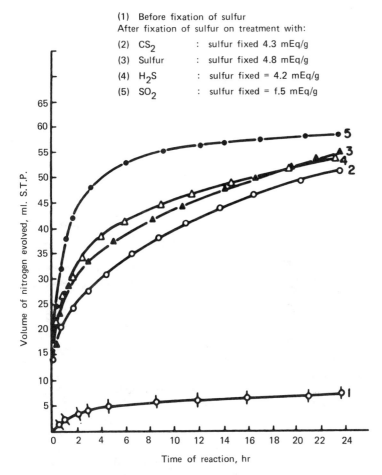

(1) Before fixation of sulfur
After fixation of sulfur on treatment with:
(2) CS_2 : sulfur fixed 4.3 mEq/g
(3) Sulfur   : sulfur fixed 4.8 mEq/g
(4) H_2S : sulfur fixed = 4.2 mEq/g
(5) SO_2 : sulfur fixed = f.5 mEq/g

Figure 24   Rate of the sodium azide—iodine reaction in the presence of 700°-outgassed sugar charcoal before and after fixation of sulfur.

was prepared by bubbling air through a mercury bath maintained at 24°C and containing 2.2 ppm by volume of mercury. The breakthrough profile of mercury from beds of charcoal containing different sulfur contents (Fig. 23) showed that increased sulfur content decreased the breakthrough time. Once breakthrough occurred, the rate of mercury buildup in the effluent from carbons containing up to 4% sulfur was independent of the amount of sulfur in the carbon. Introduction of moisture into the contaminated stream to the extent of 1.5% by volume markedly reduced the breakthrough time. However, when the contaminated stream was passed through the carbon

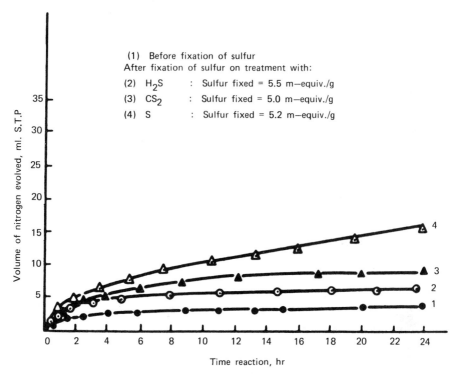

(1) Before fixation of sulfur
After fixation of sulfur on treatment with:
(2) H_2S : Sulfur fixed = 5.5 m—equiv./g
(3) CS_2 : Sulfur fixed = 5.0 m—equiv./g
(4) S   : Sulfur fixed = 5.2 m—equiv./g

Figure 25 Sulfurized carbons as catalysts for sodium azide-iodine reaction. [From Puri and Hazra (105). Reproduced with permission from Pergamon Press.]

bed at 150°C, the breakthrough time of the sulfurized carbons in-creased and the mercury buildup in the effluent stream was also very slow compared to the unsulfurized carbon. This greater capac-ity of the sulfurized charcoal at 150°C for mercury was attributed to its reaction with sulfur on the carbon surface forming mercuric sulfide. However, attempts to recover Hg^{2+} ions from aqueous solu-tions by sulfurized Saran carbons did not show any promise.

The sulfurized carbons also catalyze the sodium azide—iodine re-action (86,91,105). The results of one such study by Puri and Hazra (105) are shown in Fig. 24. The sulfurized product obtained on reacting 1000°-outgassed sugar charcoal with sulfur, hydrogen sulfide, and carbon disulfide (Fig. 25) was a much less effective cata-lyst. This was attributed to the fact that in this charcoal the sulfur was present in the form of heterocyclic ring structures and only sulfur present as sulfides can catalyze this reaction.

5.5 MODIFICATION OF CARBONS BY NITROGEN SURFACE COMPOUNDS

The formation of carbon—nitrogen surface compounds and their influence on surface modification of carbons has been only sparsely reported, although it has been observed that the treatment of carbons with ammonia gas results in the formation of basic carbons which may have some applications. Boehm et al. (62) and Hofmann and Ohlerich (87) found that when an oxidized charcoal was heated with dry ammonia, nitrogen was bound to the surface. At low temperatures the fixation of nitrogen was equivalent to the number of acid groups as determined by sodium hydroxide neutralization capacity. This was attributed to the formation of ammonium salts. However, at high-temperature treatment a substitution of the hydroxyl groups by amono groups was postulated. The carbon became hydrophobic and markedly changed the adsorption capacity of the carbon for methylene blue, which is a basic dye (Table 27).

Puri and Mahajan (111) observed that the interaction of a sugar charcoal with dry ammonia gas involved neutralization of the surface acidic complexes and fixation of some additional amounts in nonhydrolyzable form. But they could not attribute this fixation of ammonia to any particular group on the carbon surface. However, Puri and Bansal (77) observed that when chlorinated sugar and coconut charcoals were treated with ammonia gas at 300°C, a part of the chemisorbed chlorine was evolved and substituted by amino groups.

Table 27 Reaction of Acid Surface Oxides with Ammonia (activated charcoal Carboraffin oxidized with HNO_3)

Reaction temperature (°C)	Nitrogen uptake (mg atom/ 100g)	Neutralization of:		Adsorption of methylene blue (mg/g)
		KOH (alcoholic) (mEq/100 g)	NaOH (in H_2O) (mEq/100 g)	
(untreated)	6	750	400	535
110	390	400	—	252
255	540	230	—	158
330	620	170	—	136
410	710	120	—	—

Source: Boehm et al. (62).

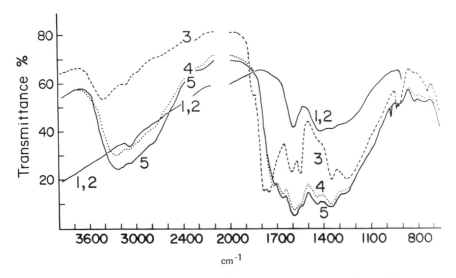

Figure 26 I.R. Spectra of ammonia sorbed on the surface of a carbonaceous film prepared from cellulose. (1) Cellulose carbonized at 400°C in CO_2 atmosphere (after desorption at 600°C). (2) Sorption of NH_3 at 1.5 kPa (1500 Pascal). (3) After oxidation in 63% aqueous HNO_3 (1 hr at 1000°C) and desorption at 200°C. (4) Sorption of NH_3 at 0.18 kPa. (5) Sorption of NH_3 at 1.5 kPa. (All spectra recorded in a vacuum cell, expansion 1.11.) [From Zawadski (112). Reproduced with permission from Pergamon Press.]

The resulting charcoals were basic in character and showed enhanced adsorption for acids. The increase in the acid adsorption corresponded to nitrogen fixed on treatment of the chlorinated charcoals with ammonia, indicating an exchange of C—Cl bond by a C—NH_2 bond (Table 21).

Zawadzki (112) treated a carbon film with ammonia gas before and after oxidation with nitric acid and using IR spectroscopy observed no adsorption of ammonia before oxidation of the film. However, the oxidation of the carbon film with nitric acid caused adsorption of ammonia, resulting in a sharp IR absorption in the N=H stretching vibration region. The three bands at 1840, 1780, and 1740 cm^{-1}, which were present in the oxidized film, disappeared on adsorption of ammonia and a single band in the region of C=O stretching vibrations at 1720 cm^{-1} was formed (Fig. 26). This band was attributed to imide structures. The disappearance of the doublet of bands 1840 and 1780 cm^{-1} was attributed to the reaction of ammonia with the cyclic anhydride structures and the disappearance of

Table 28 Irreversible Sorption of Ammonia on Surface
of Carbonaceous Films

Kind of carbonaceous film	Quantity of NH_3 sorbed irreversibly after the desorption at room temperature	
	(mmol/g)	(mg/g)
Desorbed at 600°C	0.18	3.1
Desorbed at 600°C and oxidized with HNO_3	2.06	35.0
Desorbed at 800°C	0.37	6.3
Desorbed at 800°C and oxidized with HNO_3	1.75	29.7

Source: Zawadski (112). Reproduced with permission
from Pergamon Press.

the 1740 cm^{-1} band to the interaction of ammonia with the lactone
structures. The desorption experiments on the ammonia-treated oxidized carbon film showed that the bonded nitrogen was very stable.
Oxidized carbon film could bind more than 10 times the amount of
ammonia than could the unoxidized film (Table 28).

REFERENCES

1. Donnet, J. B. and Papirer, E., *Rev. Gen. Caoutchouc Plastiques, 42:* 889 (1965).
2. Donnet, J. B. and Papirer, E., *Bull. Soc. Chim. France, 1965:* 1912.
3. Puri, B. R., Aggarwal, V. K., Bhardwaj, S. S., and Bansal, R. C., *Indian J. Chem., 11:* 1020 (1973).
4. Dietz, V. R. and Bitner, J. L., *Carbon, 11:* 393 (1973).
5. Smith, R. N., Swinehard, J., and Lessini, D., *J. Phys. Chem., 63:* 544 (1959).
6. Puri, B. R., Bansal, R. C., and Bhardwaj. S. S., *Indian J. Chem., 11:* 1168 (1973).
7. Puri, B. R., Singh, S., and Mahajan, O. P., *J. Indian Chem. Soc., 42:* 427 (1965).
8. Behraman, A. S. and Gustafson, H., *Ind. Eng. Chem., 27:* 426 (1935).

9. Puri, B. R., Mahajan, O. P., and Singh, D. D., *J. Indian chem. Soc.*, *37:* 171 (1960).
10. Puri, B. R., Singh, D. D., Sharma, L. R., and Chander, J., *J. Indian Chem. Soc.*, *35:* 130 (1958).
11. Puri, B. R. and Mahajan, O. P., *J. Indian Chem. Soc.*, *39:* 292 (1962).
12. Puri, B. R. and Bansal, R. C., *Carbon*, *1:* 457 (1964).
13. Puri, B. R. and Sharma, S. K., *Chem. Ind. London*, *1966:* 160.
14. Donnet, J. B., Bouland, J. C., and Jaeger, J., *Compt. Rendu*, *256:* 5340 (1963).
15. Donnet, J. B., French Patent 1,164,786, Oct. 14, 1958.
16. Puri, B. R., Sharma, A. K., and Mahajan, O. P., *Res. Bull. Panjab University Chandigarh*, *15:* 285 (1964).
17. Donnet, J. B., Heuber, F., Reitzer, C., Odoux, J., and Riess, G., *Bull. Soc. Chim. France*, *1962:* 1927.
18. Ghandi, D. L., Ph.D dissertation, Panjab University, Candigarh, 1976.
19. Puri, B. R. and Kalra, K. C., *Carbon*, *9:* 313 (1971).
20. Puri, B. R., Sharma, L. R., and Singh, D. D., *J. Indian Chem. Soc.*, *35:* 770 (1958).
21. Donnet, J. B. and Henrich, G., *Bull. Soc. Chim. France*, *1960:* 1609.
22. Puri, B. R. and Mahajan, O. P., *J. Indian Chem. Soc.*, *39:* 292 (1962).
23. Boehm, H. P., in *Advances in Catalysis*, Vol. XVI, Academic Press, New York, 1966, p. 179.
24. Garten, V. A. and Weiss, D. E., *Rev. Pure Appl. Chem.*, *7:* 69 (1957).
25. Puri, B. R., in *Chemistry and Physics of Carbon*, P. L. Walker, Jr., Ed., Marcel Dekker, New York, 1970, p. 191.
26. Billings, B. H. M., Docherty, J. B., and Bervan, M. J., *Carbon*, *22:* 83 (1984).
27. Baker, J. A. and Poziomek, E. J., *Carbon*, *12:* 45 (1974).
28. Collins, D. A., Taylor, L. R., and Taylor, R., in *Proc. 9th AEC Air Cleaning Conf.*, CONF. 660904 Vol. I, 1957, p. 59.
29. Grabenstetter, R. J. and Blace, F. E., in *Summary of Technical Report of Div. No. 10 NDRC Military Problems with Aerosol and Nonresistant Gases*, Vol. I, PB 158505 1964, Chapter IV.
30. Barnir, J. and Aharoni, C., *Carbon*, *13:* 363 (1975).
31. Chion, C. T. and Reucroft, P. J., *Carbon*, *15:* 49 (1977).
32. Reucroft, P. J. and Chion, C. T., *Carbon*, *15:* 285 (1977).
33. Freeman, G. B., Rao, P. B., and Reucroft, P. J., *Carbon*, *18:* 21 (1980).
34. Freeman, G. B. and Reucroft, P. J., *Carbon*, *17:* 313 (1979).

35. Capon, A., Alves, V. R., Smith, M. E., and White, M. P., *15th Bienn. Conf. Carbon*, 1981, Preprints, p. 232.
36. Jonas, L. A., *Carbon*, *16:* 115 (1978).
37. Chihara, K., Matsui, I., and Smith, J. M., *AIChE J.*, *27:* 220 (1981).
38. Umehara, T., Harriot, P., and Smith, J. M., *AIChE J.*, *29:* 232 (1983).
39. Umehara, T., Harriot, P., and Smith, J. M., *AIChE J.*, *29:* 737 (1983).
40. Reichenberger, J., Dissertation, Technische Hochschule, Aachen, 1974.
41. Krebbs, C. and Smith, J. M., *Carbon*, *23:* 223 (1985).
42. Hall, P. G., Gittins, P. M., Winn, J. M., and Robertson, J., *Carbon*, *23:* 353 (1985).
42a. Kaistha, B. C. and Bansal, R. C., unpublished work.
43. Gandhi, D. L., Sharma, S. K., Kumar, A., and Puri, B. R., *Indian J. Chem.*, *13:* 1317 (1975).
44. Reyerson, L. H. and Wishart, A. W., *J. Phys. Chem.*, *42:* 679 (1938).
45. Puri, B. R., Mahajan, O. P., and Singh, D. D., *J. Indian Chem. Soc.*, *38:* 943 (1961).
46. Puri, B. R. and Bansal, R. C., *Indian J. Chem.*, *5:* 381 (1967).
47. Puri, B. R., Sandle, N. K., and Mahajan, O. P., *J. Chem. Soc.*, *1963:* 4880.
48. Puri, B. R. and Bansal, R. C., *Carbon*, *3:* 533 (1966).
49. Puri, B. R., Gandhi, D. L., and Mahajan, O. P., *Carbon*, *15:* 173 (1977).
50. Puri, B. R., Singh, D. D., and Arora, V. M., *J. Indian Chem. Soc.*, *55:* 488 (1978).
51. Puri, B. R., Malhotra, S. L., and Bansal, R. C., *J. Indian Chem. Soc.*, *40:* 179 (1963).
52. Puri, B. R. and Bansal, R. C., *Carbon*, *5:* 189 (1967).
53. Puri, B. R. and Sehgal, K. C., *Indian J. Chem.*, *4:* 206 (1966).
54. Puri, B. R. and Sehgal, K. C., *Indian J. Chem.*, *5:* 379 (1967).
55. Puri, B. R., Mahajan, O. P., and Gandhi, D. L., *Indian J. Chem.*, *10:* 848 (1972).
56. Bansal, R. C., Dhami, T. L., and Parkash, S., *Carbon*, *18:* 395 (1980).
57. Ryerson, L. H. and Cameron, A. E., *J. Phys. Chem.*, *39:* 181 (1935).
58. Reyerson, L. H. and Cameron, A. E., *J. Phys. Chem.*, *40:* 233 (1936).

59. Alekseeviskii, L. A. and Likharev, N. A., *J. Gen. Chem. (USSR)*, *12:* 306 (1942).
60. Emmett, P. H., *Chem. Rev.*, *43:* 69 (1948).
61. Ruff, V. T., *J. Chem. Ind. (Moscow)*, *13:* 348 (1936).
62. Boehm, H. P., Hoffman, U., and Clauss, A., in *Proc. 3rd Conf. Carbon*, Pergamon Press, New York, 1959, p. 241.
63. Rivin, D. and Aron, J., *Proc. 7th Conf. Carbon*, Cleveland, Ohio, Abstracts, 1965.
64. Puri, B. R., Tulsi, S. S., and Bansal, R. C., *Indian J. Chem.*, *4:* 7 (1966).
65. Bansal, R. C., Vastola, F. J. and Walker, P. L., Jr., *Carbon*, *9:* 185 (1971).
66. Stacy, W. O., Imperial, C. R., and Walker, P. L., Jr., *Carbon*, *4:* 343 (1966).
67. Puri, B. R., Dhingra, A. K., and Sehgal, K. C., *Indian J. Chem.*, *7:* 174 (1969).
68. Puri, B. R. and Sandle, N. K., *Indian J. Chem.*, *6:* 267 (1968).
69. Tobias, H. and Soffer, A., *Carbon*, *23:* 281 (1985).
70. Stearns, R. S. and Johnson, B. L., *Ind. Eng. Chem.*, *43:* 146 (1951).
71. Brooks, J. D. and Spotswood, T. M., *Proc. 5th Conf. Carbon*, Vol. I, Pergamon Press, New York, 1962, p. 416.
72. Watson, J. W. and Parkinson, D., *Ind. Eng. Chem.*, *47:* 1053 (1955).
73. Puri, B. R. and Bansal, R. C., *Carbon*, *3:* 227 (1965).
74. Juhola, A. J., *Carbon*, *13:* 437 (1975).
75. Hassler, J. W. and McMinn, W. E., *Ind. Eng. Chem.*, *37:* 645 (1945).
76. Hill, A. and Marsh, H., *Carbon*, *6:* 31 (1966).
77. Puri, B. R. and Bansal, R. C., *Indian J. Chem.*, *5:* 566 (1967).
78. Tobias, H. and Soffer, A., *Carbon*, *23:* 291 (1985).
79. Wibaut, J. P., *Proc. 3rd Int. Conf. Bituminous Coal*, 1932, p. 657.
80. Wibaut, J. P., *Rec. Trav. Chim.*, *38:* 159 (1919).
81. Wibaut, J. P., *Rec. Trav. Chim.*, *41:* 153 (1922).
82. Wibaut, J. P. and Bastide, G. L., *Rec. Trav. Chim.*, *43:* 731 (1924).
83. Wibaut, J. P., *Z. Agnew. Chem.*, *40:* 1136 (1927).
84. Juja, R. and Blanke, W., *Z. Anorg. Allgem. Chem.*, *210:* 81 (1933).
85. Young, T. F., *Chem. Rev.*, *43:* 69 (1948).
86. Boehm, H. P., Hoffmann, U., and Clauss, A., in *Proc. 3rd Conf. Carbon*, Vol. I., Pergamon Press, New York, p. 24.
87. Hoffmann, U. and Ohlerich, G., *Angew. Chem.*, *62:* 16 (1950).

88. Sykes, K. W. and White, P., *Trnas. Faraday Soc.*, *52:* 660 (1956).
89. Owen, A. J., Sykes, K. W., Thomas, D. J. D. and White, P., *Trans. Faraday Soc.*, *49:* 1198 (1953).
90. Studebaker, M. L. and Nabors, L. G., *Rubber Age*, *80:* 661 (1957).
91. Blayden, H. E. and Patrick, J. W., *Carbon*, *5:* 533 (1967).
92. Karpinski, K. and Swinarski, A., *Pyzem. Chem.*, *43:* 71 (1964).
93. Karpinski, K. and Swinarski, A., *Chem. Stosow. Ser.*, *A 9:* 307 (1965).
94. Karpinski, K. and Swinarski, A., *Chem. Stosow. Ser.*, *A 8:* 17 (1964).
95. Puri, B. R., Kaistha, B. C., and Hazra, R. S., *Chem. Ind. London*, *1967:* 2087.
96. Bansal, R. C. and Gupta, U., *Indian J. Technol.*, *18:* 131 (1980).
97. Blachand, L., in *C. R. 27th Cong. Int. Chim. Ind.*, 1955, p. 576.
98. Rassow, B. and Hofmann, K., *J. Prakt. Chem.*, *104:* 207 (1922).
99. Siller, C. W., *Ind. Eng. Chem.*, *40:* 1227 (1948).
100. Stacy, W. O., Vastola, F. J., and Walker, P. L., Jr., *Carbon*, *6:* 917 (1968).
101. Youg, R. T. and Steinberg, M., *Carbon*, *13:* 411 (1975).
102. Puri, B. R., Balwar, A. K., and Hazra, R. S., *J. Indian Chem. Soc.*, *44:* 975 (1967).
103. Puri, B. R., Jain, C. M., and Hazra, R. S., *J. Indian Chem. Soc.*, *43:* 67 (1966).
104. Puri, B. R., Kaistha, B., and Hazra, R. S., *J. Indian Chem. Soc.*, *45:* 1001 (1968).
105. Puri, B. R. and Hazra, R. S., *Carbon*, *9:* 123 (1971).
106. Puri, B. R., Jain, C. M., and Hazra, R. S., *J. Indian Chem. Soc.*, *43:* 554 (1966).
107. Chang, C. H., *Carbon*, *19:* 175 (1981).
108. Sinha, R. K. and Walker, P. L., Jr., *Carbon*, *10:* 754 (1972).
109. Ciersa, R., *Gazz. Chim. Ital*, *52:* 130 (1922); *55:* 385 (1925).
110. Wibaut, J. P. and Stoffel, A., *Rec. Trav. Chim.*, *38:* 132 (1919).
111. Puri, B. R. and Mahajan, O. P., *J. Indian Chem. Soc.*, *41:* 586 (1964).
112. Zawadski, J., *Carbon*, *19:* 19 (1981).

6

Active Carbon Applications

Activated carbons are unique and verstile adsorbents because of their extended surface area, microporous structure, high adsorption capacity, and high degree of surface reactivity. Their important applications relate to their use in the removal of color, odor, taste, and other undesirable organic impurities from potable waters, in the treatment of domestic and industrial wastewater, solvent recovery, air purification in inhabited spaces such as restaurants, food processing and chemical industries, for the removal of color from various types of sugar syrup, in air pollution control, in the purification of many chemical, pharmaceutical, and food products, and in a variety of gas phase applications. They are being increasingly used in the field of hydrometallurgy for the recovery of gold and silver and other inorganics and as catalysts and catalyst supports. Their use in medicine to combat certain types of bacterial ailment is well known.

6.1 PURIFICATION OF AIR

There are two types of system for purification of air. One is purification of air for immediate use in inhabited spaces such as offices, hospitals, laboratories, restaurants, and food processing plants where free and clean air is a requirement. The other system prevents air pollution of the atmosphere from exhaust airstreams emitted frcm various industrial operations such as production of gunpowder, celluloid, plastic and artificial leather, extraction process industries, rubber industries, paints and varnish industries, rayon, and adhesive plastic production.

Air purification in inhabited spaces operates at pollutant concentrations below 10 ppm, generally near 2–3 ppm, and in such systems panel-type carbon filters can be used. As the concentration of the pollutants is low, these filters can work for a long time and the spent carbon can be discarded since regeneration may be expensive. Air pollution control requires a different setup to deal with much larger concentrations of the pollutants. The spent carbon needs to be regenerated by steam, air, or nontoxic gaseous treatment. These two applications require carbons of different porous structures. The carbon required for purification of air in inhabited spaces should be highly microporous to effect greater adsorption at lower concentration. In the case of the carbon used for atmospheric pollution control, the pores should have a higher adsorption capacity in the concentration range 10–500 ppm. It is difficult to specify the pore diameters exactly, but generally pores with diameters in the supermicro and transitional range are preferred because they fill in this concentration range.

6.2 ACTIVE CARBON IN RESPIRATORS

There are two different types of respirators in which the nature of the carbon required is distinctly different. The respirators required in the chemical industry have chemicals of relatively low toxity and, usually, high molecular weight. They are strongly and quickly adsorbed by ordinary active carbons. But the respirators required for military purposes demand more complete protection against chemicals and vapors of extreme toxity. Furthermore, some of these chemicals may involve physisorption, whereas others may be chemisorbed (chloropicrin and cyanogen chloride). Thus for the latter case impregnated carbons are preferred with the impregnator having a specific reactivity with the different chemicals.

6.3 AIR PURIFICATION IN NUCLEAR PLANTS

Activated carbon filters are also used in nuclear power plants to prevent the release of radioactive vapors and gases such as iodine, organic iodides—mostly methyl iodide—and noble gases such as krypton and xenon into the atmosphere. The major concern in nuclear power reactors is the methyl iodide and noble gases under reactor failure conditions. In addition, activated carbons can be used for the decontamination of helium used as a protective gas in reactors cooled and moderated by heavy water.

6.4 RECOVERY OF GASOLINE, PROPANE, AND BUTANE FROM NATURAL GAS

Natural gas contains about 3% propane and 4-5% of higher hydrocarbons which can be recovered by adsoprtion on active carbon. About 35% of propane, 70% butane, and 98—99% pentane and higher hydrocarbons can be removed by activated carbon.

6.5 ACTIVE CARBON IN BEVERAGES

Wines and spirits are filtered through a bed of granulated activated carbon to remove any traces of fusel oil. In the case of brandies, purification by carbon serves to remove undesirable flavors which are picked up during manufacture and storage. Activated carbons also reduce the amount of aldehydes in a raw distillate and accelerate maturing. In beers, activated carbons are used to improve the quality of the defective brew, to improve the color of the beer, and to remove flavors that are attributed to phenol and coloring matter.

6.6 REFINING OF SUGAR, OILS, AND FATS

In the sugar industry active carbons are mainly used to remove coloring matter from syrups, thus giving the sugar a better appearance and also improving markedly its processing properties. Treatment with active carbons helps in removing surface-active agents and colloidal substances, raising their surface tension and decreasing their viscosity. This leads to higher rates of sugar crystallization and improved separation of syrups from crystals by centrifugation.

In the case of oils and fats, the activated carbons are used in conjunction with certain bleaching clays to remove undesirable coloring matter. A small addition of active carbon to bleaching clay substantially reduces the amount of adsorbent mixture needed to obtain the desired effect.

6.7 ACTIVE CARBON IN MEDICINE

A special type of active carbon is used in treating several ailments of the digestive system to remove bacterial toxins which are easily and quickly adsorbed by the carbon because of their high molecular weight. In addition, activated carbons are used in treating gastritis and enteritis and as an antidote in all cases of poisoning arising from mushrooms, food, alkaloids, metals, phosphorus, phenols, etc. Active carbon filters are used in cigarettes to remove nicotine and other toxic substances that are part of the smoke. Active carbons of neutral character are also being used for decaffeination of coffee beans.

6.8 ACTIVE CARBON FOR RECOVERY OF GOLD

The leaching of gold from its ores is carried out by oxidation of gold to Au(I) by oxygen. The oxygen is adsorbed and reduced cathodically on the gold surface and Au(I) is obtained in solution by complexation with cyanide ions as $Au(CN)_2^-$. However, at the mixed potential of the leach solution, which is usually between 0 and 0.5 V, several other metallic impurities present in the ore are oxidized and form complex cyanides, oxyanions, or sulfides and are leached with the gold cyanide complex. Thus a typical leach solution has the composition given in Table 1. Conventionally gold is recovered from such leach solutions by cementation with zinc. The process, although simple, is very sensitive to free cyanide concentration and to the presence of sulfides, arsenic, antimony, and copper, which make it much less effective and more laborious. Consequently, attempts have been under way to recover gold from the leach liquor by adsorption on carbons.

The history of the use of activated carbon for the recovery of gold can be traced back to 1880 when Davis in Australia (1) patented a process for the recovery of gold by adsorption on a wood charcoal from chlorination leach solutions. Soon after it was recognized that cyanide was a better solvent for gold and chlorination was gradually replaced by cyanidation as a leaching process. Johnson (2) then suggested that the wood charcoal could also be used for the adsorption of gold from cyanide leach liquors. These workers continued their investigation of the possibilities of using charcoal for the recovery of gold. However, the charcoals available in those days did not have the pore structure and the surface areas of the activated carbons which are available today, so that the adsorption capacity for gold was very small. Moreover, no suitable method was known to elute gold from the carbon surface. The only method known at that time was melting of gold after burning the charcoal. This was not economically viable because the ash formed was much larger compared to the amount of gold recovered. Therefore, carbon was never used seriously for the recovery of gold until the early 1950s when Zadra et al. (3, 4), at the U. S. Bureau of Mines, developed the direct carbon-in-pulp (CIP) process in which the carbon granules were directly added to the cyanide pulp and moved countercurrent to it. The gold-loaded carbon was recovered by screening.

CIP technology was the basis for the building of the first commercial plant in the United States in 1973 by the Homestake Mining Company (5) for large-scale recovery of gold by activated carbon. The success of the CIP process has aroused much interest in South Africa and several plants are being built at various mines. The applicability of the process is further enhanced by the availability of large quantities of more suitable activated carbons with large surface areas and varied porous structures.

Table 1 Composition of a Typical Gold Leach Solution[a]

Na	Ca	Fe	Ni	Zn	Cu	Ag	Au	Si	$S\bar{C}N$	$S_2O_3^{2-}$	SO_4^{2-}	$C\bar{l}$	$C\bar{N}$
310	588	2	1.5	15	7	1	5.4	20	90	4	1670	232	94

[a]Table based on report by D. M. Muir (personal communication).

The activated carbon process has certain definite advantages over the zinc concentration process because the adsorption of gold or silver is not appreciably affected by the presence of impurities such as copper and nickel in the leach liquors (6). Moreover, the carbon can be added directly to the cyanide pulp, avoiding the filtration and clarification stages of the zinc cementation process, making the activated carbon process more economical (Fig. 1). Furthermore, several different elution procedures have been investigated to recover gold from the carbon, regenerating the carbon, and recycling it. Since activated carbon can be used to adsorb even very small concentrations of gold (0.2 mg/L or even less), it is expected that activated carbons will be used in finishing applications such as the removal of dissolved gold cyanide from gold plant effluents and dam return waters.

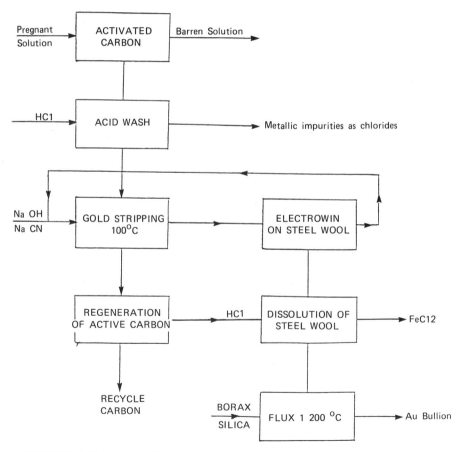

FIGURE 1 Schematic for removal of gold by activated carbon.

6.8.1 Mechanism of Gold Adsorption on Activated Carbons

The high selectivity of carbons for gold and silver in the presence of large concentrations of base metals such as copper, nickel, iron, cobalt, and antimony, and their availability in abundance and at competitive prices, is going to play a significant future role in the gold and silver recovery processes. The only difficulty at the present time is that there is no clear understanding regarding the nature of the interaction between the activated carbon surface and the gold cyanide solution. This makes the selection of an activated carbon for the recovery of gold arbitrary.

Several mechanisms have been advanced by different investigators to explain their results. The development of these mechanisms is considered under four different categories.

6.8.1.1 Reduction Theory of Gold Adsorption

Carbon is well known as a reductant. Many activated carbons prepared by activation with steam have very low reduction potential, of the order of 0.14 V against saturated calomel electrode (SCE). The chlorine leach solutions of gold have a reduction potential of + 0.8 V against SCE. Brussov (7) observed that when activated carbon was used for the recovery of gold from chlorine leach solutions, the color of the carbon changed and the gold could be seen on the surface of the carbon even at very low gold loadings. This led Brussov to suggest that the adsorption of gold in carbon occurred by reduction, which proceeded by the transfer of electrons from the interior to the surface of the carbon granule. Similarly in the case of $[AuBr_2]^-$, which have standard reduction potentials of + 0.7 V and + 0.3 V respectively against SCE, the gold was adsorbed partially or completely by the reduction mechanism. The metallic gold, according to Brussov (7), was held on the external surface of the carbon granules by dispersion forces. In the case of cyanide leach liquors, however, Brussov pointed out that since the aurocyanide complex is very strong, the adsorption of gold might not be completely by reduction.

Green (8), while investigating the loading of finely crushed wood charcoal with gold from cyanide solutions, observed that the charcoal turned yellow, which he suggested was due to the reduction of the cyanide on the carbon surface with the deposition of metallic gold. According to him this reduction was caused by the presence in charcoal of reducing gases such as CO and H_2. In support of his views he cited his observation that when these reducing gases were eliminated from the charcoal on outgassing at 500°C, the charcoal lost its capacity to load gold and that graphite, which did not contain these reducing gases, did not have the capacity to load gold. However, when he passed CO and H_2 gases through a solution of the gold cyanide, no precipitation or reduction of gold was observed. He suggested that these gases were present in the charcoal in the highly

activated state. Allen (9) rightly critized the arguments of Green
(8) that CO or H_2 which were strongly bonded chemically on the carbon surface and which could be evolved only on heat treatment at
500°C, could be available for the reduction of gold, which occurred
readily at room temperature. Allen was of the view that the decrease
in the gold-loading capacity of the charcoal on heat treatment at 500°
C was due to the elimination of carbon—oxygen surface structures
from the carbon surface. These surface structures are known to
render the carbon surface hydrophilic. The elimination of these structures rendered the carbon surface hydrophobic, so that the gold
cyanide solution could not completely wet and make contact with the
carbon surface, resulting in a decrease in gold loading.

Grabovskii and co-workers (10, 11) observed that phenol-formal-
dehyde resin-based carbons were better adsorbents and could extract
larger quantities of gold from their mixtures with silver, zinc, copper,
nickel, iron, and cadmium in cyanide solutions. These carbons adsorbed almost twice as much gold as KAD carbon and more than four
times as much as the Am-2B ion exchanger. The desorption of gold
from the carbon surface could be achieved completely only in the presence of gold complexing agents such as CN^- or SCN^- ions, which can
form strong complexes with gold. Thus the adsorption of gold, according to them, involved a reduction mechanism with carbon as the
reducing agent:

$$)C_x + NaAu (CN)_2 + 2NaOH \longrightarrow)C_xO...Na + Au + 2NaCN + H_2O$$

where $)C_x$ represent the carbon surface and $)C_xO...Na$, the oxidized
carbon surface with an adsorbed cation. However, it looks unlikely
that a carbon surface after losing an electron for reduction would be
able to adsorb a cation rather than an anion. Nevertheless, the reducing action of the carbon surface can be attributed to the presence
of delocalized π electrons of the aromatic condensed ring systems of
the activated carbon structures.

6.8.1.2 Ion-Pair Adsorption Theory

Feldtmann (12) observed that the loading of gold on wood charcoal
occurred more rapidly from a gold chloride solution than from a gold
cyanide solution. The gold loaded from the gold chloride solution
could be desorbed readily by a weak cyanide solution, whereas the
gold loaded from cyanide solution could not be desorbed significantly
in cyanide solution. This indicated that the adsorption of gold on
charcoal from a cyanide solution involved a different mechanism than
the adsorption of gold from a chloride solution. Further, the adsorption of gold decreased with increasing free cyanide ion concentration,
while increasing the sodium hydroxide concentration had no significant effect on adsorption. The adsorbed gold was soluble in solu-

tions of alkaline sulfides such as Na_2S, K_2S, NaHS, and KHS, although ammonium sulfide and alkaline polysulfides were less effective elutants. When charcoal was contacted with gold cyanide solution which contained excess of potassium cyanide, cyanogen was evolved and was retained on the carbon surface. These observations led Feldtmann to suggest that the gold cyanide was adsorbed on the charcoal surface, not in the metallic state but by a chemical precipitation mechanism involving the formation of a carbonyl aurocyanide addition compound $AuCN.CO(CN)_2$, which could be eluted with alkaline sulfides as sodium aurocyanide complex:

$$AuCN.CO(CN)_2 + NaS = NaAu(CN)_2 + NaSCN + CO$$

The elute solution, in fact, was found to contain appreciable amounts of thiocyanate and no gold sulfide. The carbon monoxide evolved in the reaction was readsorbed on the carbon and restored its activity to adsorb gold. Edmands (13) observed that the gold-loading capacity of wood charcoal was enhanced when the solution was acidic, whereas it was depressed in the presence of soluble sulfides. This observation supports Feldtmann's observation that alkaline sulfides were elutants for adsorbed gold. The addition of thiocyanate or thiosulfate to the gold cyanide solution was found to have a negligible effect on the gold-loading capacity of the charcoal. Edmands also found that the time required to reach equilibrium gold adsorption capacity depended on the particle size of the charcoal and the relative amounts of charcoal and gold present in the solution; further, time required varied when the solution was agitated. Thus although Edmands confirmed many of the observations of Feldtmann (12), he disagreed with the details of the mechanism proposed by him. Edmands still felt that carbon monoxide was involved in the adsorption of gold and suggested that it formed an additional compound with $Au(CN)_2$ such as $KAu(CN)_2CO$ or $HAu(CN)_2CO$, the latter one when the solution was acidic. This mechanism involved the oxidation of Au(I) to Au(III). In a discussion of the Edmands' paper (13), Allen (9) critized the formation of both carbonyl structures $AuCN.CO(CN)_2$ and $KAu(CN)_2CO$ involving chemical reaction with carbon monoxide and proposed that the $Au(CN)_2$ was physically adsorbed as $NaAu(CN)_2$ on charcoal without undergoing any chemical change. Allen's proposed theory was based on the fact the Feldtmann's results for the loading of carbon with gold from gold cyanide solutions could be accurately described by the empirical Freundlich equation, which was applicable to true adsorption phenomenan.

Gross and Scott (14), while investigating the factors that influence the adsorption of gold on a steam-activated wood charcoal, found that the adsorption process was very slow and equilibrium could not be obtained in 24 hr. The maximum adsorption of gold corresponded to about 20—25% of the monolayer coverage. The adsorption capacity of gold was enhanced by the addition of an acid and reduced appreciably by the addition of excess free hydroxide, sulfide, or cyanide ions. The adsorption capacity was higher from chloride solutions than from cyanide solutions (569 kg against 68 kg of gold per ton of carbon). The adsorption capacity of the charcoal also decreased with increase in the temperature of the cyanide solution. This was attributed to the increased solubility of the cyanide with the increased temperature. Analysis of the residual cyanide solution after adsorption showed that the solution contained larger quantities of bicarbonate and that the potassium ions remained in solution. On the basis of these observations these workers proposed that the loading of gold on pine wood charcoal, which contains appreciable amounts of CaO as ash, involved a chemical interaction between the Ca ions and the $Au(CN)_2^-$ anion. The cations of the gold complex $KAu(CN)_2$ remaining in solution as part of the bicarbonate which was formed with the CO_2 from the carbon:

$$2K[Au(CN)_2] + Ca(OH)_2 + CO_2 = Ca[Au(CN)_2] + 2KHCO_3$$

In the case of sugar charcoals, which contain little ash, a different reaction mechanism was proposed:

$$K[Au(CN)_2] + CO_2 + H_2O \rightleftharpoons H[Au(CN)_2] + KHCO_3$$

The increase in the loading of gold in acidic solution was attributed to the decomposition of the $Au(CN)_2^-$ complex anions in the presence of acid into insoluble AuCN:

$$Au(CN)_2^- + H^+ \longrightarrow AuCN + HCN$$

The AuCN was precipitated and retained in the pores of the carbon. Thus these workers made a distinction between the gold loading in acidic and alkaline solutions. Whereas in the former the loading was due to adsorption, in the latter precipitation was the main cause of loading of gold.

Kuz'minykh and Tyurin (15) observed that the adsorption capacity of birch and aspen carbons for gold cyanide was not influenced by the presence of simple anions such as Cl^-, I^- ions in the cyanide solution even when present in concentrations on the order of 1.5 M. Furthermore, the presence of neutral organic molecules such as kerosene and octyl alcohol decreased the gold-loading capacity of the carbon. These workers argued that the adsoprtion cannot be electrostatic in nature as proposed by Garten and Weiss (16), who proposed that the gold was adsorbed on the carbon surface as a neutral molecule, the nature of which depended on the pH of the solution. Under acidic conditions the adsorbate was a neutral molecule $HAu(CN)_2$, which involved a capillary condensation type of mechanism. However, under neutral or alkaline conditions the adsorbate was a salt of the type $NaAu(CN)_2$ and the adsorption involved dispersion forces only. Thus according to their mechanism the nature of the adsorbing species was different in alkaline and acidic solutions.

Davidson (17) examined some of the factors that influence the adsorption of gold using granulated activated coconut shell charcoal. A known weight of the charcoal was shaken with 100 ml of approximately 100 ppm solution of potassium aurocyanide for 17 hr at room temperature. The slurry was filtered using a Whatman No. 540 filter paper and the filtrate was analyzed for gold. The adsorption of gold obeyed the Freundlich equation. The adsorption of the carbon for gold increased significantly by increasing the ionic strength of the medium with sodium or calcium ions but the effect was more marked by the calcium ions (Fig. 2). The addition of 1000 ppm of Ca^{2+} showed a greater effect on rates and amount of gold adsorbed than the addition of 2000 ppm of Na ions. (Fig. 3). These complementary small cations were considered to stabilize the charge on the relatively large aurocyanide ion. The markedly greater influence of the calcium ion was attributed to the greater stability of calcium aurocyanide then sodium aurocyanide. Thus the spector ions constituting the charged double layer played an important role in adsorption. The increase in the adsorption capacity of the carbon for gold from 18 mg/g at pH 10 to 44 mg/g at pH 4 was attributed to the adsorption competition between H^+ and OH^- ions on the carbon surface sites. The rate of adsorption also increased with decrease in the pH of the solution (Fig. 3).

These differences in the behavior of sodium and calcium ions in enhancing the adsorption of gold, which cannot be completely explained on the basis of the ionic strength alone, and the fact that the calcium addition resulted in faster adsorption rates within the pH range 3—11 and the slopes and the shapes of the isotherms (Figs. 2 and 3) indicated that the adsorption mechanism was different when calcium was present as the spectator ion. Furthermore, it was found that there was no adsorption of gold in the complete absence of the stabilizing cations Na^+ and Ca^+ and that no gold could be desorbed from the charcoal surface that had been exposed to Ca^{2+} ions. These results indicated that the gold was adsorbed strongly as either calcium aurocyanide or hydrogen aurocyanide, depending on the pH of the solution (adsorption medium). Thus Davidson postulated that these spectator ions, the concentration and the character of which determined the adsorption of gold on the charcoal, formed on adsorption complex on the carbon surface. The strength of the complex was in the order

$$Ca^+ > Mg^{2+} > H^+ > Li^+ > Na^+ > K^+$$

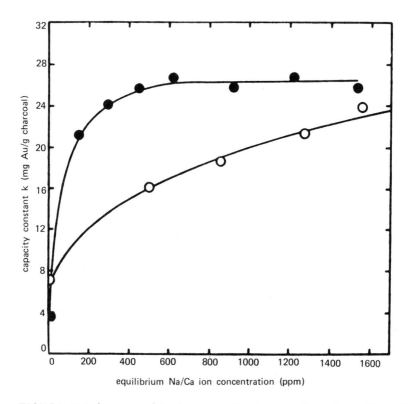

FIGURE 2 Influence of ionic concentration on the adsorption of gold by carbon at pH 7.2. Closed circles, Ca^{2+}; open circles, Na^+. [From Davidson (17). Reproduced with permission from *South African Inst. Min. Metall.*]

the calcium aurocyanide being the most stable complex and potassium aurocyanide the least firmly bound. In other words, these observations pointed out that the alkaline earth metal aurocyanide complexes were more stable than the alkali metal complexes. The elution of ad-sorbed gold, which was very much enhanced on treatment of the gold-loaded carbon with alkali metal carbonates, was due to the exchange of the calcium ions of the adsorbed complex with the alkali ion:

$$Ca[Au(CN)_2]_2 + K_2CO_3 \longrightarrow 2K[Au(CN)_2] + CaCO_3$$

and converting the firmly bound adsorbed complex into a more readily desorbable adsorption complex so that it could be more easily eluted with hot water.

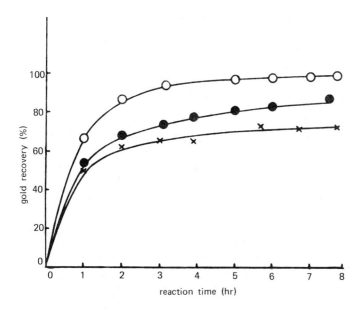

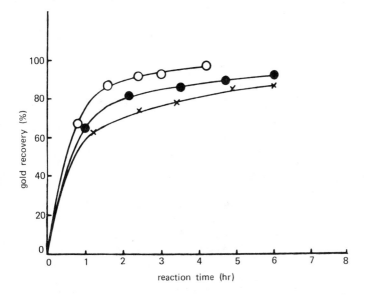

FIGURE 3 Influence of pH on the rates of adsorption of gold in the presence of (2000 ppm) sodium and (1000 ppm) calcium ions. Open circles, pH 3.1; closed circles, pH 7.2; -x-x-, pH 10.6. [From Davidson (17). Reproduced with permission from *South African Inst. Min. Metall.*]

Cho (18), however, pointed out that the enhancement of the desorption of gold from the carbon surface by treatment with alkali carbonates, as suggested by Davidson, did not necessarily imply that the exchange of the calcium ions with potassium ions was the primary driving force for the observed desorption effect. The fact that the solutions of alkali carbonate in water were alkaline in nature and would raise the pH of the medium could equally well account for the increased elution of the gold. Gross and Scott (14) had shown that the best eluants for adsorbed gold are those substances that raise the pH of the solution.

6.8.1.3 Adsorption of Gold as Aurocyanide Anion [Au(CN)$_2$]$^-$

Garten and Weiss (16) suggested that the gold was adsorbed from the cyanide solution on carbons as $Au(CN)_2^-$ anion by an anion exchange mechanism which involved simple electrostatic interaction between positive and negative charges. The positive sites on carbons were the carbonium ion sites, which were formed in alkaline solutions by the oxidation of chromene groups that existed on carbon surfaces. The adsorption of gold according to these workers occurs as

However, their mechanism cannot explain the increase in the adsorption of gold from acidic cyanide solutions.

Dixon et al. (19) invoked the Frumkin electrochemical theory (20−22) of acid adsorption to explain the mechanism of gold adsorption on carbons. According to this theory, the carbon−oxygen surface structures present on activated carbons undergo interaction with water, producing positive sites on the carbon surface

$$C_x O + H_2O \rightarrow C_x^{2+} + 2OH^-$$

The gold cyanide anion $Au(CN)_2^-$ was adsorbed on these positive sites by electrostatic interaction forces. When the solution was acidified, the equilibrium in the preceding equation was shifted to the right, resulting in the creation of more positive sites. This caused an increase in the adsorption of gold at low pH values. The increase in the adsorption on bubbling oxygen through the solution could be explained by this mechanism. Clauss and Weiss (23), however, found that the activity of carbon to adsorb gold was destroyed almost completely on oxidation of the carbon by boiling with a mixture of sulfuric acid and nitric acid. Furthermore, the addition of reducing agents such as hydrazine and hydroquinone to the cyanide solution decreased the adsorption of gold, whereas the addition of oxidizing agents such as hydrogen peroxide enhanced the adsorption. These workers also calculated the surface area covered by adsorbed gold from the data of Davidson (17) and found that only a small fraction of the surface area was involved in the adsorption process. These observations led them to suggest that the active sites provided by surface oxide structures were not involved in the adsorption of gold. Instead the gold was adsorbed partly in some special type of micropores or on certain quinonic sites. McDougall and Hancock (24), however, critized this view of Clauss and Weiss and suggested that the loss of activity of carbon to adsorb gold on treatment with the mixture of sulfuric and nitric acid may be due to a complete modification of the structure of the carbon, whereas the decrease in gold adsorption capacity by the addition of hydrazine and hydroquinone may be attributed to the competition between the gold cyanide and these neutral molecules because carbons are known to adsorb such molecular species (25). In fact the addition of several organic molecules such as ethanol, acetone, and methanol has been found to enhance the rate and efficiency of gold elution from a carbon surface because of the preferential adsorption of these molecules.

Cho et al. (26) studied the kinetics of adsorption of gold cyanide on a coconut shell—activated carbon as a function of temperature, free cyanide concentration, and charcoal content and observed that the adsorption involved simultaneously a pore diffusion process and adsorption on active sites. The adsorption isotherms using cyanide solutions containing 25−500 ppm of free cyanide at two different levels of initial gold concentration (90 and 180 ppm) and the temperatures of 25, 40, and 55°C obeyed the Freundlich equation. The adsorption rates were calculated from a diffusion control model developed by Crank (27) with the effective diffusion having an activation energy of 8−13 kj/mol. Good agreement was obtained between the experimental rate data and the rate predicted by the diffusion model (Fig. 4) at different dosages of carbon, different free cyanide concentrations, and different temperatures. The initial rates of adsorption were also a function of the size of the carbon granules, smaller size fractions being faster. But the size of the carbon granules had no effect on the equilibrium value of adsorption of gold cyanide, which was a function of the pH of the cyanide solution, being twice as high in the pH range 4−7 than in the pH range 8−11.

Cho and Pitt (28), while investigating the adsorption of silver cyanide on activated coconut shell charcoal from solutions containing varying concentrations of Na^+, Ca^{2+}, free cyanide CN^-, and hydrogen ions, observed that the adsorption of silver cyanide, $Au(CN)_2^-$, increased with increase in the Na^+ or Ca^{2+} ion concentrations while it decreased with increase in the concentration of free CN^- ions (Fig. 5). The positive effect of the Na^+ ion concentration exceeded the negative

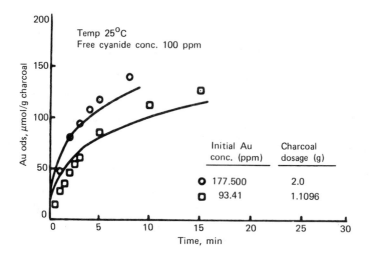

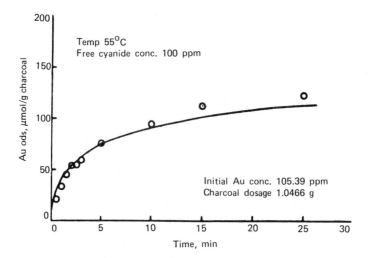

FIGURE 4 Comparison of adsorption rates obtained from a diffusion model with experimental data. [From Cho et al. (26). Reproduced with permission from *Metallurigal Transactions*.]

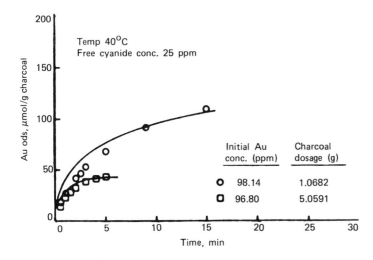

FIGURE 4 Continued.

effect of free CN^- ion concentration at low concentrations of Na^+ ions by a factor of 1:2, although the effect was reversed at higher concentrations of sodium ions. The maximum loading of silver cyanide anion, $Ag(CN)_2^-$, occurred at $\cong 45$ mmol/L $Na(CN)_2$ concentration. The increased adsorption of silver cyanide in the first half of the curve in Figure 6 was attriduted to the fact that more sodium ions were adsorbed in this region than the cyanide ions, the difference in the amounts adsorbed between the two accounting for the increased adsorption of silver cyanide. It was found that sodium or calcium ions were not adsorbed from solutions of their salts in the absence of silver cyanide ions. This indicated that these ions (Na^+ or Ca^{2+} ions) were adsorbed on the silver cyanide complex already adsorbed on the charcoal surface.

The adsorption of silver syanide on charcoal increased with increasing acidity of the solution except in the intermediate regions of acidity where the adsorption remained more or less unchanged (Fig. 7). This could be attributed to pH changes of the solution as a result of adsorption. However, the results of pH measurements of the initial and equilibrium solutions (Table 2) showed clearly that there was only a negligible pH change when the adsorption was carried out at higher pH values, although an appreciable amount of silver cyanide was being adsorbed (at initial pH values of 10.98 and 10.36, the change in pH value due to adsorption was only 0.08 and 0.05 respectively) and the pH of the solution continue to change even when no appreciable amount of silver cyanide was being adsorbed

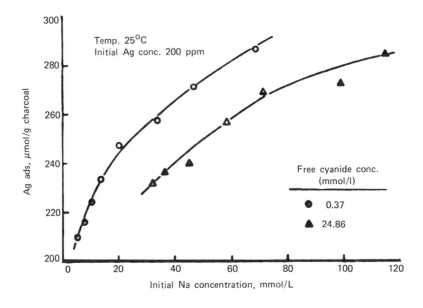

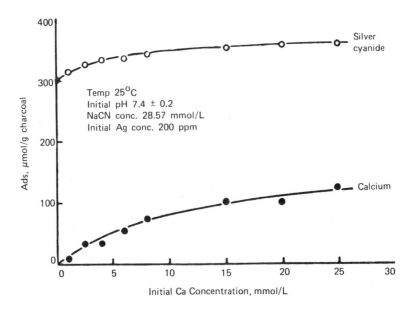

FIGURE 5 Influence of sodium, calcium, and free cyanide ions on the adsorption of silver cyanide. [From Cho and Pitt (28). Reproduced with permission from *Metallurgical Transactions*.]

(Fig. 7 and Table 2). These workers also measured the zeta potential of the carbon at different pH values and observed that the variation of zeta potential with pH showed behavior similar to the adsorption of silver cyanide. This indicated that the adsorption of silver cyanide was more influenced by the zeta potential of the charcoal particles. The influence of zeta potential on the adsorption of gold was studied by Gupta (27a) who suggested that the gold cyanide adsorption on carbons depended on the zeta potential of the carbon at a particular pH. A comparative study of the adsorption of gold and silver cyanides from their solutions containing similar concentrations of the two cyanides (28) showed that gold cyanide adsorbed three times more than silver cyanide. This could not be explained on the basis of the electrostatic forces alone, as suggested by Garten and Weiss (16), since both $Au(CN)_2^-$ and $Ag(CN)_2^-$ are univalent negative ions. Similarly, the adsorption of silver cyanide at pH 11−11.5, where the zeta potential of the carbon is -50 to -60 mV, which

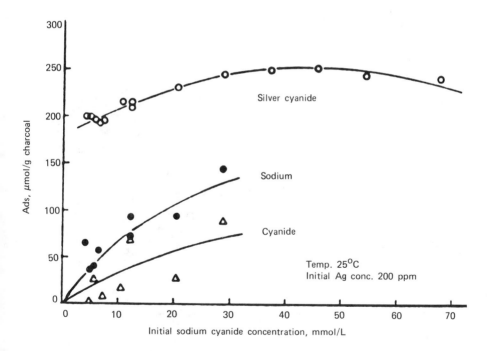

FIGURE 6 Adsorption isotherms of silver cyanide, sodium, and cyanide ions on activated carbon. [From Cho and Pitt (28). Reproduced with permission from *Metallurgical Transactions*.]

indicates that the carbon surface will have more negative sites than positive sites, could not be explained by the electrostatic forces.

On the basis of these observations Cho and Pitt (28) suggested a mechanism for the adsorption of both $Au(CN)_2^-$ and $Ag(CN)_2^-$ on carbons which was in accordance with the ionic solvation theory developed by Bockris and co-workers (29–31) to explain the specific adsorption of anions on metal electrodes. According to this mechanism, a large, weakly hydrated anion such as $Au(CN)_2^-$ or $Ag(CN)_2^-$ was adsorbed specifically on the charcoal surface while small anions such as CN^- remain in the outer part of the electrical double layer. This received support from their observation that the adsorption of Ag $(CN_2)^-$ anion was much larger than that of CN^- ions and that the hydrated Na and Ca cations were adsorbed only when they were present in the solution along with $Ag(CN)_2^-$ anions and not when present with smaller CN^- ions. These workers also invoked the multilayer formation character of the electrical double layer (double layer accord-

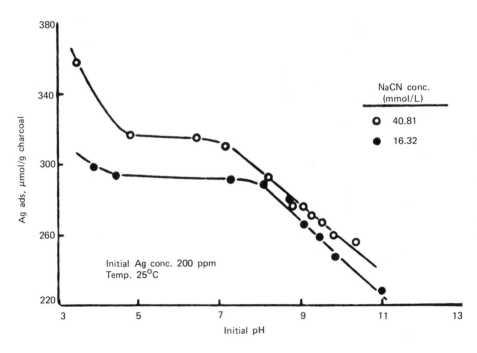

FIGURE 7 Influence of pH on the adsorption of silver cyanide ion. [From Cho and Pitt (28). Reproduced with permission from *Metallurgical Transactions*.]

Table 2 pH Change through the Adsorption Reaction

| NaCN (mmol/L) | | | | | |
| 16.32 | | | 40.81 | | |
Initial pH	Equilibrium pH	Difference	Initial pH	Equilibrium pH	Difference
3.90	8.20	4.30	3.48	6.47	2.99
4.48	8.28	3.90	4.83	7.79	2.96
7.28	8.37	1.09	6.46	7.87	1.41
8.09	8.58	6.49	7.16	7.99	0.83
8.71	8.95	0.24	8.21	8.50	0.29
9.07	9.27	0.20	8.80	8.92	0.12
9.46	9.64	0.18	9.06	9.15	0.09
9.85	10.08	0.23	9.27	9.28	0.01
10.98	11.06	0.08	9.53	9.64	0.11
			10.36	10.41	0.05

Source: Cho and Pitt (28). Reproduced with permission from *Metallurgical Transactions*.

ing to Cho and Pitt was a misnomer) according to which the electrical double layer formation also included short-range forces of chemical nature in addition to long-range electrostatic forces so that the interfacial region consisted of more than two distinct subregions of layers. Thus according to Cho and Pitt (28) both silver syanide and free cyanide ions were adsorbed on the carbon-active sites but the free cyanide ions had much less adsorbility compared to silver cyanide ions. Therefore, from a solution containing both ions [CN^- and $Ag(CN)_2^-$ or by analogy $Au(CN)_2^-$ ions], the silver cyanide or the gold cyanide anions were adsorbed preferentially and to a greater extent. The sodium or calcium ions were adsorbed in the diffuse layer on the cyanide complex already adsorbed on the carbon-active sites. These cations adsorbed in the diffuse layer provided more active sites for the adsorption of more cyanide ions in the diffuse layer when the concentration of sodium or calcium ions was increased in the solution. The decrease in the amount of $Ag(CN)_2^-$ with increase in the concentration of free CN^- ions was due to the competition between the two anions for the carbon-active sites. At higher concentrations of CN^- ions for a given concentration of $Ag(CN)_2^-$ anions, some of the free cyanide ions may attach to carbon-active sites. The larger adsorption of $Au(CN)_2^-$ anions compared to that of $Ag(CN)_2^-$ from their solutions of similar concentrations was attributed to the larger size of the $Au(CN)_2^-$ anion.

6.8.1.4 Adsorption as a Cluster Compound of Gold

McDougall et al. (32) reinvestigated the influence of such factors as ionic strength, pH, and temperature on the loading, not only of gold cyanide but also of a neutral molecular species, $Hg(CN)_2$, with a view to get some clear understanding of the adsorption of gold on carbon. These workers carried out analysis of adsorbed layer and used X-ray photoelectron spectroscopy to determine the oxidation state of the gold adsorbate. The influence of ionic strength on the adsorption of gold was examined by adding different salts such as KCl, NaCl, and $CaCl_2$ at the same ionic strength. The rate of adsorption, which was quite fast in the initial stages, was slowed down after 6 hr and the equilibrium was not obtained even after 3 months, although the rate of adsorption after 24 hr was very slow. The addition of sodium, potassium, and calcium ions enhanced the adsorption of gold, the increase being larger in the presence of calcium ions. This was in accordance with the earlier observations (17, 28). However, the increase in adsorption was not linear with the increase in the amount of the electrlyte. It increased tremendously with the first small additions but was much slower with further additions (Fig. 8) of the electrolyte. This indicated that the effect was not merely an ionic strength effect but the ions such as Ca^{2+} had a specific involvement in the adsorption of gold cyanide. The equilibrium adsorption isotherms at equal ionic

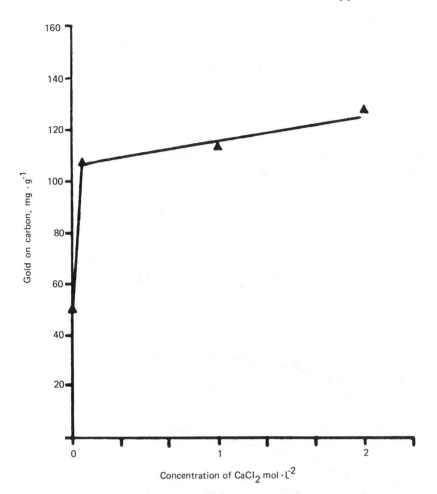

FIGURE 8 Influence of electrolyte concentration on the adsorption of gold. [From McDougall et al. (32). Reproduced with permission from *South African Inst. Min. Metall.*]

strengths of different electrolytes and in the presence of an acid (Fig. 9), showed that the highest adsorption occurred in the presence of the acid. With chloride ions as the common ion the calcium ions at the same ionic strength gave a higher gold adsorption than did potassium ions. Similarly, in the presence of the common ion K^+, the adsorption was higher when the anion was chloride than when it was cyanide or ClO_4^-. Furthermore, the adsorption of gold was about 1 1/2 times more in the presence of 0.01 M nitric acid ($\cong$ 180 mg/g) than when a much stronger calcium chloride was present ($\cong$ 120 mg/g), although the ionic strength in the latter solution was much higher. Thus McDougall et al. (32) suggested that the increase in the adsorption of gold may be attributed more to the pH effect than to ionic strength effect. The influence of pH on the adsorption of gold at different pH values of the solutions with no electrolyte additives (Fig. 10) did show an appreciable increase in adsorption at lower pH values.

A Comparison of the adsorption of gold in the presence of 8 x 10^{-2} M Cl$^-$ ions and 8 x 10^{-2} M to 2.4 x 10^{-2} M ClO$_4^-$ anions in the adsorption medium in the case of activated carbon and an anion exchange resin IRA-400 (Table 3) showed that the behavior of the carbon was completely different from that of the resin. The increase in the amount of ClO$_4^-$ anions by a factor of 3 did not affect the gold adsorption capacity of the carbon, and at an ionic strength of 8 x 10^{-2} M, the gold capacity was slightly less when the competing anion was ClO$_4^-$ than when it was Cl$^-$ anion. The gold adsorption capacity of the ion exchange resin was, however, considerably reduced in the presence of the ClO$_4^-$ anion and more so as its concentration was in-

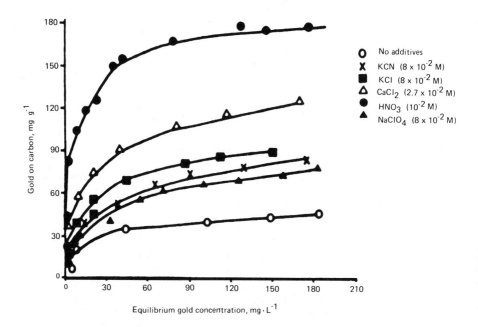

FIGURE 9 Adsorption isotherms of gold in the presence of equal strengths of different ions. [From McDougall et al. (32). Reproduced with permission from *South African Inst. Min. Metall.*]

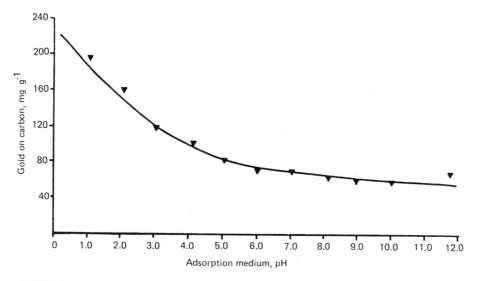

FIGURE 10 Adsorption of gold cyanide as a function of pH. [From McDougall et al. (32) Reproduced with permission from *South African Inst. Min. Metall.*]

Table 3 Influence of Cl⁻ and ClO₄⁻ Ion Concentration on the Adsorption of Gold by Activated Carbon and Resin IRA 400[a]

Concentration of ion	Gold adsorbed (mg/g)	
	Carbon	IRA 400
8×10^{-2} M NaCl	74.6	95.6
8×10^{-2} M NaClO$_4$	69.0	45.6
1.6×10^{-1} M NaClO$_4$	70.6	32.2
2.4×10^{-1} M NaClO$_4$	68.6	21.6

[a]Experimental conditions: volume of solution 50 ml; mass of carbon or resin 0.25 g; equilibrium time 24 hr. [From McDougall et al. (32). Reproduced with permission from *South African Inst. Min. Metall.*]

creased. McDougall et al. (32) inferred that the adsorption of gold on activated carbon cannot be attributed to simple electrostatic inter-actions between $Au(CN)_2^-$ anions and the positively charged sites on the carbon surface (15, 16, 18, 28). The adsorption isotherms of gold at various temperatures between 22 and 79°C (Fig. 11) showed that the adsorption of gold decreased as the temperature was increas-ed. This was attributed to the greater solubility of potassium auro-cyanide at higher temperatures. The average value of the isosteric heat adsorption of gold was found to be 42 J/mol. The fact that the adsorption of gold was sensitive to temperature to the extent of 42 J/mol indicated that the adsorption mechanism did not involve ion ex-change-type mechanisms, because the effect of temperature on the ion exchange equilibria is usually very small (17, 19).

The $4f(\frac{7}{2})$ binding energies of gold metal, $KAu(CN)_2$, and for the carbons loaded with gold were calculated from their XPS spectra (Table 4). The binding energies were almost the same for all the gold-loaded carbons irrespective of the composition of the adsorption medium, which indicated that the gold adsorbate was the same in each case. These values differ from the binding energy of metallic gold

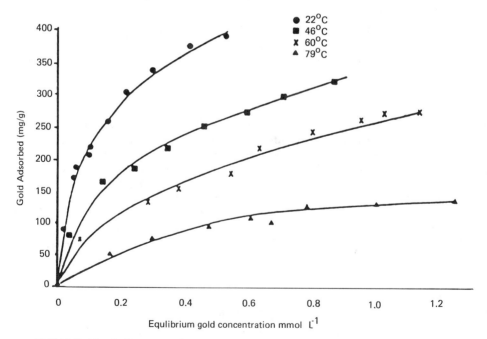

FIGURE 11 Influence of temperature on the adsorption of gold. [From McDougall et al. (32). Reproduced with permission from *South African Inst. Min. Metall.*]

Table 4 Binding Energies for Some Gold Compounds and Gold-Loaded Carbons[a]

Material	Binding energy (eV)[b]
Au metal	90.8
Au(I) in $KAu(CN)_2$	93.6
Gold-loaded carbons	
Only $KAu(CN)_2$	91.5
$KAu(CN)_2$ + 0.1 M HNO_3	91.5
$KAu(CN)_2$ + 0.1 M $CaCl_2$	91.6
$KAu(CN)_2$ + 0.1 M KCl	91.5
$KAu(CN)_2$ + 0.1 M KCN	91.6

[a] The gold loading on the carbons varied from 1 to 15%.
[b] The standard devistion on all these values is 0.3 eV.
Source: McDougall et al. (32). Reproduced with permission from
South African Inst. Min. Metall.

Au(O) by about 0.7–0.8 eV. The difference between the binding energy of metallic gold Au(O) has been found to be between 1.5 and 2.3 eV (33). Furthermore, a comparison of the XPS spectra of Au(I) in $KAu(CN)_2$, of gold metal adsorbed on carbon from a solution of $AuCl_4^-$ (where it is known that the gold is adsorbed by a reduction mechanism as metallic gold) and of gold adsorbed on carbon from gold cyanide solution (Fig. 12) showed a slight shift between the $4f(\frac{7}{2})$ levels of Au(O) and gold adsorbed from the cyanide solution. These observations led McDougall et al. (32) to suggest that the gold was adsorbed on the carbon surface neither as $Au(CN)_2^-$, where it has an oxidation state of 1, nor as metallic gold, where its oxidation state is 0. These workers postulated that the adsorption of gold on carbon involved a reduction mechanism in which a unique gold cyanide species containing gold atoms with an oxidation number 0.3 was formed and adsorbed on the carbon. This view was further supported by the correlation which these workers found between the gold adsorption capacities of a series of commercially available activated carbons, prepared from coconut shell and bituminous coal, from a solution containing initially 250 mg Au/L and their reduction potentials. The results (Fig. 13) indicated that the adsorption capacity of a carbon increased with increase in its reduction potential. A linear correlation was also found between the amount of gold cyanide adsorbed on a carbon and the amount of metallic gold adsorbed on the same carbon from a solution containing $AuCl_4^-$ anions.

McDougall et al. (32) also observed that the neutral $Hg(CN)_2$ and the anion $Au(CN)_2^-$ competed for the adsorption sites on carbon when present together in the same solution and could even displace some of

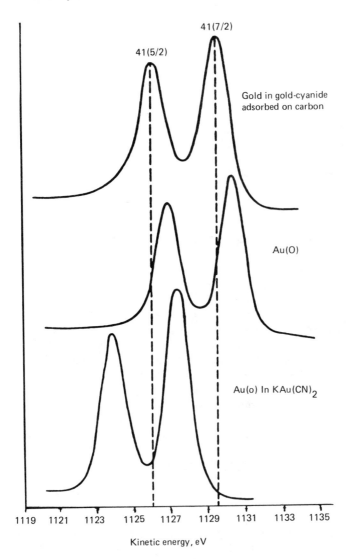

FIGURE 12 X-ray photoelectron spectra (Au 4f doublet) of gold. [From McDougall et al. (32). Reproduced with permission from *South African Inst. Min. Metall.*]

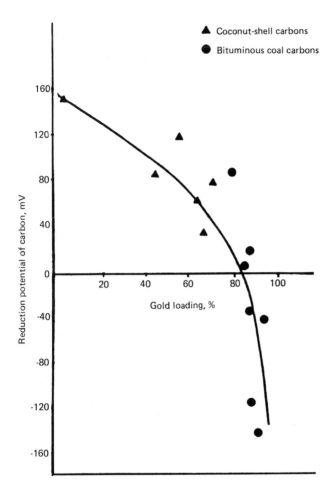

FIGURE 13 Adsorption of gold in relation to reduction potential of carbons. [From McDougall et al. (32). Reproduced with permission from *South African Inst. Min. Metall.*]

the adsorbed gold from activated carbon. This indicated clearly that the negative charge of the gold cyanide anion was not essential for the adsorption on carbon. A comparison of the nitrogen content of the gold-loaded carbons from microanalysis with the nitrogen content calculated on the basis of adsorption of $Au(CN)_2^-$ anion (Table 5) showed less nitrogen by microanalysis, which indicated again that the gold was not adsorbed as $Au(CN)_2^-$ anion on the carbon surface.

On the basis of these observations McDougall and co-workers (24) postulated that the adsorption of gold by carbons involved two steps. The first step was the adsorption of an ion pair $M^{n+}[Au(CN)_2]^{n+}$ complex where M^{n+} could be H^+ ion or a metal ion. The nature of the M^{n+} ion determined the solubility of the complex, being more when it was an alkali metal and less when it was H^+ or an alkali earth ion. This could explain the effect of pH, temperature, and the ionic strength on the adsorption of gold and some of its elution characteristics. In the second step the cyanide complex underwent reduction into a species which these workers postulated to be either a substoichiometric AuCN polymer with metallic gold in the matrix or a cluster compound of gold comprising gold atoms both in 0 and 1 oxidation states. This cluster compound was thought to be similar structurally to the well-known triphenyl phosphine compounds of gold $Au_{11}(CN)_3[P(C_6H_5)_3]_7$, as their XPS spectra were found to be almost similar to those for the gold-loaded carbons. The presence on the carbon of a gold cyanide complex which had lost cyanide also accounted for the complete desorption of gold in the presence of gold complexing agents such as CN^- or SCN^- in the eluant. These workers (32), however, pointed out that whether these compounds were loaded

Table 5 Nitrogen Content of Gold-Loaded Carbons

Gold on carbon	Nitrogen content of carbon if $Au(CN)_2^-$ adsorbed (%)	Nitrogen content from microanalysis (%)
0.0	—	0.10
3.4	0.48	0.50
5.8	0.82	0.53
7.3	1.04	0.40
14.1	2.00	1.21

Source: McDougall et al. (32). Reproduced with permission from *South African Inst. Min. Metall.*

on the carbon surface by physisorption or whether these are deposited as microcrystallites in pores in a manner similar to capillary condensation was not clear.

Tsuchida et al. (34, 35) studied the adsorption isotherms of $Au(CN)_2^-$, $Ag(CN)_2^-$, and $Hg(CN)_2$ on original and deoxygenated Norit carbon. The adsorption on both samples of the carbon followed the Freundlich equation. However, the values of the Freundlich equation constants varied with the nature of the carbon as well as with the type of cyanide complex (Fig. 14). The slopes of the isotherms for

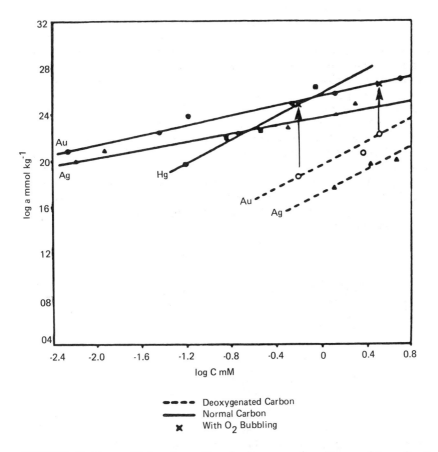

FIGURE 14 Freundlich adsorption isotherms of gold cyanide, silver cyanide, and mercuric cyanide on as-received and deoxygenated carbons. [From Tsuchida et al. (34).]

$Au(CN)_2^-$ and $Ag(CN)_2^-$ were higher for the deoxygenated sample. The original carbon loaded much larger amounts of gold compared to the deoxygenated sample at all studied concentrations. However, when the deoxygenated carbon sample was contacted with gold cyanide solution through which oxygen was being bubbled, the adsorption of gold increased to that observed in the case of the original carbon (Fig. 14). Furthermore, the molar ratio of $Au(CN)_2^-$ to K^+ on the surface of the original carbon was 2.5:1 at pH 10 while it was 1:1 in the case of the deoxygenated sample. This indicated that the original charcoal had some neutral or ion pair species adsorbed on its surface and that oxygen was involved in some way in the adsorption process. The difference in the molar ratios of $Au(CN)_2^-$ to K^+ on the two original and deoxygenated carbon samples indicated that about half of the adsorption of $Au(CN)_2^-$ involved some interaction between the gold species and the surface oxygen. The adsorption behavior of the carbon toward neutral $Hg(CN)_2$ was almost similar with or without deoxygenation. The cyclic voltamograms of pure $KAu(CN)_2$ and $KAg(CN)_2$ on a rotating platinum electrode showed an oxidation peak around 0.4 V and a reversible reduction peak at 0.5 V from $Au(CN)_2^-$ solution and at 0.2 V for $Ag(CN)_2^-$ solution (Fig. 15). Similar but less defined peaks were obtained on glassy carbon while poor peaks were observed with the activated carbon. The reversible peak around 0.4 V was found to be dependent on the $Au(CN)_2^-$ concentration, pH of the solution, and the scan rate and disappeared almost completely on addition of 0.01 M CN^- or upon raising the pH to 12 (36). This reversible peak was attributed by Tsuchida (35) to the following redox reaction of $Au(CN)_2^-$, $AuCN$, and $(CN)_2$ which were chemisorbed on the platinum surface:

$$Au(CN)_2^- + e \rightleftarrows Au(CN) + 1/2\ (CN)_2$$

$$\overline{\underset{/////////////////////////////}{CN^-\quad Au\quad CN^-\quad Au\quad Cn^-}}\ Pt\ surface$$

$$+\ 2e\ \Big\downarrow\Big\uparrow +\ 2e$$

$$\overline{\underset{/////////////////////////////}{CN^-\quad Au^I\quad (CN)_2\quad Au^I\quad CN^-}}$$

Such a layer would disappear on the addition of CN^- ions or on increasing the pH of the solution since these factors promote the decomposition of cyanogen $(CN)_2$. Thus these workers (34,35,36) inferred that the adsorption of Au or Ag on carbons involved oxidation of Au $(CN)_2^-$ or $Ag(CN)_2^-$ anions to neutral AuCN species. According to these workers, the active sites on carbon surfaces which may be oxygen-

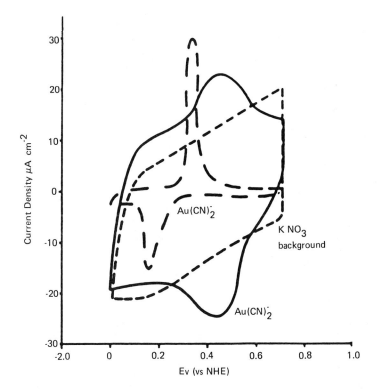

FIGURE 15 Cyclic voltamograms of pure $KAu(CN)_2$ and $KAg(CN)_2$ on a rotating platinum electrode. (0.1 M KNO_3, 0.01 M $Au(CN)_2^-$ scan rate 1 Vs^{-1}; 0.1M $Ag(CN)_2^-$ scan rate 10 mVs^{-1}). [From Tsuchida et al. (34).]

containing structures have a multifunctional character. These act as charge-separated groups and allow the adsorption of $Au(CN)_2^-$ and cations. The acidic or oxygen-containing surface groups then oxidize $Au(CN)_2^-$ to neutral $AuCN$ species with further adsorption of $Au(CN)_2^-$ anions. The mechanism is presented in Figure 16. Desorption of gold required the reformation of $Au(CN)_2^-$ anion and charge reunion either through change in temperature or solvent or through removal of ions. The carbonate ion caused the precipitation of calcium and magnesium on the carbon surface and deactivated the carbon sites, inhibiting the adsorption process.

Fleming and Nicol (36) recently observed that the factors such as mixing efficiency and the particle size of the activated carbon also influence the loading of gold on carbons. The rate of adsorption was related directly to the stirring speed (Fig. 17) and inversely to the

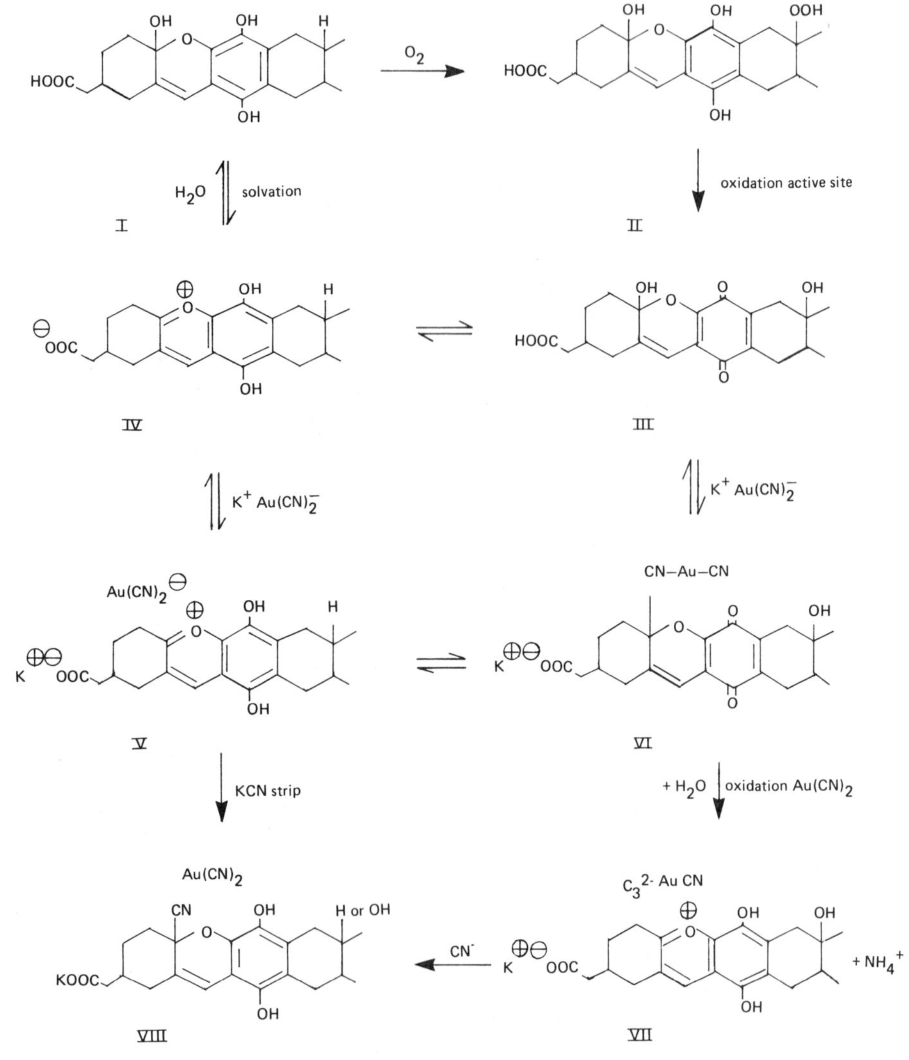

FIGURE 16 Mechanism of gold adsorption on carbons. [From Tsuchida et al. (34).]

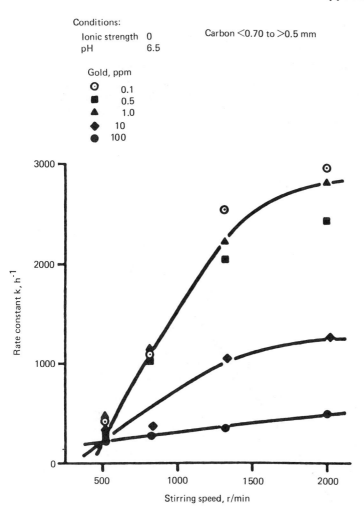

FIGURE 17 Influence of stirring speed on the adsorption of gold by carbons. [From Fleming and Nicol (36). Reproduced with permission from *South African Inst. Min. Metall.*]

particle size (Fig. 18). This was attributed to the predominance of film diffusion in limiting the rate of extraction of gold cyanide. In addition, the presence of certain organic species such as flotation reagents, lubricating oils, and kerosene or inorganic species such as calcium carbonate or hemite decrease the adsorption rate and the capacity of the carbon to load gold (Table 6a). These species act as poisons and are adsorbed strongly on the carbon surface. However,

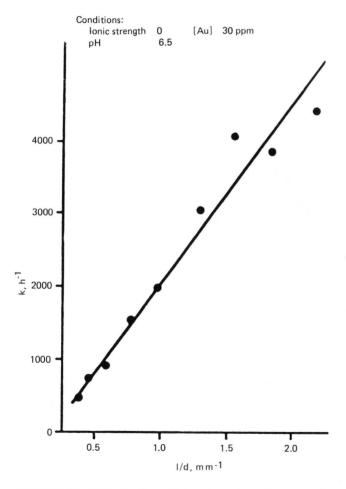

FIGURE 18 Effect of particle size of carbon on the adsorption of gold.

Table 6a Effect of Organic Solvents on Loading of Gold

Solvent concentration (%)	Date constant k (h^{-1})					
	Heptane	Benzene	Acetone	Acetonitrile	Ethanol	Methanol
0	2070	2070	2070	2070	2070	2070
0.005	1479	1169				
0.02	896	1023				
0.05	136	317	461	810	1140	1095
1.00	—	—	286	384	852	1245
2.0			174	245	327	1086

their influence was felt more in the initial stages. Consequently, it was suggested that more activated carbon should be added in the initial stages to eliminate these species by adsorption.

The rate of loading gold was also influenced by the temperature as well as by the concentration of the cyanide solution (36). The results of some such experiments in the temperature range between 20 and 80°C and at free cyanide concentrations between 0 and 1300 ppm (Table 6b) indicate that the equilibrium loading is exothermic. This observation forms the basis of the high-temperature elution procedures that are adopted at different gold recovery plants. The increase in loading rate with temperature was fairly small compared to the diffusion-controlled processes. The Arrhenius plot showing the effect of temperature on gold extraction is linear (Fig. 19) and gives an activation energy of 2.6 Kcal mole^{-1}, which agrees with the value obtained by Dixon et al. (19). An increase in the concentration of free cyanide ions decreases the rate of loading and the equilibrium capacity, and this feature is used in the elution of gold from activated carbons.

Table 6b Effect of Temperature and Sodium Cyanide Concentration on the Adsorption of Gold

Temperature (°C)	Free cyanide concentration (ppm)	Adsorption capacity (ppm)
20	0	73000
25	130	62000
24	260	57000
23	1300	59000
44	0	48000
43	130	47000
42	260	42000
43	1300	33000
62	0	35000
62	130	29000
62	260	29000
62	1300	26000
81.5	260	20000

Polania et al. (44) examined the influence of surface chemical structures and the porous structure of active carbons on the adsorption of gold cyanide using coconut shell, wood, and peat charcoals which were oxidized and heat-treated in nitrogen to vary the nature and the amount of the surface chemical structures. Among the three carbons, the adsorption was maximum in the case of original coconut charcoal, although the rate of adsorption was maximum in the case of peat charcoal. This was attributed to the existence of larger pores in the case of the peat charcoal which resulted in faster rates of diffusion of the gold cyanide into the pores.

The kinetics of gold cyanide on the three carbons, after modifying then by oxidation with air, with nitric acid, and with air and water vapor mixture followed by thermal desorption in nitrogen at 500°C, showed that the rate of adsorption was lowest in the case of untreated coconut charcoals, but increased appreciably when the charcoal was voidized in air and then degassed in nitrogen. However, the adsorption value, which was very small in the case of untreated coconut charcoal for shorter equilibrium times, approached the value for the treated sample after 120 hr of equilibrium time. The coconut and peat charcoals when oxidized with nitric acid showed a decreased adsorption capacity for gold cyanide.

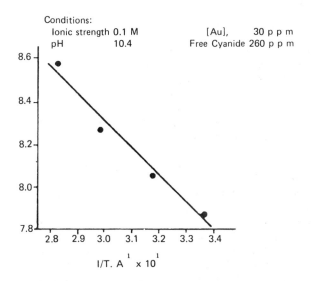

FIGURE 19 Effect of temperature on the rate of gold extraction.

6.9 DESORPTION OF GOLD FROM ACTIVE CARBON

Use of activated carbon for the recovery of gold was limited until the 1970s because of the difficulty of stripping it off the carbon surface. Several methods have now been developed which can desorb gold from the carbon surface almost completely. These methods are based on elution at high temperatures using dilute solutions of certain electrolytes or their mixtures, or they depend on the selective adsorption of certain organic solvents at the carbon-active sites at lower temperatures. The more important of these processes are summarized in Table 7.

6.9.1 Desorption of Gold by Elution

Gross and Scott (14) tested a range of inorganic salts for the elution of gold from charcoal surface and found that only potassium cyanide and sodium sulfide were effective. More than half of the adsorbed gold could be desorbed from the surface of the activated charcoal by elution with a solution of sodium sulfide. The desorbing action of sodium sulfide was explained differently by different workers. While Allen (9) suggested that the desorption of gold was due to changes in the surface energy at the charcoal solution interface in the presence of sodium sulfide, Feldtmann (12) ascribed it to the interaction of sodium sulfide with the carbonyl addition compound of gold cyanide formed on the surface of the carbon:

$$AuCN \cdot CO(CN)_2 + Na_2S \longrightarrow NaAu(CN)_2 + NaCNS + CO$$

Kuz'minykh et al. (37) suggested that the desorption was due to the formation of adsorptively inactive colloidal aurous sulfide.

The homestake process used a mixture of sodium cyanide and sodium hydroxide in dilute solutions for the elution of gold from activated carbons. However, the elute obtained was very dilute. The National Institute of Metallurgy Johannesberg, South Africa, modified the method further to obtain concentrated elutes by using higher temperatures.

Davidson (17) observed that gold could be eluted from activated carbon after pretreatment of the gold-loaded carbon with certain metal carbonates before elution with hot water low in calcium content. The alkali carbonates such as those of potassium, lithium, and sodium were found to be most effective (Fig. 20). When a gold-loaded carbon from a gold plant effluent was eluted with deionized water at 90°C, only 2% of the loaded gold was recovered. But when the same gold-loaded carbon was first pretreated with a 10% solution of potassium carbonate and then eluted with deionized water at 90°C the recovery of gold was as much as 84%. Davidson optimized conditions for the maximum recovery of gold from the carbon surface and found that a pretreatment with a solution containing 5% potassium carbonate and 10% potassium hydroxide followed by elution with hot water at 90°C could recover about 99% of loaded gold in 12 bed volumes and almost all of it in 22 bed volumes (Table 8). The elution of gold was further enhanced at higher temperatures so that concentrated gold elutes could be obtained on shorter elution times using faster flow rates (Fig. 21).

6.9.2 Desorption of Gold by Elution in the Presence of Organic Solvents

Heinen et al. (38) and Martin et al. (39) observed that addition of ethanol and other water-soluble organic solvents to aqueous cyanide solutions enhanced the desoprtion of gold from the activated carbon

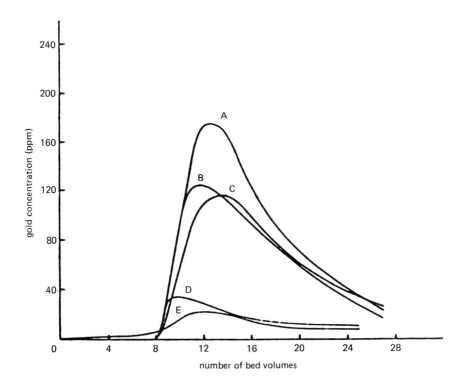

FIGURE 20 Effect of pretreatment with carbonates on the elution of gold with hot water. [From Davidson (17). Reproduced with permission from *South African Inst. Min. Metall.*]

Table 7 Desorption of Gold from Carbon Surface

Process	Eluant	Gold on carbon (ppm)	Temperature (°C)	Gold desorbed (%)
Homestake (1970)	1% NaOH 0.2% NaCN	5,000–15,000	90	Very low
Davidson (1974)	Pretreated + water 5% K_2CO_3	30,000	90	91% in 19 bed volumes
	5% K_2CO_3 + 3% KOH	30,000	90	96% in 16 bed volumes
	10% K_2CO_3 + 5% KOH	30,000	90	94% in 7 bed volumes; 98% in 12 bed volumes
	5% K_2CO_3 + 10% KOH	30,000	90	94% in 7 bed volumes; 99% in 12 bed volumes; 100% in 22 bed volumes
Anglo-American (1976)	90% acetone ethanol methanol	2,900	70–90	30–70%
NIM (1978)	5% NaCN 1% NaOH Presoak and water	20,000	110	25%
Murdoch (1981)	20–40% organics	5,500	25	≈ Complete

Table 8 Adsorption of Gold from Carbon by Pretreatment Followed by Elution with Deionized Water[a]

Pretreatment	Flow velocity (cm/min)	Number of bed volumes (carbonate pretreatment)	Temperature of pretreatment (°C)	Temperature of elution (°C)	Gold recovery
1% K_2CO_3	2.37	4	50	90	70% in 24 bed volumes
5% K_2CO_3	0.74	8	50	90	91% in 19 bed volumes
10% K_2CO_3	2.37	8	50	90	48% in 20 bed volumes
0.5% K_2CO_3 + 0.5% KOH	0.74	≈ 1/2	90	90	35% in 24 bed volumes
1% K_2CO_3 + 1% KOH	0.74	≈ 1/2	90	90	73% in 16 bed volumes; 82% in 24 bed volumes
3% K_2CO_3 + 1% KOH	0.74	≈ 1/2	90	90	88% in 16 bed volumes; 93% in 24 bed volumes
5% K_2CO_3 + 3% KOH	0.74	≈ 1/2	90	90	96% in 16 bed volumes; 98% in 25 bed volumes
5% K_2CO_3 + 3% KOH	0.74	1	90	90	98% in 16 bed volumes; 99% in 25 bed volumes
10% K_2CO_3 + 5% KOH	0.74	1	90	90	94% in 7 bed volumes; 98% in 12 bed volumes; 99% in 24 bed volumes
5% K_2CO_3 + 10% KOH	0.74	1	90	90	94% in 7 bed volumes; 99% in 12 bed volumes; 100% in 22 bed volumes

[a]Gold adsorbed on the carbon = 30,000 ppm.
Source: Davidson (17). Reproduced with permission from *South African Inst. Min. Metall.*

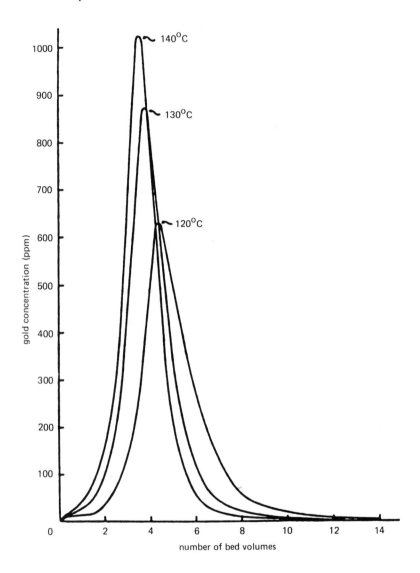

FIGURE 21 Effect of temperature of water on the elution of gold. [From Davidson (17). Reproduced with permission from *South African Inst. Min. Metall.*]

surface. It was found that the gold desorption was much more rapid and efficient at 60°C using 2% ethanol solutions compared to the use of aqueous cyanide alone (3,4,34). Tsuchida et al. (34) used a variety of organic solvents with different functional groups such as ketones, higher alcohols, dimethyl formamide, and methanol and observed that, in general, the gold desorption capacity was in the order

ketones > higher alcohols ≥ dipolar aprotics ≥ simple alcohols > water

These workers found that aqueous acetonitrile was the best solvent for desorbing gold (35,40). When a solution containing 40% V/V acetonitrile and 0.25 M sodium cyanide was contacted with gold-loaded carbon at room temperature, more than 80% of the gold was recovered in less than 10 hr. Acetonitrile and acetone had almost the same potential to desorb gold and both were better than dimethylformamide and ethanol as solvents at the same concentration. Muir and colleagues (41,42), during their further studies in the desorption of gold from carbons, found that aqueous solutions of acetonitrile could be effectively used to desorb gold in 1−2 bed volumes from activated carbon columns which compared favorably with the Zadra (42) and Anglo-American methods (43). The larger desorption capacity of the acetonitrile was attributed to the fact that it was adsorbed more strongly on the surface of the carbon from dilute solutions, whereas the other solvents were adsorbed equally strongly only from stronger solutions. The elution of gold-loaded carbons with organic solvents in the absence of cyanide ions could not desorb any gold. Thus it was suggested that these solvents changed the activity of cyanide and gold cyanide anions. Tsuchida (35) measured the activity coefficients for several organic solvents and the distribution coefficients of gold between carbon and these solvents (D_s) and between carbon and water (D_w) and found them to be significantly different on transfer from water to 40% V/V organic mixtures (Table 9). These workers represented the desorption equilibria of gold as

Carbon surface | $AuCN + CN^- \rightleftharpoons Au(CN)_2^-$ aq.

Carbon surface | $Au(CN)_2^- + CN^- \rightleftharpoons$ | $CN^- + Au(CN)_2^-$ aq.

This equilibria will be more favorable to the right in the presence of aqueous acetonitrile due to the combined increase in activity of CN^- and decrease in activity of $Au(CN)_2^-$. This change in the activity of anions in the presence of aqueous acetonitrile promoted the desorption of gold using salts which was otherwise ineffective in water (Table 10). It was evident that both cyanide and lauryl sulfate

$(n \cdot C_{11}H_{23}CH_2 \cdot O \cdot SO_2)^{2-}$ anions desorbed gold readily and more efficiently and that the thiocyanate ion was more effective than the OH^-, NO_3^-, and Cl^- anions. Thus these workers suggested that it was not essential that the ion should be able to form a complex with gold cyanide in order to be able to desorb the gold anion from the surface of carbons. They are rather of the opinion that the eluting anion displaces the gold cyanide anion from the carbon-active sites. These workers also observed a good correlation between the nucleophilicity of the anions and their ability to desorb gold. This indicated that the active sites were able to undergo nucleophilic substitution reactions.

6.10 ACTIVE CARBON FOR REMOVAL OF CADMIUM FROM AQUEOUS SOLUTIONS

Several investigators have used several different carbons modified by different treatments for the removal of heavy metals including cadmium from aqueous solutions (45,50). However, systematic studies were first reported by Huang and Ostovic (51) using several different activated carbons. These workers investigated the influence of such factors as pH, the nature of the carbon surface, the carbon dose, and the influence of addition of chelating agents on the adsorption of Cd(II). The amount of Cd(II) adsorbed increased with increase in the pH of the solution. The amount of Cd(II) adsorbed was very small at pH values less than 3 but increased appreciably at higher pH values, and became more or less constant at pH values greater then 8. But the range of pH values where maximum adsorption

Table 9 Activity and Distribution and Coefficients of $Au(CN)_2^-$ and CN^- in Different Organic Solvents

Organic solvent 40% (v/v)	Log D_s/D_w	γ_{tr} $[Au(CN)_2^-]^a$	γ_{tr} $(CN^-)^b$	Log K^c
Acetonitrile	−3.36	0.03	3.5	2.05
Acetone	−3.14	0.02	1.5	1.85
DMF	−2.42	0.32	4.3	1.14
Ethanol	−1.92	0.09	2.5	1.44

^{a}Measured distribution coefficient of gold between carbon and solvent and between carbon and water.
^{b}Measured activity coefficient in solvent relative to water.
^{c}Calculated change in equilibrium $(AuCN) + CN^- \rightleftharpoons Au(CN)_2^-$ upon change from water to solvent based upon γ_{tr}.
Source: Tsuchida (35).

Table 10 Effect of Electrolytes on the Desorption of Gold in 40% Acetonitrile-Water Solution[a]

Salt (0.25 M)	Recovery of gold (%)
NaCN	77.0
n-$C_{11}H_{23}CH_2OSO_3Na$ (sodium lauryl sulfate)	66.8
NaSCN	17.1
NaOH	5
$NaNO_3$	1.7
NaCl	1.2
Blank	1.4

[a]Time, 24 hr; temperature 28°C.
Source: Tsuchida (35).

occurred depended on the nature of the carbon. The basic carbons showed a much larger adsorption even at lower pH values compared to to the acidic charcoals (Fig. 22). Thus nuclear C-190-N—activated carbon, which is basic in character, adsorbed about three times as much cadmium as nuclear 722 or Filtrasorb 400 carbons at pH 5 from the same concentration of sadmium ions in the solution. The amount of Cd(II) adsorbed was also a function of the amount of Cd(II) in the solution and the carbon to cadmium ratio. By increasing the carbon content 100 times, the removal of cadmium was increased by three times. The efficiency of the Filtrasorb 400—activated carbon increased appreciably in the presence of chelating agents such as nitrilotriacetate (NTA) and ethylene diamine tetracetate (EDTA). With the addition of 1% of the chelating agent on the weight of the cadmium in solution, the efficiency of the carbon to remove Cd(II) increased from 20 to 40% (Fig. 23) at a pH of about 7. This increased adsorption was attributed to the formation of univalent cadmium complex anions which were adsorbed on the carbon positive sites. This was followed by the association of the Cd(II) ions with the adsorbed anions.

Dobrowolski et al. (52) studied the adsorption of Cd(II) ions from aqueous solutions on a Merck deashed activated charcoal (sample A) modified by oxidation with hydrogen peroxide (sample B) and by heat treatment in argon at 1400 K (sample C). These workers also carried out the isotopic ion exchange reactions with the carbon having adsorb-

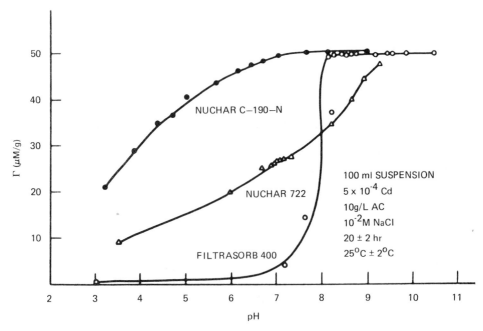

FIGURE 22 Influence of pH on the adsorption of Cd(II) by different activated carbons. (Courtesy C. P. Huang.)

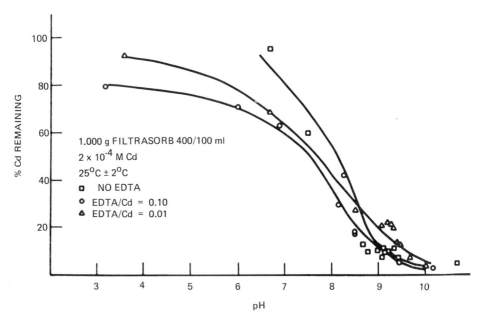

FIGURE 23 Effect of addition of chelating agents (EDTA) on the adsorption of Cd(II) by activated carbon. (Courtesy C. P. Huang.)

ed cadmium ions. The kinetics of isotopic exchange reactions showed that the exchange was very fast in the initial stages at high pH values (Fig. 24). This was attributed to exchange of the Cd(II) adsorbed at easily accessible surface sites and micropores. The slow exchange reactions at low pH values was due to exchance of physisorbed Cd(II) into the deep micropores. Thus these workers inferred that the adsorption of Cd(II) on carbon involved physisorption in the micropores and exchange mechanism with the surface groups. These workers also observed that the amount of Cd(II) adsorbed on carbon B which was oxidized with H_2O_2 was much higher than that of carbon A, indicating that the concentration of the ion exchanging groups on the surface of carbon B was higher compared to carbon A. In the case of carbon C, which was basic in character and showed a solution pH of about 8, the adsorption decreased rapidly with decrease in pH and at pH 7 practically no cadmium was adsorbed. The adsorption in the case of carbon C was thus attributed to the precipitation of Cd(II), which occurred at pH 8 first on the carbon surface.

The adsorption isotherms of Cd(II) from aqueous solutions obeyed the Langmuir equation. The adsorption of Cd(II) was maximum in the

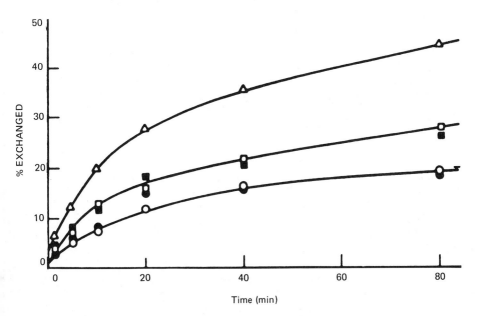

FIGURE 24 Isotopic exchange kinetics as a function of pH (carbon 1 g + 50 ml 10^{-3} M $CdCl_2$ solution). Open circles, pH 1; closed circles, pH 2.1; open squares, pH 2.85; closed squares, pH 3.15; and open triangles, pH 4.05. [From Dobrowolski et al. (52). Reproduced with permission from Pergamon Press.]

case of the degassed carbon (sample C) and minimum in the untreated
sample (Fig. 25). The higher adsorption in the case of the oxidized
sample (sample B) was attributed to a larger pH change caused by
the exchange of cadmium and hydrogen ions between the solution and
the surface acidic groups and the oxidized sample naturally contained
larger amounts of these exchangeable groups. In the case of the degassed sample (sample C), the higher adsorption was due to the fact
that it had already a pH of about 8. The relative adsorption isotherms (Fig. 26) did show that the degassed carbon adsorbed smaller
amounts of cadmium compared to carbons A and B and at lower concentrations the isotherm for the carbon B was above the isotherm for
carbon A because of the presence of large concentration of exchangeable surface acidic groups. The charge density versus pH curves
for the three carbons showed that carbon C had a positive charge
over most of the pH range. This positive charge was stabilized by

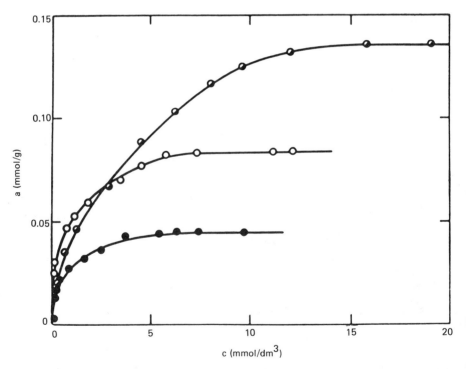

FIGURE 25 Adsorption isotherms of Cd(II) from aqueous solutions on
activated carbon at 25°C. Closed circles, ash-free carbon; open circles,
oxidized with H_2O_2; half-open circles, degassed in argon at 1000°C.
[From Dobrowolski et al. (52). Reproduced with permission from
Pergamon Press.]

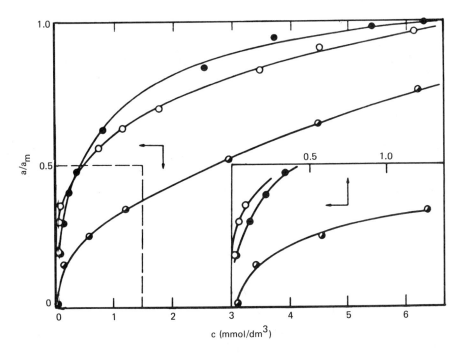

FIGURE 26 Relative adsorption isotherms of Cd(II) from aqueous solutions at 25°C. Closed circles, ash-free carbon; open circles, oxidized with H_2O_2; half-open circles, degassed in argon at 1000°C. [From Dobrowolski et al. (52). Reproduced with permission from Pergamon Press.]

the excess of OH^- ions in the diffuse external part of the double layer. This caused the solubility product of $Cd(OH)_2$ to be exceeded easily, resulting in the precipitation of cadmium. Carbons A and B showed a positive charge only over very low pH values and it changed from positive to negative as the pH increased. These workers suggested that this change of charge may be attributed to the participation of different surface groups in the adsorption process at different pH ranges. However, they did not define the ranges of pH and the nature of surface groups which contribute to this phenomenon.

6.11 REACTION OF CARBONS WITH FERRIC IONS

Vasatko (53) and Heymann et al. (54) observed that the reducing ability of active carbon for ferric ions depended on the surface acidity of the carbon and decreased with increase in the acidity. Garten and Weiss (55), while agreeing with this view, observed that the reduction of ferric ions was the result of two reactions, one of which reached completion rapidly whereas the other continued at a slow rate. Puri and Mahajan (56) investigated the reduction of ferric chloride in the presence of charcoals with acid, neutral, and alkaline surface. The rate measurements showed the reaction to be bimolecular. The reaction was very sensitive to temperature with the temperature coefficient being about 2 (Table 11). The magnitude of the reaction increased with increase in the concentration of the solution in each charcoal but ultimately reached a limiting value (Table 12), indicating that the carbon was participating in the reaction and its surface was being modified during the course of the reaction making it ineffective for further reaction. The efficiency of the carbon to reduce ferric chloride decreased with decrease in the acidity of the carbon. The comparative efficiency of the three carbons was different in the initial stages of the reaction. However, when the solution was kept in contact with the charcoal samples for 48 hr, the comparative efficiency of the alkaline charcoal sample was more than 1 1/2 times that of the acidic charcoal (Table 12, bottom). This was explained by the reaction

$$2FeCl_3 + H_2O \longrightarrow 2FeCl_2 + 2HCl + (O)$$

Since basic charcoals such as those outgassed at 750 and 1000°C can adsorb larger amounts of the acid, these will be in a better position to shift the equilibrium to the right in the initial stages of the reaction. However, the ultimate capacity of the charcoal will depend on the capacity of the charcoal to utilize the evolved oxygen and in this original charcoal was found to be more reactive (Table 13) than the degassed materials. Thus the reduction of ferric chloride very much depended on the nature of the carbon used.

Ford and Boyer (57) reported that activated carbons can be used to remove ferrous ions from solution. The presence of carbon catalyzes the oxidation of ferrous to ferric. The kinetics of ferrous oxidation in the presence of activated carbons showed that the rate of

Table 11 Reduction of N/10 Ferric Chloride Solution by Original Charcoal at Different Temperatures after 12 Hours

Temperature (°C)	$FeCl_3$ reduced (mg/g)
0	0
25	13.2
35	25.9
50	74.8
70	238.14

Source: Puri and Mahajan (56).

Table 12 Reduction of Ferric Chloride in Solutions of Different Concentrations at 25° by Various Samples of Charcoal

Concentration of solution	Original charcoal (pH 2.85)	750°-evacuated charcoal (pH 6.95)	1200°-evacuated charcoal (pH 8.80)
	Amount of ferric chloride reduced after 31 days (mg/g)		
0.10 N	356.3	340.2	210.6
0.30 N	542.6	429.3	226.8
0.50 N	656.1	469.8	259.2
0.75 N	737.0	499.1	299.7
1.00 N	748.1	510.3	299.7
	Amount of ferric chloride reduced after 48 hr		
0.10 N	48.6	65.3	77.7

Source: Puri and Mahajan (56).

Table 13 Oxygen Rendered Available by the Reduction of Ferric Chloride and Chemisorbed by the Different Carbons

Concentration of the solution	FeCl$_3$ reduced (mg/g)	O$_2$ rendered available (mg/g)	O$_2$ chemisorbed by charcoal (mg/g)
Original charcoal			
0.10 N	356.3	17.54	4.74
0.30 N	542.6	26.71	10.40
0.50 N	656.1	32.30	16.11
0.75 N	737.0	36.28	20.20
750° evacuated charcoal			
0.10 N	340.2	16.75	10.45
0.30 N	429.3	21.14	12.60
0.50 N	469.8	23.13	14.30
1200° evacuated charcoal			
0.10 N	210.6	10.37	4.82
0.30 N	226.8	11.16	5.04
0.50 N	259.2	12.76	5.92

Source: Puri and Mahajan (56).

oxidation was dependent on the oxygen pressure, the square of the hydroxyl ion concentration, and the concentration of Fe(II) in solution. These workers also observed that ferrous ions can be adsorbed on the carbon surface at pH close to 3. George and Chandhuri (58) also observed that natural bituminous coal had a marked efficiency for the removal of Fe-soluble Fe(II) by the oxidation process. Huang and Bowers (59) noticed that the oxidation of ferrous ions by carbon was not improved by aeration since air was readily adsorbed by the carbon, which modified its surface properties.

6.12 REMOVAL OF CHROMIUM FROM WATER

Puri and Satija (60) studied the interaction of charcoals, coated with different amounts of carbon—oxygen surface compounds, with acidified potassium dichromate solution and found that the charcoals associated with oxygen, which was evolved as CO_2 on evacuation, were more effective for reduction of Cr(VI) to Cr(III) than those free of any chemisorbed oxygen. The oxygen rendered available during the reaction was partly evolved as CO_2 and partly chemisorbed by the carbon. The reaction was found to occur in two stages:

$$K_2Cr_2O_7 + H_2SO_4 \longrightarrow K_2SO_4 + Cr_2O_3 + H_2O + 3(O)$$

and

$$Cr_2O_3 + H_2SO_4 \longrightarrow Cr_2(SO_4)_3 + 3H_2O$$

The reaction at lower concentrations of potassium dichromate was found to occur by the first reaction only as the amount of Cr_2O_3 formed was found to be almost exactly equivalent to that of potassium dichromate in the solution (Table 14). However, as the concentration of potassium dichromate in the solution increased, the second reaction, involving the formation of $Cr_2(SO_4)_3$ was observed to take place to increasing extents until, at concentrations higher than 0.7 there was no formation of Cr_2O_3 (Table 14).

Huang and Wu (61,62) used a calcined charcoal and Filtrasorb 400 —activated carbon for the removal of chromium ions and observed that the adsorption of Cr(VI) and Cr(III) was dependent on the pH (Fig. 27) as well as the chromium content of the solution. The adsorption of Cr(VI) was much higher than the adsorption of Cr(III) (Fig. 27). The adsorption increased with increase in pH, attained a maximum value around pH6, and declined thereafter. This was attributed to the reduction of Cr(VI) to Cr(III) on the surface of the carbon in the acidic solution. Kim (63) and Nagasaki and Terada (64) found that the reducing action of the carbon could be decreased when the proton concentration of the solution was about the same as the Cr(VI)

Table 14 Reduction of Acidified Potassium Dichromate by the Original Sugar Charcoal

Concentration of $K_2Cr_2O_7$	Amount of $K_2Cr_2O_7$ present in 100 ml of solution mixed with 1 g charcoal (mEq/g)	$K_2Cr_2O_7$ reduced (mEq/g)	Cr_2O_3 obtained (mEq/g)
0.02 N	2	2	2.0
0.03 N	3	3	2.9
0.05 N	5	4.8	4.3
0.10 N	10	8.4	4.0
0.20 N	20	12.0	2.1
0.30 N	30	17.4	1.25
0.40 N	40	22.2	1.00
0.50 N	50	23.0	0.36
0.60 N	60	25.0	0.34
0.70 N	70	25.8	0.20
0.80 N	80	26.50	0.00
0.90 N	90	28.50	0.00
1.00 N	100	29.70	0.00

Source: Puri and Satija (60).

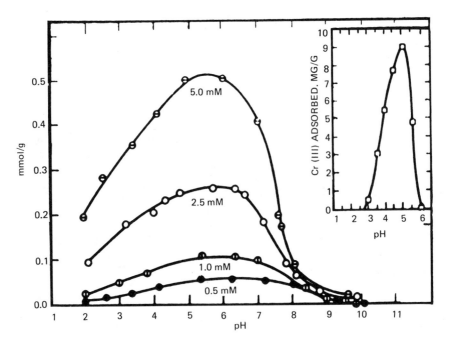

FIGURE 27 Influence of pH on the adsorption of Cr(III) and Cr(VI) from aqueous solutions. (Courtesy C. P. Huang.)

concentration in the solution. However, Huang and Bowers (59) fail-ed to confirm this observation. These workers (59) found that the adsorption and reduction were occurring simultaneously and were res-ponsible for the removal of Cr(VI). They were able to calculate the rates of the two reactions. These workers also observed that when the carbon was oxidized with nitric acid and the solution containing (Cr(VI) with chlorine water, the removal of Cr(VI) was decreased considerably. This was attributed to the fact that these oxidations resulted in the chemisorption of appreciable amounts of oxygen enhanc-ing the acidity of the carbon surface so that the carbon had greater affinity toward the reduced form of the metal Cr(III).

Yoshida et al. (65) observed that Cr(VI) was readily adsorbed by the activated carbon as anionic species such as $HCrO_4^-$ or CrO_4^{2-} while Cr(III) was scarcely adsorbed. They also found that the acidic solu-tion of Cr(VI) was easily reduced to Cr(III) in the presence of car-bon. The adsorbed Cr(VI) species could be recovered by washing with sodium hydroxide or hydrochloric acid solutions. The reduction of Cr(VI) into Cr(III) by activated carbon was also reported by Ro-ersma et al. (66). The removal of chromate ions by activated carbon from aqueous solutions (67,68) was found to be a function of pH, the best pH range being acidic range pH 3—7.

6.13 ADSORPTION OF MERCURY ON CARBONS

The adsorption of Hg(II) from aqueous solutions using activated car-bon was found to be pH dependent (69—71). The adsorption of Hg(II) increased as the pH of the solution was decreased, the acidic pH range being most suitable. Twice as much mercury was removed when the pH was decreased from 9 to the acidic range. Yoshida et al. (65,72) studied the use of several activated charcoals for the removal of Hg(II) and observed that the removal efficiency of a carbon depended on the nature of the carbon and the nature of the activation treatment that the carbon received. The carbons obtained from wood, coconut shells, and coal and activated in steam were found to have a high adsorption capacity for Hg(II) from solutions below pH 5. The ability decreased with increase in pH of the solution. The wood charcoal given a chem-ical activating treatment with zinc chloride was found to have a higher capacity for adsorption of Hg(II) even at pH greater than 5. The adsorption of $HgCl_4^{2-}$ in HCl medium was reversible both in the case of steam-activated and chemically activated carbons but at pH $>$ 7 the steam-activated carbons showed irreversible adsorption accompani-ed by reduction of Hg(II) on the surface of the carbon. The effi-ciency of an activated carbon to adsorb Hg(II) was also enhanced by the addition of certain chelating agents (71) or by sulfurization of the carbon surface (73,74). Activated carbons modified by impregnation with cartain metallic sulfides were also found to be very efficient for selectively adsorbing Hg(II) from aqueous solutions (75,76).

Thiem et al. (70) observed that mercury can be removed from water using powdered activated carbons, the amount of removal depending upon the pH and the amount of charcoal (Fig. 28). Roughly twice as much mercury was removed at pH 7 as at pH 9. Increasing the hydroxyl ion concentration evidently reduced the adsorption of mercury by the carbon. The addition of chelating agents such as tannic acid or EDTA improved the adsorption of mercury. The addition of as little as 0.02 mg/L of the chelating agent increased the adsorption of mercury from 10 to 30% depending on the pH of the solution and the amount of the carbon used. Tannic acid was more effective (Fig. 29), whereas nitric acid showed minimum effect. The presence of calcium ions in water also improved the adsorption of mercury (Fig. 30). The mercury adsorption increased from 10 to 20% as

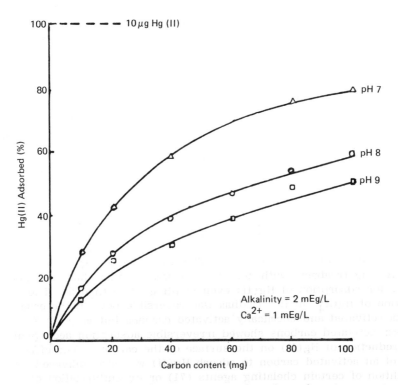

FIGURE 28 Removal of mercury as a function of carbon content at different pH values. [From Thiem et al. (70). Reproduced with permission from American Water Works Association.]

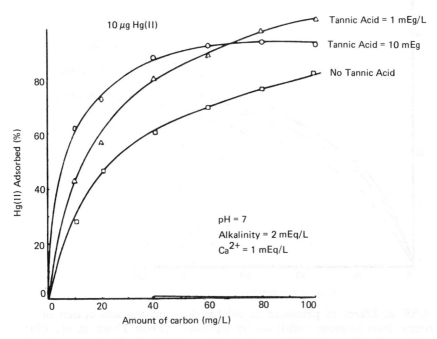

FIGURE 29 Influence of addition of tannic acid on the removal of mercury from aqueous solutions. [From Thiem et al. (70). Reproduced with permission from American Water Works Association.]

the calcium ion concentration was increased from 50 to 200 mg/L. When both calcium ions and tannic acid were present, the removal of mercury was almost doubled, even with smaller amounts of carbon.

Sinha and Walker (74) observed that sulfurized carbons could be used for the removal of mercury vapors from contaminated streams at 150°C. They sulfurized a Saran carbon to sulfur loadings varying between 1.0 and 11.8% by weight of sulfur by the oxidation of hydrogen sulfide on the carbon surface at 140°C in a fluidized bed. The mercury-contaminated airstream was passed through a bed of 1 g of the sulfur-loaded carbon with a contact time of 0.05 sec. The rate of mercury buildup in the effluent was extremely low when the mercury-contaminated stream was passed through the bed at 150°C (Fig. 31). This large capacity of the sulfurized carbon was attributed to the interaction between the mercury and the sulfur present on the carbon surface.

Ammons et al. (77) studied the adsorption of methyl mercuric chloride from aqueous solutions on Filtrasorb 200—activated carbon at room temperature and observed that the adsorption isotherms were nonlinear and did not fit either the Freundlich or the Langmuir isotherm. The adsorption was, however, reversible in nature.

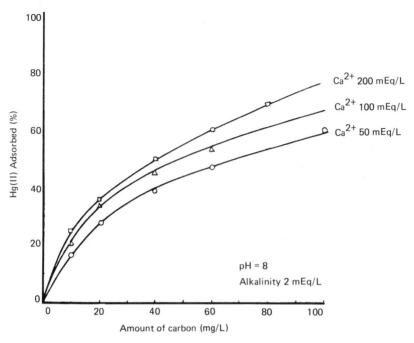

FIGURE 30 Effect of presence of calcium ions on the adsorption of mercury from aqueous solutions by carbon. [From Thiem et al. (70). Reproduced with permission from American Water Works Association.]

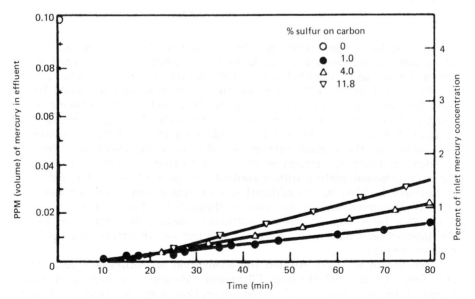

FIGURE 31 Effect of sulfur content in carbon on the removal of mercury vapors from air. [From Sinha and Walker (74). Reproduced with permission from Pergamon Press.]

6.14 REMOVAL OF COPPER BY ACTIVE CARBONS

Moore (78) studied the adsorption of Cu(II) from seawater desalination plant effluent using several different activated carbons. The adsorption isotherms were generally type I of the BET classification. The extent of adsorption was a function of the total Cu(II) concentration and pH of the solution as well as the nature of the carbon. The rate of adsorption at pH < 3 was very high in the initial stages, reached a maximum value, and then declined steeply. However, the adsorption at pH 3 increases very fast and then becomes asymptotic after a few minutes (Fig. 32). The maximum in the adsorption at pH < 3 was attributed to the instantaneous precipitation of the solid copper hydroxide on the carbon surface during which the pH of the system became alkaline. The adsorption then decreased as a result of the slow release of Cu(II) ions into the solution. The adsorption of Cu(II), however, was small, never reaching more than 6%. The treatment of the carbons with organic chelating agents increased adsorption but only slightly. Nelson et al. (79) observed that the adsorption of Cu(II) from solutions containing varying concentrations of sodium chloride depended on the ionic strength of the solution. The adsorption increased by about seven times when the concentration of sodium chloride in the solution was increased from 0.1 to 6 M. This rapid increase in adsorption was attributed to the adsorption of the negatively charged CU(II) chloride complexes.

6.15 ADSORPTION OF VANADIUM ON ACTIVE CARBONS

The adsorption of V(IV) and V(V) by activated carbons has also been reported (80,81). The adsorption has been found to be sensitive to the pH of the solution and the nature of the carbon surface. The adsorption of both V(IV) and V(V) increased with pH and was maximum in the pH range between 2.5 and 3 and was decreased thereafter. The oxidized charcoals generally showed more adsorption capacity toward both V(IV) and V(V). Kunz et al. (80) studied the adsorption from sodium metavanadate solution using Filtrasorb 400—activated carbon and observed that about 90% of the vanadium could be removed from a solution containing 50 mg/L. The efficiency of the carbon was improved when larger quantities of the carbon were used (Fig. 33).

6.16 REMOVAL OF ORGANICS FROM WATER BY ACTIVE CARBONS

More than 700 specific organic chemicals have been identified in drinking waters. These organics are derived from industrial and municipal discharge, urban and rural runoff, natural decomposition of vegetable and animal matter, as well as from water and wastewater chlorination practices. Considerable amounts of halomethanes—specifically—chloro-

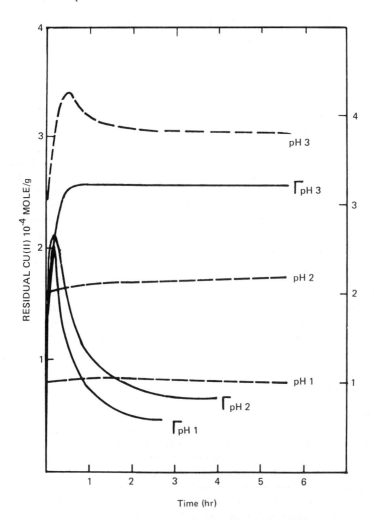

FIGURE 32 Rate of adsorption of Cu(II) from aqueous solutions at
different pH values. (Courtesy R. H. Moore.)

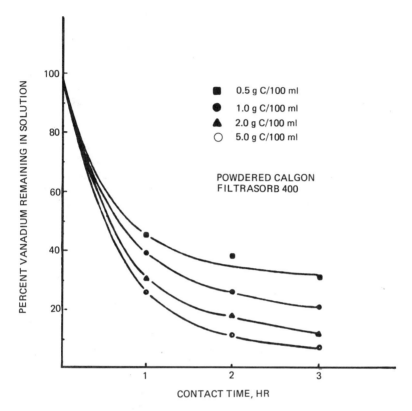

FIGURE 33 Adsorption of vanadium from sodium metavanadate solution on activated carbon as a function of time. [From Kunz et al. (80). Courtesy R. G. Kunz.]

form—can form in water due to chlorination after entering the distribution system and on its route to the consumer. Many of these organic chemicals are carcinogenic and cause many others ailments of varying intensity and character. It is essential, therefore, that water before distribution to the general public be given some treatment to avoid these health hazards. Effluent water from the chemical industries also needs to be treated before it is discharged into the ground or some watershed to avoid harm to the aquatic animals.

Several methods have been used with varying degrees of success for the control of organic chemicals in water. These include coagulation, filtration, oxidation, aeration, ion exchange, and activated carbon treatment. However, many studies including laboratory tests, pilot scale experiments, and actual field operations in several different countries have indicated that the use of activated carbons is perhaps the best broad spectrum control technology available at the present moment.

Activated carbons have a high degree of porosity, an extensive surface area, and a high degree of sruface reactivity. Consequently, they are effective adsorbents for many organic compounds of concern in water and wastewater treatment. The efficient use of activated carbon, however, depends on basic knowledge of the adsorption mechanism. The adsorption capacity of an active carbon is determined by the nature of the carbon surface, i.e., the amount and the concentration of the surface oxygen structures, its surface area and pore structure, the nature of the aqueous phase, and the nature and structure of the organic compound. The nature, the surface area, and the microporosity of a carbon are determined by the history of its preparation and the conditions of activation. Thus a wide variety of active carbons have been obtained by varying these conditions.

The choice of the activated carbon for the removal of organics has been arbitrary, probably because a wide spectrum of organic compounds exists in waters. Thus a through study of the surface chemistry of the activated carbons is needed so that carbons suited for a particular application can be prepared. For example, when the organic compounds to be removed from water are polar in character, the carbons associated with polar oxygen surface structures should be preferable, whereas oxygen-free carbons should be used for removing nonpolar organic compounds. Similarly, when the organic compounds have a large size, carbons with a greater percentage of large pores should be used.

The importance of the removal of organics from the aqueous phase can be gauged by the number of conferences that are being organized in different countries and from the number of monographs that have appeared on the subject (82–84). Since several monographs are already available on the subject, we discuss only very briefly the adsorption of some of these groups of organic compounds in order to emphasize the importance of the surface chemistry of activated carbons for this removal.

6.16.1 Removal of Phenolic Compounds

The wastewater from coal conversion industries contains substantial amounts of phenolic compounds, which include methyl phenols, ethyl phenols, and dimethyl phenols. These compounds cause several harmful effects in living beings. Therefore, their removal is part of water pollution control. Singer and Yen (85) investigated the adsorption of a number of alkyl phenols on powdered activated carbon in the concentration range 20–60 mg/l. The solution was contacted with the carbon for 4 hr, after which an aliquot was examined for phenol concentration

using UV spectrophotometry. The adsorption data obeyed the Langmuir adsorption equation. The adsorption isotherms for different
phenols showed (Fig. 34) that the alkyl-substituted phenols were adsorbed more strongly than phenol itself, and the adsorption increased
with increase in the length of the alkyl chain. The adsorption remained unaffected by the position of the alkyl group but was increased con-
considerably by an increase in the number of substituents. The increase or decrease in adsorption was attributed to the variation in the
solubility of the phenol. The substitution of an alkyl group made phenol less polar and its solubility was reduced and adsorption increased.
An ideal solution adsorption model was tried to explain the results
from a two-component system but deviations were obtained, probably
because of the heterogenous nature of the carbon surface.

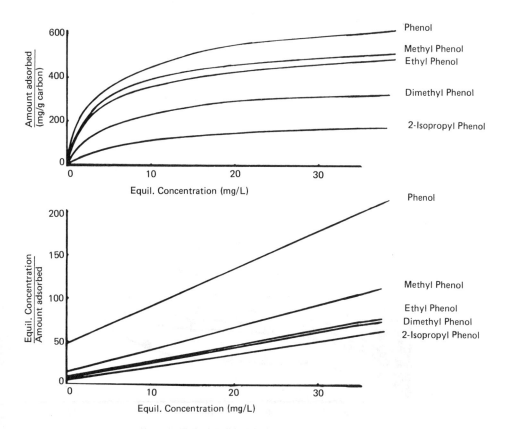

FIGURE 34 Adsorption isotherms of different phenols on activated carbon.
[From Singer and Yen (85). With permission from Ann Arbor Science
Publishers.]

Puri et al. (86), while studying the adsorption of phenol and P-nitrophenol on active carbons and carbon blacks containing varying amounts of associated oxygen, observed that the adsorption was a function of the surface area when carried out at moderate concentrations. The adsorption isotherms were type II of the BET classification (Fig. 35), with well-defined knee, and could be used to calculate the surface area of the carbons using the parallel orientation of the phenol molecules on the carbon surface (molecular area of phenol 0.522 nm^2). However, when the adsorption of the phenol was studied using very dilute solutions of phenol, the adsorption increased with decrease in the amount of acidic oxygen complexes on outgassing at temperatures up to 600°C while retaining the more stable oxygen complex. This was attributed to the interaction of the electrons of the benzene ring with the partial positive charge on the carbonyl groups, which became

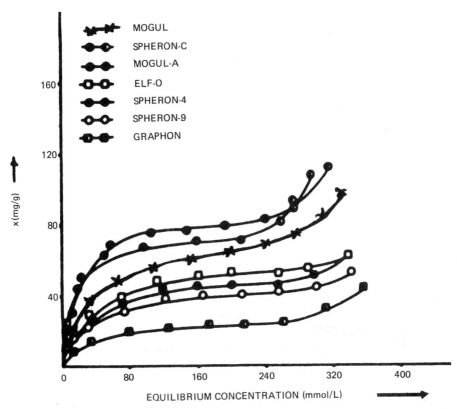

FIGURE 35 Adsorption isotherms of phenol on carbon blacks. [From Puri et al. (86).]

predominant after the removal of acidic oxygen groups. Coughlin et al. (87), however, observed that the presence of chemisorbed oxygen considerably reduced the adsorption of phenol, particularly from dilute solutions. This was attributed to the clustering of water molecules around the polar oxygen groups reducing the adsorption capacity. These workers, however, made no distinction between the acidic and nonacidic oxygen groups.

Marsh and Campbell (88) observed that the adsorption of P-nitro-phenol at low concentrations on polyfurfuryl alcohol and polyvinylidene chloride—activated carbons was extremely sensitive to the microporous structure of these carbons and showed adsorption isotherms similar to those of nitrogen or CO_2 (Fig. 36). The adsorption of P-nitrophenol was so strong at low C/C_o values that the adsorption could be used to compare the microporosities of the carbons (excluding ultramicro-porosity). The surface areas calculated were smaller than the nitrogen surface areas. This was attributed to differences in the efficiencies of packing of micropore and transitional pore volumes and to the effect of the solvent during building up of layers in transitional porosity of the adsorbate molecules.

Thus it is apparent from these studies that the adsorption of phe-nols is a function of the surface area as well as the oxygen-containing surface structures. While the presence of an acidic group decreases adsorption of phenols, the presence of carbonyl oxygen enhances the adsorption. The choice of the carbons for the removal of phenols from aqueous solutions should favor activated carbons having large surface areas but low acidic oxygen content. Carbon prepared at high tem-peratures should thus be preferable.

6.16.2 Removal of Halomethanes

The presence in drinking waters of light halogenated hydrocarbons in general and of trihalomethanes in particular has caused considerable concern because these compounds can cause cancer of the bladder. Chloroform is produced by the reaction of chlorine with the natural constituents of organic color in surface water such as humates and fulvates. Youssefi and Faust (89) studied the formation of these halo-methanes by the chlorination of naturally occurring organic compounds in water such as tannic acid, humic acid, D-glucose, vannillic acid, and gallic acid. These workers found that appreciable amounts of chloroform were actually formed when 10 mg/L of these precursors were contacted with 10 mg/L of chlorine (Fig. 37). The reaction was quite fast and about 90% of the chloroform was formed in the first two hours. The amount of the precursor reacted was extremely small.

These workers (89) also studied the adsorption of chloroform, bromoform, bromodichloromethane, dibromochloromethane, and carbon tetrachloride on a granulated activated carbon. The pH of the solu-

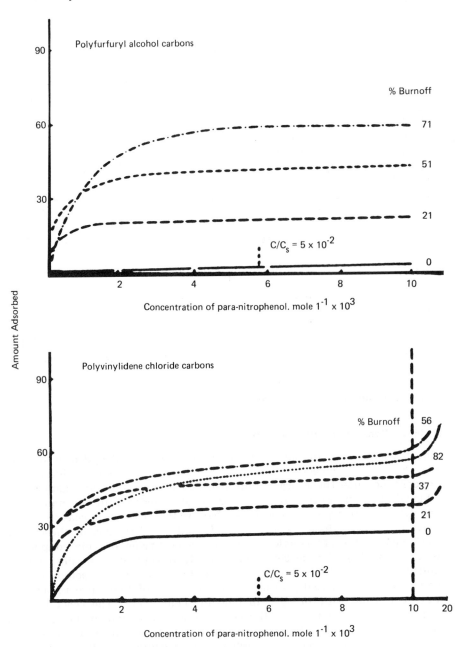

FIGURE 36 Adsorption isotherms of P-nitrophenol on polyfurfuryl alcohol and polyvinylidene chloride carbons. [From Marsh and Campbell (88). Reproduced with permission from Pergamon Press.]

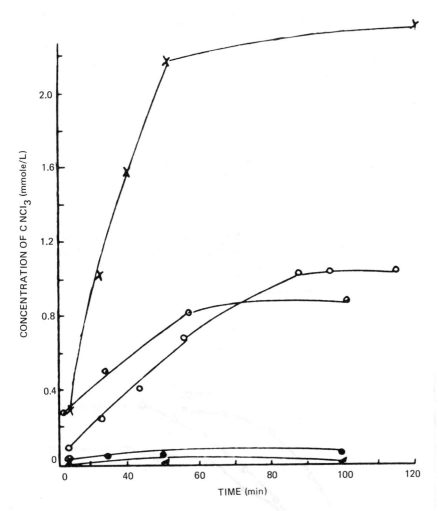

FIGURE 37 Formation of chloroform by the chlorination of water containing tannic acid, humic acid, D-glucose vanillic acid, and gallic acid. [From Youssefi and Faust (89). With permission from Ann Arbor Science Publishers.]

tion was maintained at 7 by using a phosphate buffer and the time of contact was 24 hr. The adsorption was the greatest in the case of bromoform and least for chloroform, the adsorption capacity being 185 mg/g for $CHBr_3$ and only 16.5 mg/g for $CHCl_3$ (Table 15). The breakthrough studies with these halomethanes showed that the breakthrough followed the order $CHCl_3 > CHCl_2Br > CHClBr_2 > CHBr_3$.

Table 15 Adsorption of Different Halomethanes on Activated Carbon[a]

Halomethane	Amount adsorbed ($\mu g/g$)
$CHCl_3$	16.5
$CHClBr_2$	130.0
$CHCl_2Br$	150.0
$CHBr_3$	185.0

[a]pH 7.0; temperature, 24°C; contact time, 24 hr.
Source: Youssefi and Faust (89). With permission from Ann Arbor Science Publishers.

Mullins et al. (90) evaluated the performance of several brands of commercially available granulated activated carbons for the removal of trihalomethanes from aqueous solutions using vertical columns operated in continuous downmode. The effluent contained 0.22−02 mg/L of the trihalomethanes (THM). Each brand of activated carbon was found to be capable of removing appreciable amounts of THM, although the amounts varied with the carbon (Fig. 38). The rate of removal was

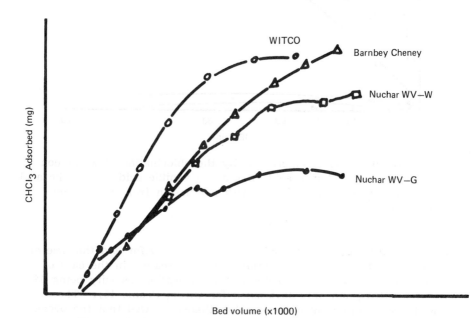

FIGURE 38 Adsorption of chloroform by different activated carbons.

also rapid, but the ultimate adsorption capacity of the carbons was very low, between 0.05 and 0.12% by weight of the carbon for chloroform and between 0.02 and 0.04% for bromodichloromethane. Thus although the removal of THM could be effected rapidly and efficiently, the carbon needed more frequent regenerations because of its lower adsorption capacity.

Ishizaki et al. (91) studied the adsorption of chloroform from aqueous solutions on as-received Filtrasorb 200 and after removing the surface oxides by evacuation at 1000°C in the equilibrium concentration range 10—200 μg/L. Both the carbons adsorbed appreciable amounts of chloroform but the outgassed sample showed a stronger affinity and larger capacity compared to the as-received sample (Fig. 39). The data obeyed the Langmuir equation. The Langmuir equation

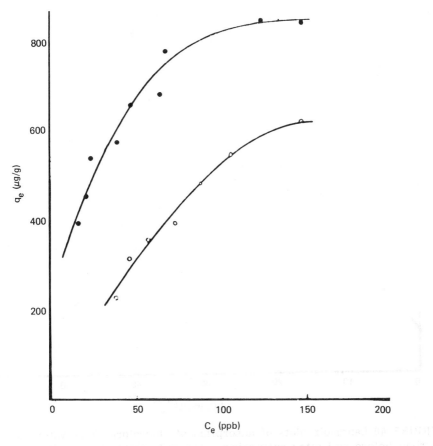

FIGURE 39 Adsorption isotherms of chloroform from aqueous solution on carbon before and after outgassing. [From Mullins et al. (90). With permission from Ann Arbor Science Publishers.]

constants calculated from linearized plots gave adsorption energies which were different in the two carbons. The adsorption energy on the outgassed sample was about 2.6 times the adsorption energy on the as-received carbon sample. As the pore size of the two carbon samples were almost the same, the difference in the adsorption energy was attributed to the presence of oxygen complexes in case of the as-received sample. When the monolayer capacities were calculated from the extrapolated linear Langmuir plots, the two lines crossed each other at higher equilibrium concentrations (Fig. 40), indicating that the adsorption capacity of the as-received sample was larger at equilibrium concentrations greater than 270 µg/L. Thus these workers suggested that the adsorption of chloroform from aqueous solutions by carbons was a function of the nature of the carbon surface as well as the equilibrium concentration.

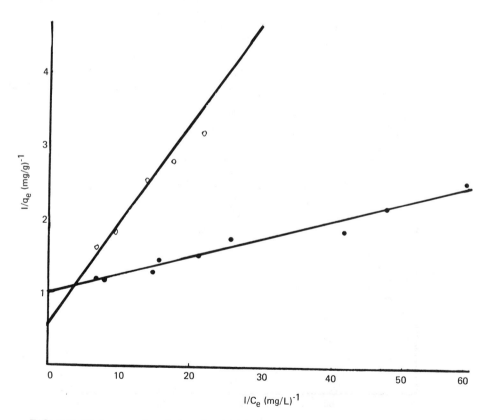

FIGURE 40 Langmuir plots of adsorption of chloroform from water on carbon before and after outgassing. [From Ishizaki et al. (91). With permission from Ann Arbor Science Publishers.]

6.16.3 Removal of Polycyclic Aromatic Hydrocarbons

Bornehoff (92) studied the adsorption of a number of polycyclic aromatic hydrocarbons (PAH) using a variety of activated carbons under laboratory and field conditions and found that activated carbons were good adsorbents for the removal of PAH from water. These workers also considered the possibility of the presence of certain polyaromatic hydrocarbons in different activated carbons by extracting 25 g of an activated carbon with 300 ml benzene in a soxhlet extractor for 8 hr. It was found that 10 different PAHs were present in the extract varying between 1 µg/kg and 227 µg/kg with a mean value of 63.4 µg/kg. However, in spite of the presence of PAH, the capacity of the carbon to adsorb PAH from water did not decrease. Furthermore, it was found in laboratory experiments that the amounts of these PAHs leached with water from most of the carbons were small—of the order of 1 µg/kg—so that the use of activated carbons did not cause any detectable contamination of the water.

Bornehoff (92) examined the adsorption capacity of 10 different activated carbons by adding 1 ml of acetone solution containing the PAH to 1 ml of sewage water which was equilibrated with 1 g of activated carbon for 30 min at 15°C. The adsorption of the PAH was greater than 99% in all cases. The adsorption experiments carried out using 10 and 100 mg of Norit-active carbon with contact times varying between 0.5 and 45 hr for eight different PAHs showed (Table 16) that some were removed to the extent of 90% in first 30 min of contact time, whereas others required a much larger contact time. Furthermore, the PAHs with lower molecular weights were adsorbed to a larger extent than the higher molecular weight PAHs. For example, for experiments carried out under similar conditions the adsorption of anthracene (molecular weight 178.4), benzo(a)anthracene (molecular weight 228.3), and dibenzopyrene (molecular weight 302) after 30 min of contact time were 98, 83.7, and 70% respectively (Table 17). Similarly, when the experiments were carried out by using polycyclic aromatic hydrocarbons of comparable molecular weights but differing in their structural configurations, the adsorption depended on the pore size distribution of the activated carbon. The carbons showed a certain degree of molecular sieving behavior toward larger molecules.

When the removal of PAH was carried out on actual plant scale, only 10% of the PAH present could be removed by the carbon when the effluent contained about 10 µg/L of the hydrocarbons. However, this efficiency was increased when the hydrocarbon content of the effluent was larger. The removal efficiency of these polycyclic aromatic hydrocarbons was also enhanced by pretreatment of water with chlorine or zone.

Table 16 Adsorption of Polycyclic Aromatic Hydrocarbons by Norit-Activated Carbon

PAH	Amount of carbon (mg/L)	Amount adsorbed, contact time of 0.5-45 hr (%)				
		0.5 hr	4 hr	22 hr	28 hr	45 hr
Fluoranthene	100	98.8	99.2	99.5	99.2	99.4
	10	81.8	92.8	98.4	98.7	99.2
Benzo[a]anthracene	100	97.2	99.0	97.7	99.3	98.8
	10	79.3	84.5	97.6	98.4	99.1
Benzo[b]fluoranthene	100	80.9	97.8	99.1	99.0	98.3
	10	99.8	66.1	95.0	97.3	98.9
Benzo[a]pyrene	100	80.7	96.9	99.5	98.8	98.6
	10	68.6	73.2	96.2	98.1	99.0
Benzo[g,h,i]perylene	100	57.6	81.8	97.7	97.3	98.7
	10	37.8	49.0	93.7	96.6	98.1
Benzo[j]fluoranthene	100	85.4	96.8	99.4	99.0	98.4
	10	59.1	74.9	96.1	98.0	98.6
Pyrene	100		98.4		99.6	98.5
	10		93.6		99.3	99.5

Source: Data based on Bornehoff (92).

Table 17 Adsorption of Polycyclic Aromatic Hydrocarbons by Activated Carbon

PAH	Molecular weight	Amount of active carbon added (mg/L)	Amount adsorbed in 30 min (%)
Anthracene	178	100	98
		10	89
Benzo(a)anthracene	228	100	83.7
		10	67.5
Dibenzopyrene	302	100	70
		10	51.3

Source: Data based on Bornehoff (92).

6.16.4 Removal of Nitrosamines

Nitrosamines are dangerous health hazards discharged into wastewater by certain chemical processing industries. Bornehoff (92) investigated the removal of several nitrosamines from water using activated carbons and observed that when 1 L of the dosed water was filtered through 23 g of an activated carbon at 40—50 L/kg/hr the removal was over 99% in the case of dimethyl, diethyl, and diphenyl nitrosamines. Fochtman and Dobbs (93) used Darco- and Filtrasorb-activated carbons for the adsorption of several carcinogenic hydrocarbons such as naphthalene, benzidine, and nitrosamines from water and observed that appreciable amounts of these substances could be adsorbed. The amount adsorbed was dependent on the nature, the molecular weight, and the chemical structure of the compound (Table 18). The adsorption data obeyed Freundlich isotherms. About 20—30 mg of carbon was sufficient to reduce the concentration of these organic compounds from 1.0 to 0.1 mg/L, which is considered to be the safe limit.

6.16.5 Adsorption of Some Miscellaneous Organic Compounds

Ishizaki and Cookson (94) selected four typical organic compounds—n-butyl mercaptan, butyl disulfide, decane, and p-hydroxybenzaldehyde—for their adsorption studies on Filtrasorb-granulated carbon from aqueous phase. The mercaptan and the butyl disulfide impart an unpleasant odor to water, decane represented the nonpolar parafinic molecules, and p-hydroxybenzaldehyde represented degradation products of larger organic molecules identified as natural color—producing compounds in water. The adsorption isotherms for all the adsorbates obeyed the Langmuir adsorption equation. The adsorption capacity of the carbon for butyl disulfide increased on outgassing the carbon at 900°C. When the outgassed carbon was oxidized in dry oxygen in the temperature range 200—300°C, the adsorption capacity remained unchanged at 200°C oxidation but increased at 300°C oxidation temperature (Table 19). This increase in adsorption of butyl sulfide was attributed to an increase in surface rather than to surface functionality because the values when normalized with respect to surface area showed no change. Furthermore, it was supported by the fact that when surface functionality was enhanced by oxidation with ammonium persulfate, there was no change in the adsorption capacity of the carbon for butyl disulfide. The adsorption of the mercaptan and disulfide also decreased on methylation of the carbon. This could be attributed to a decrease of surface functionality due to methylation, but these authors attributed it to a decrease in surface area as a result of pore blocking.

Table 18 Adsorption of Different Carcinogenic Compounds on Activated Carbons

Compounds	Molecular weight	Equilibrium concentration range (mg/L)	Adsorption capacity	Carbon dose to reduce 1.0 mg/L to 0.1 mg/L (mg/L)
Naphthalene	128			
Darco		0.002-9.3	6.3	29
Filtrasorb		0.06-5.3	16.8	19
1,1-Diphenylhydrazine	184			
Darco		0.2-1.5	9.4	18
Filtrasorb		0.4-9.1	14.9	10
Naphthyl amine	143			
Darco		0.2-1.5	7.7	27
Filtrasorb		0.1-1.5	16.6	10
4,4'-Mathylene-bis (2-choroaniline)	264			
Darco		0.08-1.5	12.1	27
Filtrasorb		0.05-1.2	24.2	15
Benzidine	184			
Darco		0.2-10	8.4	19
Filtrasorb		0.003-8.2	17.2	10
Dimethylnitrosamine	74			
Darco		6.5-9.5	6×10^5	$>10^5$
Filtrasorb		6.5-9.0	6×10^7	$>10^7$
3,3'-Dichlorobenzidine	253.1			
Darco		0.8-2.5	12.6	13
Filtrasorb		0.6-2.5	24.1	5

Source: Fochtman and Dobbs (93). With permission from Ann Arbor Science Publishers.

In the case of p-hydroxybenzaldehyde, the adsorption on a per unit area basis decreased on oxidation, indicating that the surface oxygen groups played a role. This was in agreement with the earlier observation that the surface oxygen groups decrease the adsorption of phenol, p-nitrophenol, nitrobenzene, etc.

Kinetic studies showed that the rate of adsoprtion of butyl disulfide remained unchanged with degassing termperature up to 700°C but decreased when the outgassing temperature was raised above 900°C. Similarly, the rate of adsorption remained unchanged on oxidation but decreased when the oxidation was followed by outgassing at 900°C (Table 20). This was attributed to the presence of metallic impurities such as copper and iron, which were removed on outgassing at 900°C and which enhanced the adsorption rate. However, the presence of these metallic impurities did not influence the rate of adsoprtion of decane, although the presence of surface oxides hindered their adsorption. The rate of adsorption of decane increased when the surface oxides were removed and decreased when the surface oxides were formed on oxidation (Table 20). This is probably due to the nonpolar character of the decane molecule, which will be preferred by a nonpolar carbon surface. In the case of p-hydroxybenzaldehyde, neither degassing nor oxidation caused any change in the rate of adsorption (Table 20), indicating that pore diffusion rather than the presence of surface oxide or metallic impurities was the limiting step in adsorption.

Table 19 Adsorption of Butyl Disulfide as a Function of Surface Treatment

Carbon	Equilibrium concentration range (mmol/L)	Adsorption	
		mmol/g	$\mu mol/m^2$
Untreated	0.008-0.45	2.40	3.37
Outgassed at 900°	0.014-0.52	2.85	—
Outgassed at 900° and oxidized 100°	0.025-0.51	2.93	—
Outgassed at 900° and oxidized at 300°	0.044-2.10	3.07	3.17
Outgassed at 900° and oxidized between 300° and 500°C	0.026-0.47	3.64	—
Chemically oxidized	0.022-0.32	2.54	—

Source: Ishizaki and Cookson (94). With permission from Ann Arbor Science Publishers.

Table 20 Rates of Adsorption of Butyl Disulfide, Decane, and p-Hydroxybenzaldehyde on Treated and Untreated Carbons

Carbon	Rate constant (L/mmol-min x 10^3)		
	Butyl disulfide	Decane	p-Hydroxybenzaldehyde
Untreated	78.4	3.80	12.6
Outgassed at 900°C	6.65	16.6	11.5
Outgassed at 900°C and oxidized at 100°C	2.21	—	—
Outgassed at 900°C and oxidized at 200°C	0.95	—	—
Outgassed at 900°C and oxidized at 300–500°C	0.29	1.18	—
Chemically oxidized	0.37	—	11.1

Source: Ishizaki and Cookson (94). With permission from Ann Arbor Science Publishers.

6.17 CATALYTIC REACTIONS OF CARBONS

In addition to being good adsorbents, activated carbons can promote a variety of surface reactions. The role of the carbon as a catalyst is, however, different in different reactions. Whereas the nature of the carbon surface and the presence of carbon—oxygen surface structures play dominant roles in some reactions, the extent of the surface area and the availability of the active sites are important factors in certain other reactions. The details of many catalytic reactions of carbons are a part of the patented literature and their actual mechanisms are closely guarded secrets. In this section we discuss some of the catalytic reactions that tend to elucidate broadly the factors that influence the catalytic activity of carbons. Although carbon is also widely used as a catalyst support, this is not included as part of discussion, nor is the reaction where carbon may be acting as a poison.

The catalytic reactions have been classified into oxidation reactions, combination reactions, decomposition reactions, and elimination reactions. A few reactions that cannot be included in these catagories are also discussed.

6.17.1 Catalytic Oxidation of Hydrogen Sulfide

Removal of hydrogen sulfide from waste gases obtained from coal conversion processes and from biogas obtained during the microbial conversion of sulfur-containing wastes is an important aspect of air pollution control. The method that has been frequently proposed and studied is the selective oxidation of hydrogen sulfide to elemental sulfur over an activated carbon catalyst. The elemental sulfur then can be recovered by extraction of the activated carbon with organic solvents or by steam distillation.

Several active carbons obtained from coconut shells and coals have been used for the catalytic oxidation but because of the large number of impurities present in them, the mechanism of the oxidation reaction is not very well understood.

When hydrogen sulfide is oxidized in the presence of oxygen, the major reaction that occurs is the formation of water and elemental sulfur:

$$H_2S + 1/2O_2 \longrightarrow H_2O + 1/2S_2$$

However, depending on the reaction conditions and the nature of the activated carbon surface, several side reactions can occur which give rise to certain unwanted products such as SO_2 and H_2SO_4. Therefore, the reaction conditions or the active carbons are chosen as to eliminate nate or at least reduce the formation of these unwanted products to a negligible amount. Many times cartain types of reaction promoters have been used.

Puri and co-workers (95) investigated the catalytic oxidation of H_2S by passing mixtures of H_2S and O_2 (2 L over a period of 1 hr on 3 g of a carbon) over a bed of sugar and coconut charcoal and a few commercial-grade carbon blacks associated with varying amounts of chemisorbed oxygen in the temperature range 120–240°C. An appreciable amount of H_2S was oxidized over the carbons into S, a part of which was chemisorbed on the carbon surface, another part deposited as free S, and only a very small part converted into SO_2. The conversion into SO_2 was very small and that occurred only when the oxygen content of the gaseous mixture was large (1:6 or more). The oxidation of H_2S increased from 25 to 30% when the oxygen content of the mixture was raised from 1:1 to 1:6. The extent of oxidation was dependent on the amount of charcoal used in the bed. The oxidation of H_2S was more than 90% when the amount of charcoal in the bed was increased to 10 g (Table 21). However, the capacity of the charcoal to fix sulfur per gram of charcoal remained more or less unchanged irrespective of the amount of the carbon used (Table 21).

The oxidation of hydrogen sulfide, whigh was negligible in the absence of a carbon, proceeded fairly rapidly even at 120°C but the rate of oxidation increased with increase in the temperature. The extent of oxidation was also found to depend on the nature of the carbon surface (Table 22), being much larger when the carbon was an outgassed sample. The disposition of sulfur obtained by the oxidation of hydrogen sulfide was more as free sulfur when carbon blacks were used and more as chemisorbed sulfur when charcoals were used (Table 23). Furthermore, the outgassed carbons showed a higher catalytic activity than the original carbons, the difference being related to the capacity of a carbon to chemisorb sulfur, which in turn depended on the availability of active sites on the carbon surface. Original charcoals and carbon blacks contained appreciable amounts of chemisorbed oxygen, which was removed on outgassing, thus making the active sites available for chemisorption of sulfur. The catalytic activity of a carbon, however, was independent of the total BET surface area because the outgassed carbons, which although having about the same surface, were more efficient catalysts. Furthermore, these workers also observed that when the active sites were blocked by chemisorption of oxygen, bromine, or sulfur, the efficiency of the carbon for the oxidation was considerably reduced (Table 24), indicating clearly that the active sites were responsible for the catalytic activity.

Cariaso and Walker (96) used saran charcoals activated by carbon dioxide at 900°C, oxygen at 300°C, and in air at 400°C for the catalytic oxidation of hydrogen sulfide in the temperature range 100–160°C at hydrogen sulfide pressures usually available in flue gases. These

Table 21 Oxidation of Hydrogen Sulfide on 1000° Outgassed Sugar Charcoal as a Catalyst[a]

H_2S/O_2 ratio	Weight of catalyst (g)	Sulfur fixed irreversibly on charcoal (mmol)	Free sulfur formed (mmol)	SO_2 formed (mmol)	Total H_2S oxidized (mmol)	Amount of H_2S oxidized (%)
1:1	3	2.71	0.80	nil	3.51	26
1:3	3	2.97	0.78	nil	3.75	27.8
1:6	1	0.95	0.46	nil	1.41	10.4
1:6	2	1.98	0.76	nil	2.74	20.3
1:6	3	3.21	0.86	0.04	4.11	30.4
1:6	7	7.66	1.23	0.10	9.09	67.3
1:6	10	10.67	1.79	0.09	12.55	93.0
1:12	3	3.27	0.89	0.10	4.26	31.6

[a]Total H_2S passed = 13.5 mmol.
Source: Puri et al. (95).

Table 22 Oxidation of H_2S from H_2S-O_2 Mixtures (1:6) over Different Charcoals and Carbon Blacks at Different Temperatures[a]

Sample	Surface area (m^2/g)	Amount of H_2S oxidized (%) at 120°	180°	240°
Blank	—	0.3	0.5	0.9
Charcoals				
Sugar charcoal				
Original	402	14.1	35.1	63.0
Degassed at 750°	426	48.2	87.1	95.2
Degassed at 1000°	421	55.9	93.0	97.0
Coconut charcoal				
Original	320	23.0	50.1	74.1
Degassed at 750°	315	49.3	89.4	96.5
Degassed at 1000°	319	55.6	96.2	99.6
Carbon blacks				
Philblack-A				
Original	45.8	26.7	92.6	94.1
Degassed at 1000°	44.3	60.0	96.5	98.2
Statex-B				
Original	48.3	21.4	38.8	50.3
Degassed at 1000°	46.2	32.5	46.4	54.6
Pelletex				
Original	27.1	18.8	31.5	40.4
Degassed at 1000°	25.4	25.4	36.9	46.8

[a]Total H_2S passed = 13.5 mmol; catalyst = 10 g.
Source: Puri et al. (95).

carbons were associated with different amounts of the active surface area. The samples were cleared of any chemisorbed species by heat treatment in vacuum at 900°C before being used for the oxidation of hydrogen sulfide. When the experiments were carried out for long periods of time (5 days) by passing a mixture of H_2S and O_2 (in excess of stoichiometric amounts) and using H as a carrier gas at 140° C, over oxygen-activated carbon, about 90% of the H_2S was oxidized and the total sulfur content of the carbon was about 18%. The instantaneous rate of sulfur loading of the carbon was directly proportional to the amount of hydrogen sulfide in the gaseous mixture (Fig. 41). Furthermore, the oxidation rates during the second run were almost the same as in the first run, although there was a further buildup of sulfur on the carbon to about 36% by weight. The reaction was found to be first order with respect to H_2S and zero order in O_2 concentration. The loading of sulfur on the carbon resulted in a loss of surface area of the carbon, the nitrogen area falling more rapidly than the carbon dioxide area (Table 25). This was attributed to the fixation of sulfur in some of the micropores and their blocking.

In the case of short runs (5 hr) the rate of hydrogen sulfide oxidation decreased rapidly with increasing reaction time during the early stages and then became more or less constant. The carbons activated in oxygen or in air were found to have a higher initial catalytic activity than the carbon activated in carbon dioxide (Fig. 42). The rates of hydrogen sulfide oxidation showed behavior almost similar to the rates of oxygen chemisorption in the case of oxygen and air-activated carbons; i.e., the rate of oxidation increased or remained constant over a reaction period in which the rate of chemisorption of oxygen

Table 23 Oxidation of H_2S on Passing H_2S-O_2 Mixtures (1:6) over Different Charcoals and Carbon Blacks[a]

Sample	H_2S removed (mmol)	Sulfur fixed irreversibly (mmol)	Free sulfur formed (mmol)
Sugar charcoal			
Original	4.74	3.06	0.83
Degassed at 750°	11.76	9.90	1.32
Degassed at 1000°	12.55	10.67	1.79
Coconut charcoal			
Original	6.77	4.29	1.24
Degassed at 750°	12.07	10.31	1.25
Degassed at 1000°	12.99	10.95	1.52
Philblack-A			
Original	12.50	4.41	7.12
Degassed at 1000°	13.03	4.51	7.72
Statex-B			
Original	5.24	0.73	4.31
Degassed at 1000°	6.26	1.88	4.11
Pelletex			
Original	4.25	0.11	3.91
Degassed at 1000°	4.98	0.55	4.12

[a]Total H_2S passed = 13.5 mmol; temperature = 180°.
Source: Puri et al. (95).

Table 24 Oxidation of H_2S and Its Breakup on Passing H_2S-O_2 Mixtures (1:6) over Variously Treated 1000° Degassed Sugar Charcoal[a]

Sample	H_2S removed (mmol)	Sulfur fixed irreversibly (mmol)	H_2S oxidized (%)
Control	12.55	10.67	93.0
Sugar charcoal pretreated with $K_2S_2O_8$	6.41	4.78	47.5
Sugar charcoal pretreated with aqueous Br_2	5.93	4.31	43.9
Sugar charcoal (reused)	6.31	3.91	46.7
Sugar charcoal reused after treatment in hydrogen at 800°	11.62	9.98	86.1

[a]Total H_2S passed = 13.5 mmol; temperature = 180°.
Source: Puri et al. (95).

increased or remained constant. But in the case of the carbon dioxide—activated or unactivated carbons the two rates differed appreciably. Thus the catalytic activity of the carbon was related to the
carbon activity for the dissociative chemisorption of oxygen (97—100),
which in turn was determined by the presence of active sites on the
carbon surfaces. This view was further supported by the fact that
when these active sites were covered by pretreatment with oxygen or
hydrogen, the extent of catalytic activity was considerably reduced.
A decrease of 54% in the steady-state rate was observed on treatment
of the carbon dioxide—activated sample with oxygen at 300°C and a
decrease of 36% on treatment with hydrogen at 140°C. This was in
agreement with the earlier observations of Bansal et al. (97), Hart
et al. (98), and several other workers (99,100) that these treatments
result in the chemisoprtion of oxygen or hydrogen on the most active
sites and render them unavailable for the catalytic oxidation of hydrogen sulfide. The loss in the catalytic activity was less in the case of
the hydrogen pretreatment because hydrogen at 140°C could cover
only part of the active sites at this temperature. Thus Cariaso and
Walker (96) suggested that the reaction between hydrogen sulfide and
oxygen on the carbon surface involved the chemisorption of oxygen
on the active sites and its reaction with the gas phase of hydrogen
sulfide. The reaction caused the removal of oxygen, creating nascent

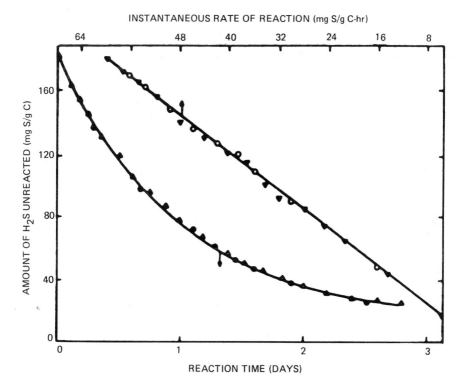

FIGURE 41 Rates of catalytic oxidation of hydrogen sulfide by carbon.
[From Cariaso and Walker (96). Reproduced with permission from
Pergamon Press.]

Table 25   Surface Areas of Sulfurized Carbons

| Sulfur content (%) | Surface areas (m^2/g) | |
	N_2	CO_2
Oxygen-activated carbon		
0	1160	850
10.5	890	620
14.6	730	—
18.5	520	530
30.0	44	160
Unactivated Saran carbon		
0	770	880
12.1	2	480

Source: Cariaso and Walker (96). Reproduced with permission from
Pergamon Press.

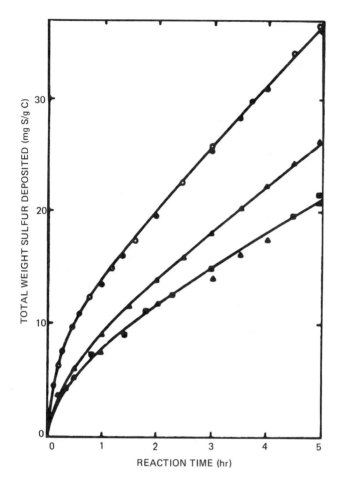

FIGURE 42 Loading of sulfur on carbon as a function of time of oxidation of hydrogen sulfide at 140°C. Closed squares, unactivated carbon; closed circles, activated in oxygen; open circles, activated in air; closed triangles, activated in CO_2 (60% burnoff); open triangles, activated in CO_2 (20% burnoff). [From Cariaso and Walker (96). Reproduced with permission from Pergamon Press.]

active sites where more oxygen could be adsorbed for the reaction to proceed. The elemental sulfur formed as a result of the oxidation of hydrogen sulfide was deposited on the carbon. This could explain the loss of surface area of the carbon as more and more sulfur was deposited. The oxidation of hydrogen sulfide using some commercial active carbons showed that catalytic activity was dependent on their

activity toward oxygen as well as on the presence of certain metallic
impurities which could catalyze the oxidation of hydrogen sulfide.
When these metallic impurities were removed by washing the active
carbons with acid, the catalytic activity was reduced in those carbons
which contained sodium or iron impurities, which are known to catalyze the oxidation of hydrogen sulfide.

Klein and Henning (101), while studying the kinetics of the catalytic oxidation of hydrogen sulfide on activated carbons in the temperature range 40–150°C, observed that the order of the reaction was
dependent on the molar ratio of H_2S and O_2 in the gaseous mixture.
The reaction order was 0.5 when the molar ratio was greater than
unity and zero when the molar ratio was less than unity. Both the
activation energy and the frequency factor increased with sulfur loading of the carbon (Fig. 43). The addition of a promoter such as

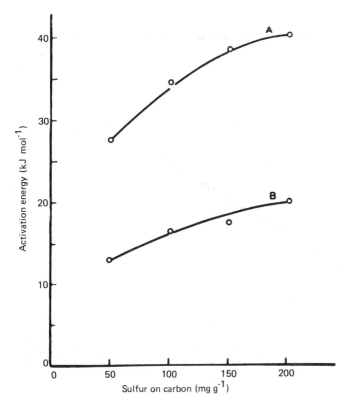

FIGURE 43 Variation in activation energy for the catalytic oxidation of
hydrogen sulfide by carbon with sulfur loading of carbon. (Courtesy
J. Klein.)

potassium iodide, however, decreased both the activation energy and the frequency factor. The frequency factor was found to be a linear function of the activation energy (Fig. 44). This compensation effect was explained by these workers (101) by postulating that the oxidation of hydrogen sulfide was catalyzed by carbon-active sites which were associated with different activation energies. The addition of the promoter enhanced the efficiency of all the active sites. The reaction first occurred on the most active sites, which had a low concentration so that low activation energy and low frequency factor was observed in the initial stages of the reaction. The sulfur formed as a result of the oxidation was deposited on the larger number of active sites with lower activity. This caused an increase in activation energy and the frequency factor as the sulfur loading on the carbon increased. The loading of the carbon with sulfur also decreased the reaction rate and the BET surface area of the carbon.

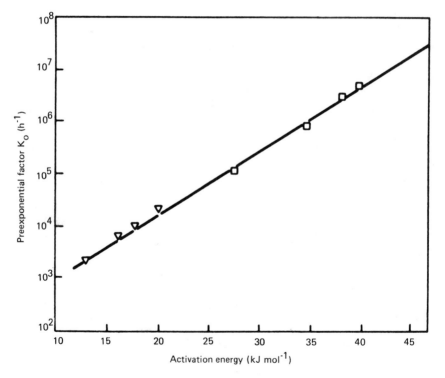

FIGURE 44 Relationship between activation energy and preexponential factor for the catalytic oxidation of hydrogen sulfide by carbon. (Courtesy J. Klein. Reproduced with permission of Butterworths.)

The addition of potassium iodide as a promotor increased the re-action rate as well as decreased the extent of the side reactions which produce sulfuric acid. The addition of 2% potassium iodide in-creased the rate of hydrogen sulfide oxidation by about 40–50% (Fig. 45).

Mechanism of the Reaction

There is general agreement that the catalytic oxidation of hydrogen sulfide in the presence of oxygen by a carbon occurs on the carbon-active sites and that the rate of the reaction depends on the concen-tration of these sites as well as the concentration of hydrogen sulfide in the gaseous mixture. There are, however, different views regard-ing the mechanism of the process. Puri et al. (95) and Cariaso and Walker (96) are of the view that the reaction occurs by the chemicorp-tion of oxygen on these active sites, which then reacts with hydrogen sulfide in the gaseous phase. Steijns et al. (102) postulate that the oxidation of hydrogen sulfide in the temperature range 20–250°C in-volves reaction between the chemisorbed oxygen and the dissociatively adsorbed hydrogen sulfide producing sulfur and water. The molecular oxygen in the gas phase then reacts with the adsorbed sulfur to pro-duce sulfur dioxide as an intermediate product which then oxidizes hydrogen sulfide. It may, however, be mentioned that Puri et al. (95) made quantitative measurements of the formation of SO_2 during the oxidation of H_2S of oxidation on several charcoals and carbon blacks and found very small amounts of SO_2. This small amount of SO_2 may not contribute substantially to the H_2S oxidation. Hedden et al. (103) proposed (Fig. 46) that the oxidation occurs through the formation of a water film on the carbon surface in which both H_2S and O_2 dissolve. The oxygen gets chemisorbed on the carbon surface and breaks into active species and reacts with the hydrosulfide species, produced by the dissociation of hydrogen sulfide, to form hydroxyl ions and sulfur, which is deposited on the carbon surface. Thus the formation of a water film on the carbon surface is an essential condi-tion for the reaction to occur (103).

6.17.2 Catalytic Oxidation of Ferrous Ions

The catalytic oxidation of ferrous ions by molecular oxygen in the pre-sence of powdered gas mask charcoal was first studied by Lamb and Elder (104), who found that the rate of oxidation was about 200 times faster in the presence than in the absence of charcoal. The oxidation was attributed to the surface peroxide groups. Posner (105) investi-gated the reaction in the presence of an activated carbon and suggest-ed that the oxidation involved adsorption of Fe^{2+} and H^+ ions on the carbon surface and subsequent oxidation of the complex with oxygen to Fe^{3+} ions, which desorbed immediately. The catalytic activity was

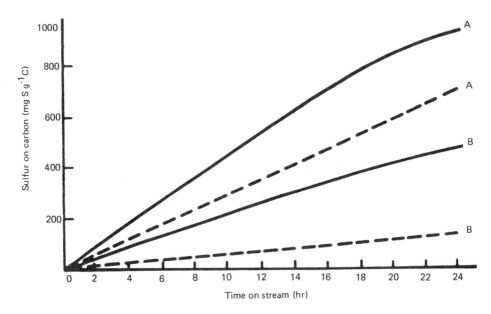

FIGURE 45 Influence of temperature and addition of potassium iodide (promotor) on catalytic activity of carbon. (Courtesy J. Klein. Reproduced with permission of Butterworths.)

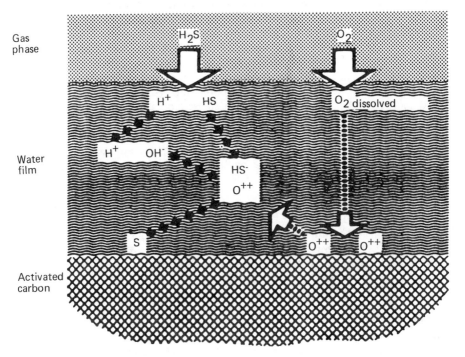

FIGURE 46 Mechanism of catalytic oxidation of hydrogen sulfide by carbon. [From Hedden et al. (103).]

not attributed to any specific surface structure or any active site but was related to the surface area of the charcoal. Garten and Weiss (106) explained the oxidation on the basis of the chromene groups. Thomas and Ingram (107) also investigated the catalytic oxidation of ferrous ions in the presence of activated carbons and proposed that the mechanism involved the adsorption of Fe^{2+} ions and oxygen, which then reacted to give Fe^{3+}, which could also be adsorbed on the surface. The catalytic activity was related to the surface area of the carbon.

Puri and Kalra (108) observed that charcoals and carbon blacks that had been outgassed at sufficiently high temperatures, and thus devoid of oxygen, evolved as CO_2 (CO_2-complex), catalyzed the oxidation of Fe^{2+} to Fe^{3+} by molecular oxygen in acid solution. The reaction was much faster in the presence of charcoals, the oxidation increasing from 1.9% in the absence of charcoal to about 25% in the presence of charcoal. The reaction was first order in ferrous ion concentration as expected because oxygen was being continuously bubbled through the solution. The activation energy of the reaction was found to be 5.7 kcal/mol. The catalytic efficiency increased with an increase in the amount of charcoal. The catalytic activity of the original charcoal was smaller but increased appreciably when the carbons were outgassed at increasing temperatures (Table 26). The catalytic performance of the original charcoals increased more than threefold on degassing the samples at 750°C and above. Since the magnitude of the surface area did not change appreciably on outgassing at these temperatures, it was suggested that the nature and not the magnitude of the surface was of primary importance in determining the catalytic efficiency of the carbon for this reaction. Further, since the 1000°-outgassed charcoals, which were free of any chemisorbed oxygen, showed maximum catalytic activity, this property could not be attributed to the peroxide (104,106) or chromene groups. The increase in the catalytic activity on outgassing was attributed to the increase in the number of unsaturated active sites, although it was not related linearly to the number of active sites available. However, when these unsaturated sites, made available by the desorption of surface oxygen structures, were covered with bromine, oxygen, or sulfur, the catalytic activity of the carbon in each case decreased appreciably (Table 27) and the catalytic activity was restored back to a considerable extent when the bromine was desorbed on treatment with hydrogen at 900°C. Thus Puri and Kalra were of the view that the carbon catalyst provided suitable active sites for the dissociative chemisorption of oxygen, which then reacted with Fe^{2+} ions and a proton to yield Fe^{3+} ions:

$$2Fe^{2+} + (O) + 2H^{+} \longrightarrow 2Fe^{3+} + H_2O$$

Table 26 Extent of Ferrous Sulfate Oxidation in the Presence of Charcoals and Carbon Blacks[a]

Carbon	Surface area (m^2/g)	Unsaturated sites as determined by bromine adsorption (mmol/100 g)	$FeSO_4 \cdot 7H_2O$ oxidized (mmol)
Charcoals			
Sugar charcoal			
Original	402	15	0.31
Degassed at 400°	401	118	0.80
Degassed at 750°	426	186	1.23
Degassed at 1000°	421	195	1.30
Coconut charcoal			
Original	320	19	0.34
Degassed at 400°	319	81	0.82
Degassed at 750°	315	145	1.31
Degassed at 1000°	319	147	1.39
Carbon blacks			
Mogul			
Original	308	4	0.31
Degassed at 1000°	305	38	1.12
Mogul-A			
Original	228	3	0.32
Degassed at 1000°	226	28	1.04
ELF-O			
Original	171	6	0.41
Degassed at 1000°	169	25	1.01

[a]Charcoal taken = 3 g and 100 ml of 0.1 M $FeSO_2 \cdot 7H_2O$.
Source: Puri and Kalra (108).

This view was further supported by the observation that when the residual carbons left after the reaction were washed, dried, and examined for their oxygen content, there was an appreciable increase in the oxygen content, which came off as CO_2 on evacuation. There was also a corresponding fall in the number of unsaturated active sites as determined by bromine adsorption (109). It was also observed that the catalytic activity of the carbons declined when carbon was reused for oxidation after two cycles. This was due to the fact that some of the oxygen remained strongly bonded at some of the active sites, which were therefore not available for the reaction.

6.17.3 Catalytic Oxidation of Sulfurous Acid

Boehm and colleagues (110–112) studied the catalytic oxidation of dilute aqueous solutions of sulfurous acid by oxygen in the presence of several active carbons and carbon blacks as catalysts. The progress of the reaction was monitored by measuring the electrical conductance of the solution. The reaction rate was found to be proportional to the sulfurous acid concentration and the square root of the oxygen pressure. Similar carbons showed a wide variation of catalytic activity. Activated carbons obtained from peat were the most active; wood charcoal was less active. Sugar charcoal and carbon blacks showed very feeble activity and that too after heat treatment (Table 28). The catalytic activity of a given carbon increased generally after heat treatment being highest near 800°C for charcoals and near 900°C for carbon blacks. The catalytic activity of carbon blacks was con-

Table 27 Effect of Presaturating the Unsaturated Sites and Their Regeneration on Catalytic Activity of 1000° Degassed Sugar Charcoal

Nature of pretreatment	Amount of O_2, Br_2, or S fixed at the unsaturated sites (mmol/100 g)	Surface unsaturation of treated product (mmol/100 g)	Ferrous sulfate oxidized (mmol)
None	0	195	1.30
Treated with aqueous H_2O	150	55	0.39
Treated with aqueous bromine	195	0	0.24
Treated with CS_2 vapor	191	0	0.15
Br-fixed sample treated with H_2 at 900°	0	178	1.14

Source: Puri and Kalra (108).

Table 28 Catalytic Activity of Different Carbons for Oxidation of Sulfurous Acid by Oxygen[a]

Carbon	Specific surface (m^2/g)	Catalytic activity $(mol/g/sec)$
Norit BRX outgassed at 200°C	1800	0.85
Norit BRX heat treated at 800°C	1800	1.93
Sugar charcoal outgassed at 200°C	165	0
Sugar charcoal heat treated at 800°C	188	0
Anthralur Sta outgassed at 200°C	560	1.16
Anthralur Sta heat treated at 800°C	545	1.78
Anthralur Sta extracted with HCL	670	1.04
Eponite extracted with HCL and then outgassed at 200°C	750	0.28
Eponite extracted with HCL and then heat treated at 800°C	755	0.40
Corax-3 (carbon black) outgassed at 200°C	83	0.008
Corax-3 heat treated at 1000°C	81	0.030
Acetylene black heat treated at 200°C	80	0
Ck-3 extracted with xylene and outgassed at 200°C	92	0

[a]Concentration of solution = 1.65×10^{-3} M; temperature = 20°C.
Source: Boehm et al. (110). Reproduced with permission of Butterworths.

siderably enhanced by chemical surface treatment (Table 29). The effect was larger on treatment of the carbon black with chlorine than on treatment with bromine. The increase in catalytic activity of the carbon black was very large on treatment with ammonia at elevated temperatures. Even inactive acetylene black became active on treatment with ammonia. X-ray photoelectron spectroscopy showed the presence of several nitrogen species on the surface of the carbon black and the catalytic activity was related to the increase in the pyridine-like nitrogen structures (N 1_s binding energy: $\cong$ 399eV) (Fig. 47). Nitrogen in the form of amines or carboxylic amide groups (binding energy 400 eV) was found to be less active. The acidity of the solution had little effect on the catalytic activity, although addition of nitric acid or potassium nitrate was found to poison the reaction. The catalytic activity of the carbons also decreased when they were used several times. This was attributed to the decrease in surface area and the low binding energy nitrogen, which was essential for the catalytic activity. The reaction rate was postulated to involve adsorption of SO_2 or HSO_3^- ions on the carbon surface as the oxidation reaction was poisoned by substances such as acetone, phenol, p-nitrophenol, and hydroquinone, which were strongly adsorbed by the carbons. The oxidation is caused by the adsorbed oxygen which gets activated on the carbon surface by transfer of electrons from the carbon conduction band forming O_2^- ions. This could explain the enhancement of the catalytic activity of the carbon on treatment with ammonia. The treatment with ammonia results in the binding of nitrogen atoms in the carbon layers as

with the extra electrons going to the conduction band at relatively high energies, facilitating electron transfer to adsorbed species. This model was supported by an increase in catalytic activity on heat treatment of the ammonia-treated wood charcoal at 900°C under argon (Table 30). The higher catalytic activity apparently resulted from the rearrangement in the periphery of the carbon layers producing more substitutional nitrogen atoms.

6.17.4 Catalytic Oxidation of Oxalic Acid

Effluents from industries engaged in producing wood pulp, refining of sugar and olive oil contain oxalic acid. Water contaminated with oxalic acid is harmful for human consumption and may deprive aquatic animals, plants, and soils of calcium by precipitating it as calcium oxalate. Therefore, removal of oxalic acid should be of importance from a water pollution control point of view.

Table 29 Effect of Surface Treatment of Corax 3 Carbon Black on the Catalytic Oxidation of Sulfurous Acid

Treatment	Surface area (m^2/g)	Cl, Br, N content (mmol/100 g)	Oxidation rate ($\mu mol/g/sec$)
Heat treatment at 300-600°C	87	9.0 N	0
O_2 at 400°C	301	n.d.	0.036
Cl_2 at 450°C	78	230.0 Cl	0.024
Br_2 at 450°C	75	93.0 Br	0.009
NH_3 at 600°C	84	11.9 N	0.046
NH_3 at 900°C	731	20.5 N	2.14

Source: Boehm et al. (110). Reproduced with permission of Butterworths.

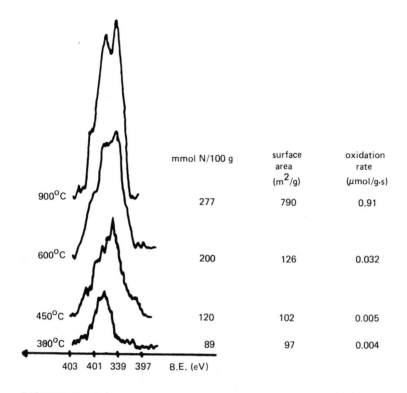

	mmol N/100 g	surface area (m^2/g)	oxidation rate (μmol/g·s)
900°C	277	790	0.91
600°C	200	126	0.032
450°C	120	102	0.005
300°C	89	97	0.004

403 401 339 397   B.E. (eV)

FIGURE 47 X-ray photoelectron spectra (N 1_s) of carbon after heat treatment of the ammonia-treated sample at different temperatures and their catalytic activity for oxidation of H_2SO_3. [From Boehm et al. (110). Reproduced with permission of Butterworths.]

Table 30 Influence of Heat Treatment (under Argon) on the Catalytic Activity of Ammonia-Treated Eponite Carbon for Oxidation of Sulfurous Acid

Catalyst	Surface area (m^2/g)	N content (mmol/g)	Catalytic activity (μmol/g)
Ammonia treated (3 hr at 870°C)	780	0.96	1.84
Ammonia treated (sample heated at 900°C)	735	0.71	2.25
Ammonia treated (sample heated at 1000°C)	720	0.69	2.26

Source: Boehm et al. (110). Reproduced with permission of Butterworths.

Kurth et al. (111) observed that dissolved oxalic acid can be oxidized with oxygen in the presence of carbons. Using a number of active carbons and carbon blacks given different surface chemical treatments, these workers found that activated carbons obtained from peat had the highest catalytic activity. Heat treatment of the carbon enhanced the catalytic activity (Fig. 48), whereas the presence of organic substances such as phenol and hydroquinone, which are adsorbed by the carbon, slowed down the activity of the carbons for oxidation. The catalytic activity was also reduced by the presence of KNO_3 in the solution.

The reaction rate was not influenced by the concentration of oxalic acid at a given pH and it was maximum for a given concentration at pH 2.6 (Fig. 49). This was attributed to the charge of the carbon surface and the ionization of the oxalic acid. The second dissociation of oxalic acid changes from 0.6 to 6% in the pH range 2−3. Higher opposite charges of the carbon surface and the reducing species lead to increased adsorption. Thus Kurth et al. (111) are of the view that the anionic reducing species were first adsorbed by basic surface oxides and then underwent oxidation by the adsorbed oxygen.

6.17.5 Catalytic Oxidation of Creatinine

Active carbons are widely used as adsorbents for the removal of toxins from blood (113−115) and for dialyrate regeneration (116). Creatinine is one such substance which is adsorbed by the active carbons. It was, however, found that creatinine reacts (117,118) on the carbon surface, releasing some oxidation products into the system, which may lead to complications during analysis. This led Smith et al. (119) to investigate the extent of oxidation of creatinine on active carbon surfaces using different carbons in the granulated and powdered form.

These workers (119) observed that when an activated carbon was placed in contact with a creatinine solution in air, the concentration of the solution continued to decrease considerably over a number of days. The decrease in concentration was comparatively much smaller when granulated carbons were used and greater when powdered carbons were used (Table 31). In the case of the control sample (i.e., no carbon), the concentration of the solution decreased for 24 hr and then became more or less constant. The decrease in the concentration of the control was attributed to the conversion of small amounts of creatinine to creatine, for which the equilibrium was established in one day (120). Further decrease in the concentration of the solution in the presence of powdered carbon was the result of the oxidation of creatinine. This was supported by the fact that when the reaction with carbon was carried out from a solution containing both creatinine and creatine, there was little conversion of creatinine into creatine because the equilibrium favored creatinine. However, the oxidation

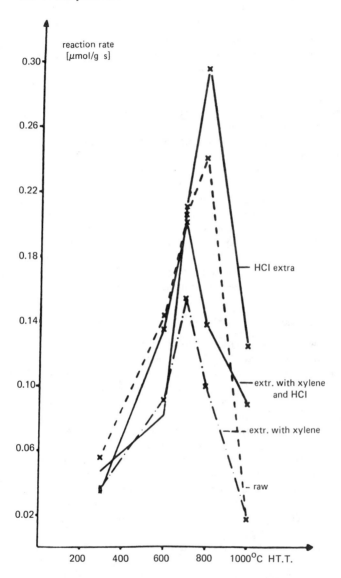

FIGURE 48 Influence of heat treatment temperature on the catalytic activity of Anthralur Sta carbon for the oxidation of oxalic acid. [From Kurth et al. (111).]

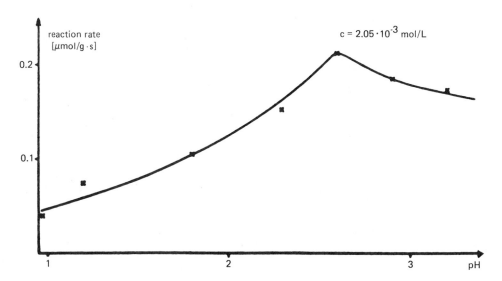

FIGURE 49 Effect of pH on the catalytic oxidation of oxalic acid at 200°C by carbon. [From Kurth et al. (111).]

of creatinine continued. Furthermore, it was found that creatinine could not be oxidized from its solution by the carbon. This showed that the dominating reaction was the oxidation of the creatinine. The rate of the reaction also increased with an increased amount of the carbon. No oxidation of creatinine was observed when the reaction was carried out in an inert atmosphere.

The kinetic measurements with granulated carbons in the low concentration range in inert atmosphere and in the presence of oxygen showed that the concentration of creatinine in the inert atmosphere decreased only very slightly after the first rapid fall, whereas it showed a steep fall in the presence of oxygen, ultimately reaching a zero concentration after about 45 hr (Fig. 50). The fall in the presence of nitrogen atmosphere, which is similar to the fall in the presence of oxygen, was attributed to the adsorption of creatinine on the carbon surface since there was no fall in concentration in the case of the control. The further continued fall in the concentration of creatinine in the presence of oxygen resulted from the catalytic oxidation of the adsorbate by the carbon. Thus these workers suggested that the oxidation of creatinine involves the adsorption on carbon-active sites followed by oxidation. It was found that the rate of oxidation from a pure creatinine solution was 1.6 times higher than the rate of oxidation from a solution containing an equimolar amount of creatine because creatine then competed for the carbon-active sites with creatinine.

Table 31 Catalytic Oxidation of Creatinine and Creatine by Powdered Charcoal[a]

Time (hr)	Creatinine oxidized using 0.1 g charcoal at 83°C (mmol/g)		Creatine oxidized using 0.5 g charcoal at 37°C	
	Control	With charcoal	Time	(mmol/g)
0	0	0	0	0
2.5	1.4	5.1		
21.5	10.5	24.4	1 hr, 5 min	0
69.5	11.4	30.3		
75	12.1	32.2	18 hr, 25 min	0.59
120	11.5	32.5		
144	11.7	32.6	24 hr, 40 min	1.12
168	11.5	32.6		
216	11.5	32.7		

[a]Solution taken = 500 ml 10^{-3} M.
Source: Smith et al. (119). Reproduced with permission from Pergamon Press.

6.17.6 Catalytic Oxidation of *n*-Butyl Mercaptan

Ishizaki and Cookson (94,121), while studying the adsorption of *n*-butyl mercaptan on active carbons, observed a secondary peak in the chromatograms, which was attributed to butyl disulfide:

$$2C_4H_9SH + 1/2O_2 \longrightarrow (C_4H_9S)_2 + H_2O$$

This butyl disulfide was produced by the disulfide catalytic oxidation of the mercaptan by the carbon. The catalytic effect of the carbon was significantly reduced by outgassing the carbon at elevated temperature (Table 32).

The catalytic oxidation of the *n*-butyl mercaptan has been explained on the basis of the existence of quinone groups and on metal ions in carbons which are known to catalyze such oxidations (122). In the case of quinones the disulfide ion was formed through redox reaction involving the formation of thiyl radicals and semiquinone anions (Fig. 51). However, no separate mechanism was proposed for the

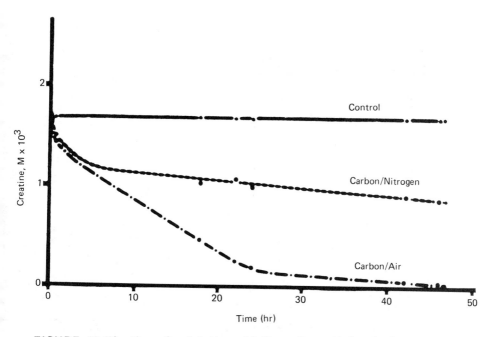

FIGURE 50 Kinetics of catalytic oxidation of creatinine in inert atmosphere and in the presence of air. [From Smith et al. (119). Reproduced with permission from Pergamon Press.]

Table 32 Butyl Disulfide Formed by the Catalytic Oxidation of
N-Butyl Mercaptan

	Carbon	
Treatment	Concentration (g/L)	Butyl disulfide formed (mmol/L)
Untreated	1.0	2.99
OG 900	1.16	1.14

Source: Ishizaki and Cookson (94). With permission from Ann Arbor
Science Publishers.

metal ions, although it was suggested that one mechanism did not necessarily exclude the other. It was likely that both occurred simultaneously. A possible combined mechanism suggested by Oswald and
Wallance (122) was the formation of a thiolate ion in a reversible step
with the quinone groups, which then reacted with the metal ion by
an electron transfer mechanism and produced the thiyl radical:

$$RS^- + M^{+n} \longrightarrow RS^\cdot + M^{+ (n-1)}$$

The thiyl radicals dimerized to the disulfide while the metal ion was
oxidized by the oxygen, producing a peroxide ion, which further
reacted with water, producing hydroxyl ions:

$$2 \ RS^\cdot \longrightarrow RSSR$$

$$2 \ M^{+(n-1)} + O_2 \longrightarrow 2 \ M^{+n} + O_2^{2-}$$

$$O_2^{2-} + H_2O \longrightarrow 2 \ OH^- + 1/2 O_2$$

The outgassing of the carbon at 900°C eliminated both the quinone
groups and the metal impurities and resulted in the reduced catalytic
activity of the carbon.

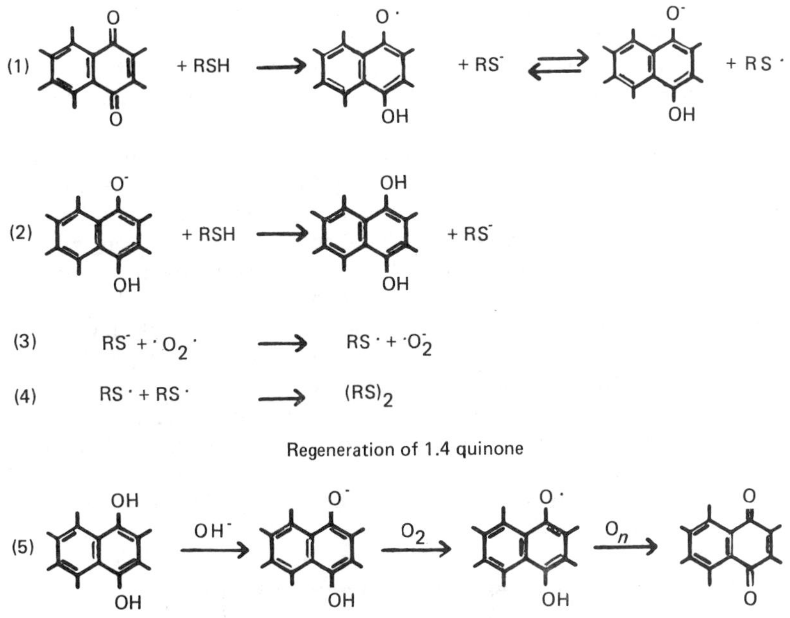

(1) [quinone structure] + RSH $\longrightarrow$ [semiquinone radical structure] + RS$^-$ $\rightleftharpoons$ [structure] + RS$\cdot$

(2) [structure] + RSH $\longrightarrow$ [hydroquinone structure] + RS$^-$

(3) $RS^- + \cdot O_2 \cdot \longrightarrow RS\cdot + \cdot O_2^-$

(4) $RS\cdot + RS\cdot \longrightarrow (RS)_2$

Regeneration of 1.4 quinone

(5) [hydroquinone structure] $\xrightarrow{OH^-}$ [structure] $\xrightarrow{O_2}$ [structure] $\xrightarrow{O_n}$ [quinone structure]

FIGURE 51 Mechanism for the catalytic oxidation of n-butyl mercaptan by quinone groups on the carbon surface.

6.17.7 Catalytic Combination of Hydrogen and Bromine

The combination of hydrogen and bromine in the presence of different varieties of carbons at different temperatures has been studied by a number of workers (123–126) and it has been shown that the reaction proceeds almost quantitatively at comparatively low temperatures. Bruns and Zarubina (123) found maximum catalytic activity with activated charcoal with a yield of about 80%. Hofmann and Höper (124), Hofmann and Ohlerich (125), and Boehm et al. (126) studied the reaction in the presence of several carbon blacks before and after activation and outgassing and observed that at 150°C the reaction efficiency was almost 100% in the presence of carbons. In the absence of carbons the reaction did not proceed. The catalytic efficiency was proportional to the BET surface area (Fig. 52) only on pure carbon surfaces containing no surface compounds. It was found that 1 m^2 of a pure carbon surface could catalyze the formation of 6.7 X 10^{-6} moles of HBr per hour at 150°C thus emphasizing the importance of surface area in the performance of the carbons as catalysts. In the case of raw carbon blacks or those that were outgassed only at 300°C, the

catalytic efficiency was considerably changed in the presence of surface compounds. Jones (127) had suggested that the carbon activates
the bromine in a manner analogous to its activation by light, and the
activated bromine then reacts with the molecular hydrogen to start a
chain reaction.

Puri and Kalra (128) carried out a detailed study of the catalytic
combination of hydrogen and bromine as a function of temperature
using different charcoals and carbon black having different surface
areas as well as surface complexes. The reaction, which was negligible in the absence of carbons even at 300°C, took place to a considerable extent even at 150°C in the presence of the carbons (Table 33).
The catalytic efficiency was almost complete in the case of degassed
charcoals even at 200°C. However, the difference in the performance
of the original and the degassed charcoals was maintained up to 250°
C. At 300°C all the carbon samples showed almost equal performance.
The rise in the catalytic activity of original charcoals at higher temperatures was attributed to the elimination of associated oxygen. The
catalytic activity of a carbon was not influenced by the surface area

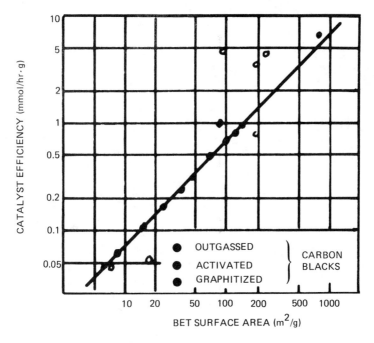

FIGURE 52 Relationship between surface area and the catalytic efficiency of carbon blacks for hydrogen-bromine combination. [From Boehm
et al. (126).]

Table 33 Conversion of Bromine into Hydrogen Bromide at Different Temperatures

Sample[a]	Conversion (%) at			
	150°	200°	250°	300°
Without Charcoal	0.50	1.0	1.4	2.2
Sugar charcoal				
Original	6.1	30.5	69.9	96.7
Degassed at 750°	85.3	96.4	99.2	99.4
Degassed at 1000°	95.8	99.5	99.4	99.7
Coconut charcoal				
Original	18.2	39.5	76.1	98.2
Degassed at 750°	90.9	99.3	99.5	99.6
Degassed at 1000°	99.2	99.4	99.5	99.7
Carbon black				
Spheron-9	23.6	98.1	99.2	99.4

[a]Amount of carbon = 4 g.
Source: Puri and Kalra (128).

because coconut charcoal (320 m^2/g) or the carbon black Spheron-9 (115 m^2/g), which had much smaller surface areas than sugar charcoals, had almost the same catalytic afficiency. The catalytic efficiency was also increased by increasing the amount of carbon catalyst. The rate studies showed that the reaction followed first-order kinetics with an activation energy of 9.4 kcal/mol, which was considerably lower than the known value of about 60 kcal/mol in the absence of the catalyst.

The efficiency of any carbon was not impaired as a result of the reaction and the same carbon could be used a number of times. There was also no significant loss of the carbon catalyst during the reaction. It is suggested that both hydrogen and bromine are adsorbed on the carbon-active sites in close proximity and then react to give HBr, which is desorbed, rendering the sites available for further chemisorption. This mechanism was in agreement with the observation that the catalytic efficiency of original charcoals which have their active sites occupied by chemisorbed oxygen showed lower efficiency at 150°C. However, when these surface groups were removed at 250°C and above in the presence of hydrogen, the active sites became available for the chemisorption of hydrogen and bromine and the catalytic efficiency improved. This was further substantiated by the poor efficiency of 1000°-outgassed sugar charcoal when the active sites were blocked by chemisorbing oxygen, sulfur, chlorine, or bromine (Table 34). Thus Puri and Kalra concluded that the catalytic efficiency of a carbon for the combination of hydrogen and chlorine was determined by the availability of the active sites and not by surface area.

6.17.8 Catalytic Combination of Phosgene and Formaldehyde

Ryan and Stacey (129) investigated the catalytic reaction between phosgene and formaldehyde for the production of dichloromethane in the temperature range 179—200°C on a variety of active carbons obtained from wood, coal, coconut shells, and also on a carbon cloth. These workers observed that the reactants were almost completely converted into dichloromethane at temperatures below 200°C (Table 35) with the evolution of carbon dioxide according to the equation

$$COCl_2 + CH_2O \longrightarrow CH_2Cl_2 + CO_2$$

However, when the reaction was carried out at temperatures greater then 200°C, the percentage conversion into dichloromethane was decreased and some carbon monoxide was also produced. The amount of carbon monoxide produced was almost double the loss of dichloromethane. The formation of carbon monoxide was attributed to the reaction

$$COCl_2 + CH_2O \longrightarrow 2CO + 2HCl$$

although the formation of hydrochloric acid was not detected by the mass spectrometer but was explained on the basis of the loss of hydrogen and chlorine from the system. The conversion into dichloro-

Table 34 Conversion of Bromine into Hydrogen Bromide at 150°C after Covering the Active Sites with Different Surface Compounds in 1000° Outgassed Sugar Charcoal

Surface compound	Amount of chemisorption of other elements (mEq/100 g)	Conversion (%)
None	0	95.8
Oxygen	404	2.8
Sulfur	499	2.2
Chlorine	309	2.1
Bromine	390	2.2

Source: Puri and Kalra (128).

methane was appreciably lower ($\cong 80\%$) when bone charcoal was used as the catalyst. This was attributed to the fact that bone charcoal contained less carbon and was associated with phosphatic impurities.

Temperature-programed desorption studies showed that both the reactants—phosgene and formaldehyde—were adsorbed on the surface of the carbon and then reacted to give an intermediate compound, chloromethyl chloroformate, which then decomposed to give dichloromethane and carbon dioxide:

$$
\begin{array}{ccc}
 & CH_2OCOCl_2 & CH_2Cl.OCOCl \\
| \ | \ | & |\ \ | & | \\
-C-C-C- + CH_2O\ (g) + COCl_2(g) \longrightarrow -C-C-C- \longrightarrow -C-C-C- \\
| \ | \ | & |\ |\ | & \\
\text{Carbon} & \text{Adsorption} & \text{Intermediate} \\
\text{surface} & \text{complex} & \text{chloromethyl} \\
 & & \text{chloroformate}
\end{array}
$$

$$\downarrow -CO_2$$

$$
\begin{array}{cc}
 & CH_2Cl_2 \\
| \ | \ | & | \\
-C-C-C- + CH_2Cl_2\ (g) \longleftarrow -C-C-C
\end{array}
$$

i.e.,

$$CH_2O\ (ads.) + COCl_2\ (ads.) \rightleftharpoons CH_2ClO.COCl\ (ads.)\ (1)$$

$$CH_2ClO.COCl\ (ads.) \longrightarrow CH_2Cl_2\ (ads.) + CO_2\ (g)\ (2)$$

The kinetic measurements and the calculations of activation energies showed that reaction (1) was slow and therefore rate determining. The catalytic activity of the carbon for the reaction was attributed to the existence of polar surface groups. Although Ryan and Stacey (129) did not speculate about the nature of the polar structures responsible for the catalytic activity, they did mention that the groups can bind the intermediate chloromethyl chloroformate ester in an unreactive form.

6.17.9 Catalytic Decomposition of Hydrogen Peroxide

The catalytic decomposition of hydrogen peroxide in the presence of activated carbons has been studied by several workers. Firth and Watson (130–132) observed an initial rapid reaction, which soon subsided, and a slow reaction, which continued for a long time. Skumburdis (133) attributed this fast reaction to the heat of immersion of the carbon in the solution. Larsen and Walton (134) found that the

Table 35 Influence of Temperature on the Catalytic Activity of a Coal Charcoal for the Reaction between Phosgene and Formaldehyde

Temperature (°C)	Reactant flow rates (dm³/hf)		Molar ratio $COCl_2:CH_2O$	Conversion[a] (%)	
	$COCl_2$	CH_2O		$COCl_2$	CH_2O
179	8.1	8.5	0.95	98.6	98.1
212	8.2	8.5	0.93	99.3	98.0
230	8.2	8.3	0.99	99.3	98.2
254	8.3	8.4	0.99	99.3	98.2

[a] % Conversion is defined as 100 x (moles of reactant consumed) / (moles of reactant fed).
Source: Ryan and Stacey (129). With permission from Butterworths.

rate of decomposition was pH dependent. It increased rapidly when pH was increased from 3 to 4, remained steady from pH 4 to 8, and increased slightly at higher pH values. Fowler and Walton (135), King (136), and Larsen and Walton (134) also observed that the catalytic activity of carbons depended on the nature of the carbon, being maximum when the carbons were heated to 850°C and least when they were heated at 400°C. Brinkmann (137,138) suggested that the decomposition of hydrogen peroxide by an activated carbon involved primary exchange of an oxonium hydroxyl group with a hydrogen peroxide anion. The adsorbed peroxide was regarded as having an increased oxidation potential and thus decomposed another hydrogen peroxide molecule and regenerated the carbon:

$$\text{Carbon surface} \quad)C-O\overset{\overset{H^+}{|}}{\underset{\underset{H}{|}}{}} ---OH^- + H^+OOH^- \longrightarrow)C-O\overset{\overset{H^+}{|}}{\underset{\underset{H}{|}}{}} ---OOH^- + H_2O$$

$$\text{Carbon surface} \quad)C-O\overset{\overset{H^+}{|}}{\underset{\underset{H}{|}}{}} ---OOH^- + H_2O_2 \longrightarrow)C-O\overset{\overset{H^+}{|}}{\underset{\underset{H}{|}}{}} ---OH^- + H_2O + O_2$$

Garten and Weiss (16), while studying the decomposition of hydrogen peroxide in the presence of carbon blacks that had been heated to different temperatures, observed a maximum catalytic activity with carbon blacks heated to 900°C. The acidity of the carbon black increased after the reaction and the basicity was almost completely lost. These observations were explained by Garten and Weiss on the basis of the existence of chromene groups, which initiated a chain mechanism for the decomposition of hydrogen peroxide. At low pH values these chromene groups were oxidized to carbonium ions and thus the carbon lost its ability to catalyze the decomposition.

Puri and Kalra (193) and Bansal (140) studied the influence of the concentration of solution, temperature, and the nature of the carbon surface on the catalytic decomposition of hydrogen peroxide using sugar and coconut charcoals. The reaction was catalyzed to an appreciable extent and the catalytic activity of the carbon increased with increase in the concentration of the hydrogen peroxide in the aqueous solution. Coconut charcoal was a better catalyst than sugar charcoal, although the BET nitrogen area of the former ($321 \ m^2/g$) was less than that of the latter ($422 \ m^2/g$). The catalytic performance of each carbon increased with increase in the amount of charcoal added to the solution. The catalytic decomposition was temperature dependent and the extent of decomposition with time at higher temperatures was greater (Fig. 53). The catalytic activity of a charcoal was considerably enhanced when the charcoal was degassed at gradually increasing temperatures. In fact the original charcoals, which were not degassed, inhibited the decomposition of hydrogen peroxide and this inhibiting effect increased with time as well as the temperature of the reaction. The sample degassed at 400°C acted as catalyst in the initial stages but as inhibitor after the reaction had proceeded for a few hours.

The catalytic performance of a charcoal increased with decrease in the pH of the solution. The 1200°-outgassed charcoals which showed a distinct alkaline pH (9.75 in case of coconut charcoal) were the most active catalysts, the coconut charcoal being more efficient because it was more alkaline although it had smaller surface area than the corresponding sugar charcoal. The inhibiting effect of the original charcoals was attributed to their acidic character. Original sugar charcoal, which showed a pH of 3.3, lowered the extent of decomposition from 9.6 mEq in the case of control to 4.0 mEq in 24 hr at 35°C. Similarly, when 1200°-outgassed charcoals were rendered acidic by oxidation with hot nitric acid (pH $\cong$ 3), they were even better inhibitors than the original charcoals. The extent of decomposition was lowered to 2.8 mEq in 24 hr at 35°C.

When the decomposition reaction was carried out in the presence of different charcoals using buffer solutions, the reaction was inhibited in the acid range buffers, the extent of inhibition increasing with increase in acidity, whereas the catalytic effect was enhanced in the alkaline range buffers, the extent of enhancement increasing with increase in the pH buffer. Thus the results pointed out that although an acidic medium was conducive to stability, an alkaline medium was conducive to decomposition of hydrogen peroxide. The function of the carbon was simply to provide an environment of a particular degree of acidity or alkalinity. This could also explain the variation of catalytic activity of carbon heat treated to different temperatures, as reported by several workers (134,136,138). These workers suggest-

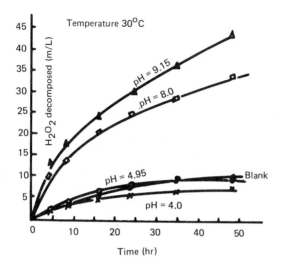

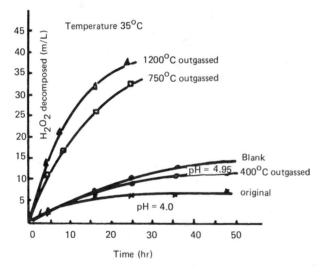

FIGURE 53 Effect of temperature on the catalytic decomposition of hydrogen peroxide by different carbons having different pH values. [From Puri and Kalra (193).]

ed that the decomposition of hydrogen peroxide proceeded through its dissociation as a weak acid (pKa 2.4×10^{-12})

$$H_2O_2 \rightleftharpoons H^+ + OOH^-$$

and not through the equation

$$H_2O_2 \longrightarrow H_2O + 1/2O_2$$

The OOH^-, being even less stable than the H_2O_2 itself, decomposed very readily. Evidently in alkaline environment the dissociation of H_2O_2 will increase, whereas it will be suppressed in the acidic environment. The increase in the decomposition of hydrogen peroxide at higher temperatures evidently resulted from increase in its dissociation. The development of increasing acidity in the case of original as well as 400°-degassed charcoals, during the reaction itself, explained the change in the behavior of the 400°-outgassed charcoal, which acted as a catalyst initially and as an inhibitor after few hours.

In their subsequent investigations using carbon blacks as catalysts for the decomposition of hydrogen peroxide, Puri and Kalra (141) observed that the catalytic activity cannot be attributed to surface acidity or alkalinity alone because Statex B and Kosmos-40 carbon blacks, which have an alkaline surface (pH >9), were poorer catalysts than Mogul or Spheron-C carbon blacks with distinct acidic pH values. Similarly, Spheron-C carbon, which had a smaller surface area (253 m^2/g) then Mogul (308 m^2/g), decomposed more hydrogen peroxide. Thus these workers concluded that both the alkalinity and the surface area of the carbon blacks determine the catalytic activity of carbon blacks toward the decomposition of hydrogen peroxide. When the amount of hydrogen peroxide decomposed was normalized with respect to surface area of the carbon black and plotted as a function of pH of the carbon black, the points could be accommodated on two straight lines (Fig. 54), indicating that the acidity or alkalinity of the carbon sample was related in some way to the activity of the carbon black. Similarly, when the catalytic decomposition was carried out in a buffer solution using different carbon blacks, the catalytic activity was related directly to the surface area (Fig. 55), indicating that at a given pH the catalytic activity was a function of the surface area.

6.17.10 Catalytic Decomposition of Benzoyl Peroxide

The catalytic decomposition of benzoyl peroxide in organic solvents has been studied by several workers using carbon blacks and activated carbons (142–145) but no mechanism for the reaction was proposed by these workers. Puri, Sud, and Kalra (146) studied the decomposition in benzene solutions using several commercial-grade carbon blacks

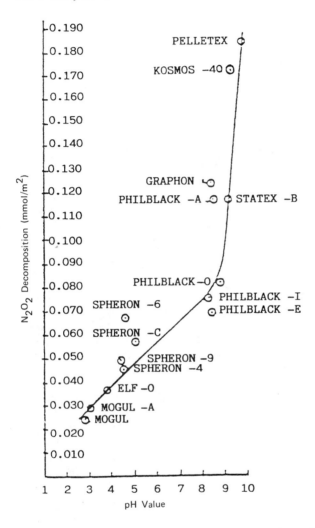

FIGURE 54 Relationship between the pH value of the carbon and the catalytic decomposition of hydrogen peroxide.

and sugar charcoals. The extent of decomposition of benzoyl peroxide increased considerably with the concentration of the solution, the temperature of the reaction, and the amount of carbon used. The kinetics of the reaction showed it to be first order in the initial stages and second order afterwards. The decomposition resulted in the formation of benzoic acid and evolution of carbon dioxide. The carbon black sample after the reaction neutralized larger amounts of alkali,

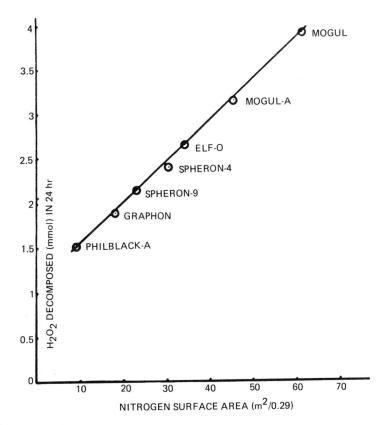

FIGURE 55 Catalytic decomposition of hydrogen peroxide as a function of surface area at a given pH.

which was attributed to the adsorption of free radical $C_6H_5COO\cdot$ on the surface of the carbon black.

It was postulated that the benzoyl peroxide underwent dissociation into two radicals:

Some of these free radicals extracted hydrogen from the solvent to produce benzoic acid, whereas others were chemisorbed on the surface of the carbon catalyst or broke down evolving carbon dioxide and changing into diphenyl. This view was based on the fact that the total free radicals obtained from the decomposition of benzoyl peroxide could be accounted for as benzoic acid, free radicals chemisorbed, and carbon dioxide evolved (Table 36). The reaction involving fragmentation of the molecule was evidently first order in the initial stages. The amount of benzoyl peroxide decomposed was a linear function of the surface area of the carbon catalyst (Fig. 56). When the same carbon catalyst was reused in a second cycle, the decomposition activity of the catalyst was considerably reduced. This was attributed to the fact that some of the surface was covered by the chemisorbed free radical species $C_6H_5COO^-$. This was checked by calculating the area occupied by these free radicals in a run (taking molecular area of $C_6H_5COO^-$ to be $52\overset{\bullet}{A}{}^2$) and comparing it to

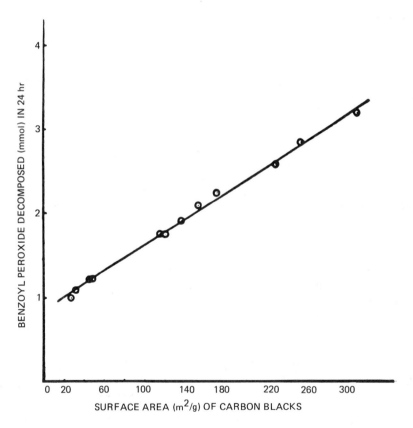

FIGURE 56 Catalytic decomposition of benzoyl peroxide as a function of surface area of carbon blacks. [From Puri et al. (146).]

Table 36 Catalytic Decomposition of Benzoyl Peroxide by Carbon Black

Carbon Black	Surface area (m^2/g)	Benzoyl peroxide decomposed (mmol)	$C_6H_5COO^·$ radicals rendered available (mmol)	Benzoic acid formed (mmol)	Free CO_2 evolved (mmol)	$C_6H_5COO^·$ radicals chemisorbed (mmol) from CO_2 evolved on evacuation at 1000°(C_1)	Barium hydroxide value (C_2)	Total C_6H_5COO accounted
Mogul	308.0	3.20	6.40	4.13	1.61	0.51	0.55	6.25
Spheron-C	253.7	2.85	5.70	3.73	1.09	0.67	0.69	5.49
Mogul-A	228.4	2.59	5.18	3.39	1.32	1.35	0.38	5.06
ELF-O	171.0	2.25	4.50	2.90	1.04	0.33	0.29	4.27
Spheron-4	152.0	2.09	4.18	2.41	1.16	0.44	0.40	4.01
Philblack-E	135.1	1.90	3.80	2.35	1.04	0.37	0.34	3.76
Spheron-6	120.0	1.77	3.54	2.48	0.71	0.33	0.30	3.52
Spheron-9	115.0	1.75	3.50	2.32	0.62	0.32	0.34	3.26
Statex-B	48.0	1.21	2.42	1.16	0.96	0.13	0.11	2.25
Philblack-A	45.8	1.20	2.40	1.25	0.90	0.12	0.13	2.27
Kosmos-40	31.2	1.10	2.20	0.96	0.91	0.09	0.08	1.96
Pelletex	27.1	1.00	2.00	1.01	0.80	0.08	0.07	1.89

Source: Puri et al. (146).

Table 37 Influence of Repeated Use of Carbon (Mogul) on the Catalytic Decomposition of Benzoyl Peroxide

Number of times sample used	Surface area (m^2/g)	Benzoyl peroxide decomposed (mmol/g)	Benzoic acid formed (mmol/g)	Free CO_2 evolved (mmol/g)	$C_6H_5COO^-$ chemisorbed (mmol/g)
First time	308	3.20	4.13	1.61	0.51
Second time	148[a]	2.10	3.02	0.78	0.21
Third time	83[a]	1.59	2.25	0.47	0.12
Fourth time	39[a]	1.15	1.75	0.26	0.03

[a]Computed surface area.
Source: Puri et al. (146).

the extent of decrease in the catalytic activity (Table 37). Thus these observations bring out clearly not only the importance of surface area of carbon black in determining the extent of the reaction brought about in their presence but also support the view that the carbon surface itself participates in the reaction.

6.17.11 Catalytic Elimination of Hydrogen Halide

It was shown by Schwab and Ulrich (147) that graphite can catalyze the elimination of hydrochloric acid from 1-butylchloride forming butene. Boehm et al. (110) used active charcoal and carbon blacks as catalysts for the elimination of hydrochloric acid from 1-butyl chloride and 1-octyl chloride by passing their vapors over the carbons using nitrogen as a carrier gas at temperatures of 350 and 400°C. The reaction resulted in the formation of corresponding olefines and the dimers of alkyl groups of the alkyl halides, e.g., *n*-octane in the case of 1-butyl chloride. The olefine to dimer product ratio was 2.7. Some polymer was also deposited on the carbon surface, which deactivated the catalyst. The catalytic activity of the peat or wood charcoals increased with heat treatment at 800°C or with ammonia treatment at 600 or 900°C.

The elimination was explained on the basis of a radical mechanism. The radicals produced return the electron to the catalyst forming carbonium ions, which decompose to olefine as

$$R - Cl + e^- \longrightarrow R^\cdot + Cl^-$$
$$R^\cdot \longrightarrow R^+ + e^-$$
$$R - CH_2 - CH_2^+ \longrightarrow R - CH = CH_2 + H^+$$

This radical mechanism was supported by the formation of the radical dimerization products during the reaction.

6.17.12 Catalytic Chlorination of Aromatic Hydrocarbons

Puri and Bansal (148) and Bansal (140) studied the chlorination of toluene in the presence of sugar and coconut charcoals associated with varying amounts of chemisorbed hydrogen by passing the hydrocarbon vapors mixed with chlorine over a bed of charcoal (2 g) heated to 300°C at a flow rate of 3 L/hr. No chlorine derivatives of toluene were obtained in the absence of the carbon. However, in the presence of the charcoals, there was appreciable conversion of toluene into chlorotoluene, benzyl chloride, and benzoyl chloride. The amount of conversion decreased appreciably with decrease in the hydrogen content of the charcoal (Table 38). It was found that the catalytic activity of the charcoal varied with the hydrogen content and disappeared alto-

Table 38 Catalytic Halogenation of Toluene in the Presence of Different Charcoals at 300°

Sample	H_2 present in sample (mEq/g)	Halogenation (%)[a]
Sugar charcoal		
Original	33.81	22
Degassed at 1200°	10.21	6
Coconut charcoal		
Original	20.31	18
Degassed at 1200°	6.20	4
Charcoal freed from H_2 on prolonged treatment with Cl_2 at 1000°	0	0

[a]No halogenation took place in the absence of charcoal.
Source: Puri and Bansal (148).

gether when the charcoal was almost free of hydrogen. It was suggested that the chemisorbed hydrogen atoms in the charcoal react with a molecule of chlorine, producing atomic chlorine which reacts with the hydrocarbon, producing chlorinated products.

6.17.13 Catalysis of Sodium Azide—Iodine Reaction

The sodium azide—iodine reaction is known to be catalyzed by sulfide, thiosulfate, and thiol groups. The reaction in the presence of sulfurized carbons was studied by Boehm et al. (126), Bladen and Patrick (149), and Puri and Hazra (150) to adduce evidence for the presence of sulfide groups in their carbons. Puri, Sud, and Kalra (151) investigated the catalysis of this reaction in the presence of sugar and coconut charcoals and several commercial-grade carbon balcks which were associated with varying amounts of acidic surface groups. The progress of the reaction was measured by the volume of nitrogen evolved. The magnitude of the reaction increased with increase in the concentration of sodium azide, the temperature of the reaction, and with the amount of carbon catalyst used. The rate of the reaction and its magnitude decreased as the charcoals were outgassed at gradually increasing temperatures up to 750°C and increased when the same charcoals were oxidized with hydrogen peroxide (Fig. 57). Similarly, the catalytic performance of a carbon black increased as the amount of acidic oxygen complex increased. The surface area of the charcoal or the carbon black was found to have no effect on the catalytic activity. Thus the catalytic activity was attributed to the

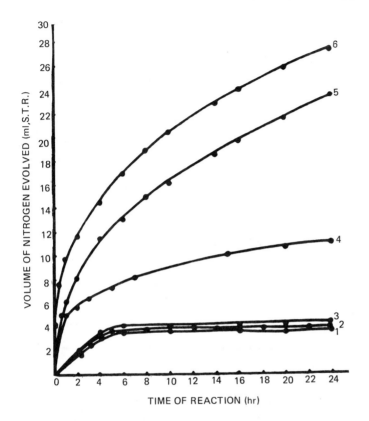

FIGURE 57 Rates of the carbon catalysis for sodium–iodine reaction on different charcoals. (1) Original sugar charcoal; (2) outgassed at 400°C; (3) outgassed at 750°C; (4) outgassed at 1000°C; (5) original oxidized with H_2O_2; (6) oxidized with H_2O_2 and outgassed at 1000°C. [From Puri et al. (151).]

presence of acidic surface structures. When these surface structures were removed on degassing at temperatures of 750°C or above, the carbons lost their capacity to catalyze the reaction. A linear relationship was observed between the acidic CO_2-evolving complex and the catalytic activity. However, two different straight lines were obtained (Fig. 58) for carbon blacks and charcoals, indicating that the catalytic activity for the same amount of acidic Co_2-complex was lower for charcoals than for carbon blacks. This lower activity in the case of charcoals was attributed to the location of some of the acidic sites within the microcapillary pores, which were not accessible to azide ions for the reaction.

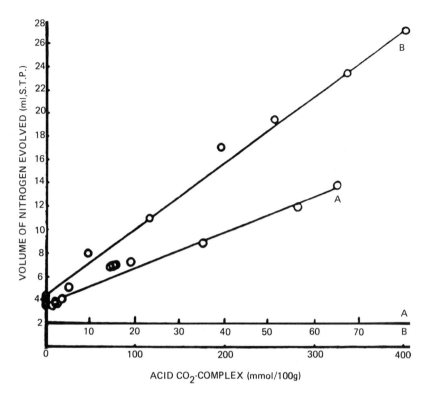

FIGURE 58 Relationship between acidic CO_2-complex and the catalytic activity of the carbon for sodium azide—iodine reaction. [From Puri et al. (151).]

REFERENCES

1. Davis, W. N., U. S. Patent 227,963, 1880.
2. Johnson, W. D., U. S. Patent 522,260, 1894.
3. Zadra, J. B., U. S. Bureau of Mines, Washington, D. C., R. I. No. 11672 (1950).
4. Zadra, J. B., Angel, A. K., and Heinen, H. J., U. S. Bureau of Mines, Washington, D. C., R. I. No. 4843 (1952).
5. Hall, K. B., *World Min.*, 27:44 (1974).
6. Muir, D. M., Private communication, 1982.
7. Brussov, G., *Z. Chem. Ind. Kolloids*, 5:137 (1909).
8. Green, M., *Trans. Inst. Min. Netall.*, 23:65 (1913—1914).
9. Allen, A. W., *Metall. Chem. Eng.*, 18:642 (1918).
10. Grabovskii, A. I., Ivanova, L. S., Korostyshevskii, N. B., Shirshov, V. M., Storozhuk, R. K., Matskevich, E. S., and Arkadakskaya, N. A., *Zh. Prikl. Khim.*, 49:1379 (1976).

11. Grabovskii, A. I., Grabshak, S. L., Ivanova, L. S., Storozhuk, R. K., and Shirshov, V. M., *Zh. Prikl. Khim.*, *50:*522 (1977).
12. Feldtmann, W. R., *Trans. Inst. Min. Metall.*, *24:*329 (1914–1915).
13. Edmands, H. R., *Trans. Inst. Min. Metall.*, *27:*277 (1917–1918).
14. Gross, J. and Scott, J. W., U. S. Bureau of Mines, Washington, D. C., Tech. Paper No. 378 (1972).
15. Kuz'minykh, V. M. and Tyurin, N. G., *Izv. Vyssh. Uchebn. Zaved. Tsved Metall.*, *11:*65 (1968).
16. Garten, V. A. and Weiss, D. E., *Rev. Pure Appl. Chem.*, *7:*69 (1957).
17. Davidson, R. J., *J. South African Inst. Min. Metall.*, *75:*67 (1974).
18. Cho, E., Ph. D. dissertation, University of Utah, Salt Lake City, 1978.
19. Dixon, S., Cho, E., and Pitt, C. H., AICRE Meeting, Chicago, Nov. 26–Dec. 2, 1976.
20. Frumkin, A., *Kolloid Z.*, *51:*123 (1930).
21. Frumkin, A., Burshtein, B., and Lewin, P., *Z. Phys. Chem.*, *157A:*422 (1931).
22. Burshtein, B. and Frumkin, A., *Z. Phys. Chem.*, *141A:*219 (1929).
23. Clauss, C. R. A. and Weiss, K., Pretoria CSIR Report No. CENG 206, Sept. 1977.
24. McDougall, G. J. and Hancock, R. D., *Gold. Bull.*, *14:*138 (1981).
25. Puri, B. R., in *Chemistry and Physics of Carbon*, Vol. 6, P. L. Walker, Jr., Ed., Marcel Dekker, New York, 1970.
26. Cho, E., Dixon, S. N., and Pitt, C. H., *Metall. Trans. B*, *10B:*185 (1979).
27. Crank, J., *Phil. Mag.*, *39:*140 (1948).
27a. Gupta, A., *AIMM Proc.*, *289:*239 (1984).
28. Cho, E. and Pitt, C. H., *Metall. Trans.*, *10B:*159 (1979).
29. Bockris, J. O'M., Devanathan, M. A. V., and Müller, K., *Proc. Roy. Soc. (London) A*, *274:*55 (1963).
30. Anderson, T. N. and Bockris, J. O'M., *Electrochem. Acta*, *9:*347 (1964).
31. Devanathan, M. A. V. and Tilak, B. V. K. S. R. A., *Chem. Rev.*, *65:*635 (1965).
32. McDougall, G. J., Hancock, R. D., Nicol, M. J., Wellington, O. L., and Copperthwaite, R. G., *J. South African Inst. Min. Metall.*, *80:*344 (1980).
33. Battistoni, C., Mattogno, G., Cariati, F., Naldini, L., and Sgamellotti, A., *Inorg. Chim. Acta*, *24:*207 1977.
34. Tsuchida, N., Ruane, M., and Muir, D. M., Private communication.
35. Tsuchida, N., Ph. D. dissertation, Murdoch University, 1984.

36. Fleming, C. A. and Nicol, M. J., *J. South African Inst. Min. Metall.*, *84:*95 (1984).
37. Kuz'minykh, V. M., Tyurin, N. G., and Nikulova, A. A., *J. Appl. Chem. USSR*, *42:*1886 (1969).
38. Heinen, H. J., Peterson, D. G., and Lindstrom, R. E., in *World Mining and Metallurgy Technology*, Vol. I, Weiss, Ed., AIME, New York, 1976, pp. 551—564.
39. Martin, J. P., Davidson, R. J., Duncanson, E., and Nkosi, N., Anglo American Res. Lab. Report No. 6, Johannesberg, South Africa, 1970.
40. Parker, A. J. and Muir, D. M., U. S. Patent Appl. 323,558; South African Patent 81/7908; Australian Patent Appl. 77505.
41. Muir, D. M. and Hinchcliffe, W., Mineral Chemistry Research Unit Report, Murdoch University, 1981.
42. Hinchcliffe, W., Muir, D. M., and Ruane, M., Paper presented at seminar on CIP Technology, Murdoch University, 1982.
43. Ruane, M., in Proc. Symp. CIP Technology for the Extraction of Gold, Aust. I. M. M. Melbourne, 1982, p. 393
44. Polania León, A., Donnet, J. B., and Papirer, E., to be published.
45. Yuki, N. and Ohsaki, K., *Japan Kokai*, *73*(95):994 (1973).
46. Tominaga, M., Mimon, M., and Okuda, K., *Japan Kokai*, *84*(39): 587 (1974).
47. Yamaguchi, O. and Nebekura, S., *Japan Kokai*, *74*(7):267 (1974).
48. Yamaguchi, O. and Nebekura, A., *Japan Kokai*, *74*(102):565 (1974).
49. Saito, I., *Kogai Shigen Kenkyasho Iho*, *5:*57 (1976).
50. Tatsuyama, K., Egawa, H., Srnmaru, H., Goto, I., Jodai, S., and Onishi, H., *Shimane Paigaku Vogakube Kenkyo Hokoku*, *9:*58 (1975).
51. Huang, C. P. and Ostovic, F., *J. Am. Soc. Civil Eng. Environ. Eng. Div.*, *1975.*
52. Dobrowolski, R., Jaronilc, M., and Kosmulski, M., *Carbon*, *24:*15 (1986).
53. Vasatko, J., *Z. Zuckerind.*, *52:*221 (1928).
54. Heymann, E., Salomon, K., and Keiffer, R., *Z. Anorg. Allgem. Chem.*, *187:*97 (1930).
55. Garten, V. A. and Weiss, D. E., *Rev. Pure Appl. Chem.*, *7:*69 (1957).
56. Puri, B. R. and Mahajan, O. P., *J. Indian Chem. Soc.*, *39:*292 (1962).
57. Ford, C. T. and Boyer, J. F., Jr., EPA report EPA-R2-73-150, 1973.
58. George, A. D. and Chaudhuri, M., *J. Am. Water Works Assoc.*, *69:*305 (1977).
59. Huang, C. P. and Bowers, A. R., *10th Int. Conf. Water Pollution Res.*, 1977.

60. Puri, B. R. and Satija, B. R., *J. Indian Chem. Soc.*, *45*:298 (1968).
61. Huang, C. P. and Wu, M. H., *J. Water Pollution Control Fed.*, *47*:2437 (1975).
62. Huang, C. P. and Wu, M. H., *Water Res.*, *11*:673 (1977).
63. Kim, J. L., *Diss. Abstr. Int.*, *B 37*, 3566 (1977). (C. A. 08620145537 R)
64. Nagasaki, Y. and Terada, A., *Japan Kokai*, *75*:721 (1975). in Japan CA. 08408049635 J
65. Yoshida, H., Kamegawa, K., and Arita, S., *Nippon Kagakn Karshi*, *3*:387 (1977).
66. Roerma, R. E., Alsema, G. L., and Anlhonissen, J. H., *Belg. Ned. Tijdschr. Oppervlakte Tech. Met. Ser.*, *19*:53 (1975).
67. Miyagawa, T., Ikeda, S., and Koyama, K., *Japan Kokai*, *76*:417 (1976). Japanese CA. 08520148621 D.
68. Nagasaki, Y., *Japan Kokai*, *76*(122):477 (1974).
69. Oppold, W. A., Paper presented at 44th Water Poll. Control Fed. Conf., San Francisco, Calif., Oct. 5, 1971.
70. Thiem, L., Badorek, D., and O'Connor, J. T., *J. Am. Water Works Ass.*, *68*:447 (1976).
71. Homenick, M. J., Jr., Sehnoor, J. L., *J. Am. Soc. Civil Eng. Environ. Eng. Div.*, *100*:1249 (1974).
72. Yoshida, H., Hamegawen, K., and Arita, S., *Nippon Kogakai Kaishi*, *5*:808 (1976).
73. Fuxeluis, K. O. H., *Ger. Offen. 1972*:14. (C.A. 07720130425 T)
74. Sinha, R. K. and Walker, P. L., Jr., *Carbon*, *10*:754 (1972).
75. Ohtsuki, S., Miyanohara, I., and Mizui, N., *Japan Kokai*, *74*:5 (1974). (Japan C. A. 08114082116 B)
76. Yamaguchi, T., *Japan Kokai*, *74*:1218 (1974). (Japan C. A. 08312102927 P)
77. Ammons, R. D., Dougharty, N. A., and Smith, J. M., *Ind. Eng. Chem. Fundam.*, *16*:263 (1977).
78. Moore, R. H., U. A. Department of the Interior Report No. 651, 1971.
79. Nelson, F., Phillips, H. O., and Kraus, K. A., Paper presented at 29th Purdue Ind. Waste Conf., Lafayette, Inc., 1974.
80. Kunz, R. G., Giannelli, J. F., and Stensel, H. D., *J. Water Poll. Control Fed.*, *48*:762 (1976).
81. Kostyuchenko, P. I., Tarkovskaya, F. A., Kononchuk, T. L., Kovalenko, T. J., and Glushankova, Z. L., in *Adsorption and Adsorbent*, D. N. Strazhesko, Ed., Wiley, New York, 1973, p. 37.
82. Suffet, I. H. and McGuire, M. J., *Activated Carbon Adsorption*, Vol. I, Ann Arbor Science Publishers, Ann Arbor, Mich., 1981.
83. Suffet, I. H. and McGuire, M. J., *Activated Carbon Adsorption*, Vol. II, Ann Arbor Science Publishers, Ann Arbor, Mich., 1981.

84. Cheremisinoff, P. N. and Ellerbusch, F., *Carbon Adsorption Handbook*, Ann Arbor Science Publishers, Ann Arbor, Mich., 1980.
85. Singer, P. C. and Yen, C. Y., in *Activated Carbon Adsorption*, Vol. I, I. H. Suffet and M. J. McGuire, Eds., Ann Arbor Science Publishers, Ann Arbor, Mich., 1981, p. 167.
86. Puri, B. R., Bhardwaj, S. S., and Gupta, U., *J. Indian Chem. Soc.*, *53*:1095 (1976).
87. Coughlin, R. W., Ezra, F. S., and Tan, R. N., *J. Colloid Interface Sci.*, *28*:383 (1968).
88. Marsh, H. and Campbell, H. G., *Carbon*, *9*:489 (1971).
89. Youssefi, M. and Faust, S. D., in *Activated Carbon Adsorption*, Vol. I, I. H. Suffet and M. J. McGuire, Eds., Ann Arbor Science Publishers, Ann Arbor, Mich., 1981, p. 133.
90. Mullins, R. L., Jr., Zogorski, J. S., Hubbs, S. A., and Allgerei, G. D., in *Activated Carbon Adsorption*, Vol. I., I. H. Suffet and M. J. McGuire, Eds., Ann Arbor Science Publishers, Ann Arbor, Mich., 1981, p. 273.
91. Ishizaki, C., Marti, I., and Ruiz, M., in *Advances in Chemistry*, No. 202, I. H. Suffet and M. J. McGuire, Eds., American Chemical Society, Washington, D. C., 1983, p. 95.
92. Bornehoff, J., in *Activated Carbon Adsorption*, Vol. I, I. H. Suffet and M. J. McGuire, Eds., Ann Arbor Science Publishers, Ann Arbor, Mich., 1981, p. 145.
93. Fochtman, E. G. and Dobbs, R. A., in *Activated Carbon Adsorption*, Vol. I, I. H. Suffet and M. J. McGuire, Eds., Ann Arbor Science Publishers, Ann Arbor, Mich., 1981, p. 157.
94. Ishizaki, C. and Cookson, J. T., in *Chemistry of Water Supply, Treatment and Distribution*, A. J. Rubin, Ed., Ann Arbor Science Publishers, Ann Arbor, Mich., 1980, p. 201.
95. Puri, B. R., Kumar, B., and Kalra, K. C., *Indian J. Chem.*, *9*:970 (1971).
96. Cariaso, O. C. and Walker, P. L., Jr., *Carbon*, *13*:233 (1975).
97. Bansal, R. C., Vastola, F. J., and Walker, P. L., Jr., *J. Colloid Interface Sci.*
98. Hart, P. J., Vastola, F. J., and Walker, P. L., Jr., *Carbon*, *5*:363 (1967).
99. Lussow, R., Vastola, F. J., and Walker, P. L., Jr., *Carbon*, *5*:591 (1967).
100. Laine, N. R., Vastola, F. J., and Walker, P. L. Jr., *J. Phys. Chem.*, *67*:2030 (1963).
101. Klein, J. and Henning, K. D., *Fuel*, *63*:1064 (1984).
102. Steijns, M., Derks, F., Verloup, A., and Mars, P. J., *Catalysis*, *42*:87 (1976).
103. Hedden, K., Huber, L., and Rao, B. R., VDI-Bericht No. 253, S. 37/42 Dusseldorf, VDI-Verlag, 1976.

104. Lamb, A. B. and Elder, L. W., *J. Am. Chem. Soc.*, *53*:137 (1931).
105. Posner, A. M., *Trans. Faraday Soc.*, *49*:389 (1953).
106. Garten, V. A. and Weiss, D. E., *Aust. J. Chem.*, *10*:309 (1957).
107. Thomas, G. and Ingram, T. R., *Unit Process Hydromett.*, *1*:67 (1965).
108. Puri, B. R. and Kalra, K. C., *Indian J. Chem.*, *10*:72 (1972).
109. Puri, B. R. and Bansal, R. C., *Carbon*, *3*:523 (1967).
110. Boehm, H. P., Muir, G., Stoehr, T., de Rincon, A. R., and Tereczki, B., *Fuel*, *63*:1061 (1984).
111. Kurth, R., Tereczki, B., and Boehm, H. P., in 15th Bienn. Conf. Carbon, Philadelphia, 1981, p. 244.
112. Zuckmantel, M., Kurth, R., and Boehm, H. P., *Z. Naturforsch.*, *34b*:188 (1979).
113. Volans, G. M., Vale, J. A., Crome, P., Widdop, B., and Goulding, R., in *Artifical Organs*, R. M. Kenedi et al., Eds., Macmillan, London, 1977, p. 178.
114. Chang, T. M. S., *Can. J. Physiol. Pharmacol.*, *47*:1043 (1969).
115. Yatzidis, H., *Boc. Eur. Dial. Transpl. Ass.*, *83*:1 (1964).
116. Gordon, A., Greenbaum, M. A., Marantz, L. B., McArthur, M. J., and Maxwell, M. H., *Trans. Am. Soc. Artif. Int. Organs*, *347*:15 (1969).
117. Tijssen, J., Kaptein, M. J. F. M., Fiejen, T., Bantjes, A., and Van Doorn, A. W. J., in *Artifical Organs*, R. M. Kenedi et al., Eds., Macmillan, London, 1977, p. 158.
118. Sharrock, P. J., Courtney, J. M., Gilchrist, T., and Afrossman, S., *Anal. Lett.*, *9*:1085 (1976).
119. Smith, E. M., Affrossman, S., and Courtney, J. M., *Carbon*, *17*:149 (1979).
120. Edgar, G. and Shiver, H. E., *J. Am. Chem. Soc.*, *47*:1179 (1925).
121. Cookson, J. T., Jr., in *Activated Carbon Adsorption*, Vol. , I. H. Suffet and M. J. McGuire, Eds., Ann Arbor Science Publishers, Ann Arbor, Mich., 1981, p. 379.
122. Oswald, A. A. and Wallance, T. J., in *The Chemistry of Organic Sulphur Compounds*, Vol. 2, N. Kharasch and C. Y. Meyers, Eds., Pergamon Press, New York, 1966, p. 205.
123. Bruns, B. and Zarubina, O., *Zh. Fiz. Khim.*, *11*:300 (1938).
124. Hofmann, U. and Höper, W., *Naturwissenschaften*, *32*:225 (1944).
125. Hofmann, U. and Ohlerich, G., *Angew. Chem.*, *62*:16 (1950).
126. Boehm, H. P., Hofmann, U., and Clauss, A., *Proc. 3rd Carbon Conf.*, Pergamon Press, Oxford, 1959, p. 241.
127. Jones, N. C., *J. Phys. Chem.*, Ithaca *33*:1415 (1929).
128. Puri, B. R. and Kalra, K. C., *Indian J. Chem.*, *5*:638 (1967).
129. Ryan, T. A. and Stacey, M. H., *Fuel*, *63*:1101 (1984).

130. Firth, J. B. and Watson, F. H., *J. Chem. Soc.*, *123:*1750 (1923).
131. Firth, J. B. and Watson, F. H., *Trans. Faraday Soc.*, *20:*370 (1924).
132. Firth, J. B. and Watson, F. H., *J. Phys. Chem.*, *29:*987 (1925).
133. Skumburdis, K., *Kolloid Z.*, *55:*156 (1931).
134. Larsen, E. C. and Walton, J. H., *J. Phys. Chem.*, *44:*70 (1940).
135. Fowler, D. and Walton, J. H., *Rec. Trav. Chim.*, *54:*476 (1935).
136. King, A., *J. Chem. Soc.*, *1936:*1688.
137. Brinkmann, G., *Kolloid Z.*, *123:*116 (1951).
138. Brinkmann, G., *Angew. Chem.*, *61:*378 (1949).
139. Puri, B. R. and Kalra, K. C., *Indian J. Chem.*, *7:*149 (1969).
140. Bansal, R. C., Ph. D. dissertation, Panjab University, Chandigarh, 1965.
141. Puri, B. R. and Kalra, K. C., *Carbon*, *9:*313 (1971).
142. Ohkita, K. and Kasahara, H., Symp. Carbon, Tokyo, Japan, 1964.
143. Latyaeva, V. N. and Nuretdinove, O. N., *Tr. Khim. Teknol.*, *3:*110 (1960).
144. Studebaker, M. L., *Rubber Chem. Technol.*, *30:*1400 (1957).
145. Breitenbach, J. W. and Preussler, H., *Osterr. Chem. Zig.*, *51:*66 (1950).
146. Puri, B. R., Sud, V. K., and Kalra, K. C., *Indian J. Chem.*, *9:*966 (1971).
147. Schwab, G. M. and Ulrich, H., *Kolloid Z. Polym.*, *190:*108 (1963).
148. Puri, B. R. and Bansal, R. C., *Indian J. Chem.*, *5:*381 (1967).
149. Blayden, H. E. and Patrick, J. W., *Carbon*, *5:*533 (1967).
150. Puri, B. R. and Hazra, R. S., *J. Indian Chem. Soc.*, *47:*651 (1970).
151. Puri, B. R., Sud, V. K., and Kalra, K. C., *Indian J. Chem.*, *10:*76 (1972).

Author Index

Numbers in parentheses are reference numbers and indicate that an author's work is referred to although the name is not cited in the text. Numbers in italics give the page on which the complete reference is listed.

Subject Index

A

Acidic carbons, 29
Activating agents, 1,2,9,11,13, 19
Activation, 1,2,7,19
  chemical, 8,9,10,11,12
  mechanism of, 18,19
  physical, 8,13
Activation temperature, 6
Active sites, 238,248,250
Active surface area, 248,249,250
Adsorption
  of alcohols, 194,195,203,206, 235
    amines, 247
    benzene, 207,212,235
    benzoic acid, 222
    carbon tetrachloride, 209
    dyes, 217
    glycol, 236
  of gold, 338
    aurocyanide anion, 348
    cluster compound, 348

[Adsorption]
  influence of,
    ionic concentration, 345, 346,351,352,357,358
    pH, 347,359
    particle size, 367
    reduction potential, 361
    temperature, 360
  ion pair theory, 342
  reduction theory, 341
  of heptane, 208,224
    hexane, 206,209
    neopentane, 209
    nitrobenzene, 220
  of oxalic acid, 222
    phenols, 220
    α-pinene, 206
    surfactants, 220,224
  of vapors, 183
    water, 151,184,195,299
Adsorption centers, 178,180, 184
Adsorption rates, 242,245
Adsorption selective, 230
Affinity coefficient, 132,134